Greening the Americas
NAFTA's Lessons for Hemispheric Trade

edited by Carolyn L. Deere and Daniel C. Esty

nstitute of Latin American Studies
31 Tavistock Square
London WC1H 9HA

The MIT Press
Cambridge, Massachusetts
London, England

A 333.7 GRE
M Coll

016995   22 MAR 2004

© 2002 Massachusetts Institute of Technology

All rights reserved. No part of this book may be reproduced in any form by any electronic or mechanical means (including photocopying, recording, or information storage and retrieval) without permission in writing from the publisher.

Set in Sabon by The MIT Press.
Printed and bound in the United States of America.

Library of Congress Cataloging-in-Publication Data

Greening the Americas : NAFTA's lessons for hemispheric trade / edited by Carolyn L. Deere and Daniel C. Esty
p. cm.
"Many of the papers included in this volume were first presented and discussed in the spring of 2000 at a conference on lessons from the NAFTA for the FTAA"—Pref.
Includes bibliographical references and index.
ISBN 0-262-04212-6 (hc. : alk. paper) — ISBN 0-262-54138-6 (pbk. : alk. paper)
1. Free trade—Environmental aspects—North America. 2. Free trade—Environmental aspects—Mexico. 3. Environmental policy—North America.
4. Green movement—North America. 5. Canada. Treaties, etc. 1992 Oct. 7.
6. Canada. Treaties, etc. 1993 Sept. 13. I. Deere, Carolyn. II. Esty, Daniel C.
HF1746 .G74 2002
333.7—dc21   2002022921

Contents

This book is dedicated to Eduardo Gitli (1947–2002)—friend, colleague, and a leader of the effort to promote economic prosperity and environmental protection in the Americas.

Foreword

Among the most important challenges facing the world community today is how to manage international trade and globalization more generally—and how to make these processes compatible with other important policy goals, such as environmental protection. This volume provides a road map for understanding how trade and environmental policies were integrated in the North American Free Trade Agreement. This volume explores in detail and from a variety of perspectives strengths and weaknesses of the NAFTA model.

As president of Costa Rica, I experienced how critical, and yet challenging, striking a balance between trade and environmental goals can be. Unfortunately, over the past decade very little analysis was available about how to make trade liberalization and environmental protection mutually supportive. Today we know a great deal more. This volume provides a primer for all those interested in the trade-environment linkage, and more specifically in how this relationship might be managed in the context of the Free Trade Area of the Americas.

Not only do these essays build an explicit bridge between the NAFTA experience and the FTAA; they also provide creative thinking and a fresh set of perspectives on the key issues at the trade-environment interface. As a whole, the volume makes clear that the hemispheric journey toward economic integration and better environmental protection stands as a challenge for all 34 countries involved, but especially for Latin American and Caribbean governments. As the Doha Declaration from the World Trade Organization makes clear, environmental issues cannot be ducked in the context of trade policy making. Calls for "greening" the trade regime are growing in intensity. The prospect of an FTAA that does not address these environmental pressures is unimaginable.

In my experience, the developing world has played too passive a role in the trade and development debate. Rather than resist discussions on the trade-environment relationship, developing countries should make their own case for a more environmentally sensitive regime on terms that meet the needs of the South. This book also provides the analytic underpinnings for developing country governments, NGO officials, and others interested in the FTAA process to define their own trade and environment agendas. There really is no alternative. The authors make a convincing case for why the challenge for the Americas is no longer *whether* trade and environment issues need to be addressed. Instead, the question is *how* to address this nexus in a way that incorporates the main priorities and concerns of Latin American and Caribbean countries.

An appropriate approach to trade and environment issues in the FTAA context will require dialogue and negotiation. NAFTA's experience shows that unilateral imposition of one nation's environmental agenda does not work. Indeed, a key finding from this volume—with important implications for both governmental officials and the environmental community— is that the use of environmentally related trade sanctions should explicitly be avoided. I strongly support the call for a more cooperative approach to addressing environmental issues.

Greening the Americas provides a solid foundation for the critical discussion—engaging government officials, NGOs, business leaders, and others—that needs to take place over how best to fold environmental sensitivity into the FTAA. Success in this endeavor is essential. The hemisphere stands to benefit from deeper economic integration, but not at the expense of environmental degradation. There are tremendous business opportunities to be explored and built upon by incorporating the environmental dimension into trade agreements. We must therefore find ways to manage our interdependence—economic as well as ecological.

José María Figueres-Olsen
Managing Director, World Economic Forum, 1994–1998
Former president of Costa Rica
April 2, 2002

Preface

This volume seeks to present to governments across the Americas, and to trade officials in particular, lessons from the NAFTA environmental experience and analytically rigorous environmental proposals for the FTAA context. Its goal is to ensure that economic integration in the western hemisphere proceeds in an environmentally sustainable and politically sensible manner. We editors do not always agree on how best to do this. Our different individual perspectives on aspects of the trade-environment interface reflect, to an extent, the divergent views that animate the broader discussion of trade and the environment. We agree, however, that environmental issues should be taken seriously by negotiators. With this objective in mind, we have worked to incorporate contributions from the diverse actors that have formulated practical policy proposals for the region. The voices we have included range from the broadly optimistic about trade and trade liberalization to the more pessimistic. What unites all the contributions is a commitment to engaging constructively in the policy dialogue concerning the environmental, economic, and social future of the Americas.

We thank the contributors for their commitment to this project. For feedback and comments, we thank Gustavo Alanis-Ortega, Mónica Araya, John Audley, Robert Bach, Greg Block, Jake Caldwell, Caroline Dommen, Kevin Gallaghar, Jasper Goss, Ana Karina González-Lutzenkirchen, Catherine Monagle, Juliette Moussa, Ricardo Mélendez-Oritz, Robin Rosenberg, Erin Rossitto, Marie-Claire Cordonnier Segger, Mathew Stilwell, Blanca Torres, and Scott Vaughan. Many of the chapters were first presented and discussed in the spring of 2000 at "The FTAA and the Environment: What Can We Learn from the NAFTA Model?" (a conference hosted jointly by the National Wildlife Federation and the Yale Center for Environmental Law and Policy and supported by the Ford Foundation and the C. S. Mott Foundation).

We thank our MIT Press editor, Clay Morgan, who saw the project through with great patience and dedication. We also appreciate the efforts of our personal assistants, Marguerite Camera and Huma Mody. Special thanks go to Rachel Spector, without whose efforts—researching, synthesizing information, proofreading—the book would not have come to fruition. Mónica Araya, Janet McGarry, and Gwen Busby also made important substantive contributions. We are grateful to Elizabeth Esty, Erion Veliaj, Catherine Monagle, Kate Raworth, and Virinder Singh for their support and good humor.

Greening the Americas

1

Trade and the Environment in the Americas: Overview of Key Issues

Carolyn L. Deere and Daniel C. Esty

A debate about the interaction of trade policy and environmental objectives has raged since the early 1990s. The 1991 GATT ruling against a US ban on imports of tuna caught in dolphin-killing nets triggered a public outcry, particularly in the US environmental community.[1] Suddenly, groups never before interested in trade mobilized against further trade agreements (Esty 1994a). Similarly, recent protests, such as the uproar that attended the World Trade Organization's attempt to launch a new round of global trade talks in Seattle in December 1999 and the 2001 Summit of the Americas Trade Ministerial in Quebec City, demonstrate that the environment remains a central concern in the public debate over the direction of trade policy.

International trade offers the potential for economic growth and greater market efficiency. Likewise, environmental protection is crucial to sustained economic growth, the efficient use of scarce resources, the protection of human health, and ecological values. Thus, the challenge is: Can we simultaneously pursue both international trade policy objectives[2] and environmental protection? Intellectually, technically, and politically, the task is complex. And those negotiating the FTAA have not even tried. Many in the environmental community complain that trade liberalization runs roughshod over efforts to control pollution and to manage natural resources. At the same time, many advocates of free trade see "green" groups as protectionist. While dialogue between these groups has improved, mutual suspicion continues to hinder the emergence of truly innovative and constructive efforts to substantively address the intersection of trade issues and environmental issues (Esty 2001). But a carefully constructed trade and environment agenda, respecting the legitimate goals of both trade policy and environmental policy, can be developed. In this regard, models for progress do exist and deserve attention. The 1994 North American Free Trade Agreement directly addressed environmental concerns both in critical elements of the agreement and in an unprecedented "side agreement" on the environment (Johnson and Beaulieu 1996): the North American Agreement for Environmental Cooperation (NAAEC), which not only provides a plan for US-Mexico-Canada environmental cooperation but also establishes the North American Commission for Environmental Cooperation (NACEC), an ongoing mechanism for addressing trade and environment issues.[3]

The NAFTA's environmental approach is not without flaws (Independent Review Committee 1998; McFayden 1998). In fact, while the NAFTA's environmental provisions won over some members of the environmental community, others remain skeptical (RMALC 2002; Sierra Club 2001).[4] And the NAFTA environmental package enraged a number of Mexican government officials who felt forced to "swallow" the environmental provisions advanced by the United States despite serious misgivings. Infuriated by the perceived slights of the NAFTA experience (some real and some imagined), officials of the Zedillo administration initiated the ongoing opposition to discussions on linking environmental considerations with trade commitments and rules in the context of the WTO, and subsequently the FTAA (de la Calle 1999).

Mexico now has a new president and a new policy team. President Vicente Fox's appointment of Victor Lichtinger (a former executive director of the NACEC) to head the Secretaría de Medio Ambiente y Recursos Naturales (SEMARNAT) bodes well for a more thoughtful future Mexican position on trade and environment issues. Upon assuming office, Lichtinger swiftly affirmed the interface of trade and environmental policy objectives and expressed an interest previously unheard of in the Mexican government.[5] The question remains whether trade officials at Mexico's Secretaría de Economía (hereafter Ministry of Education) will ultimately be convinced by Lichtinger's logic and advance a forward-looking trade and environment agenda.[6] In the meantime, the Zedillo administration's hostility to consideration of environmental issues in the trade context has set the tone across the Americas. Nearly all Latin American governments have adopted the Mexican stance against developing an environment agenda as part of the FTAA negotiations.[7]

With an overarching goal of trying to find ways to make trade arrangements and environmental protection more mutually reinforcing, this volume explores and evaluates the traditional hostility of the Mexican government under the Salinas and Zedillo administrations to trade-environment linkages generally and to the environmental provisions of the NAFTA in particular. This inquiry raises a number of critical questions: What is at the heart of the Mexican complaints? Do the concerns raised hold up to scrutiny? Have the NAFTA environmental provisions diminished US-Canada-Mexico trade or otherwise hurt Mexico economically? If so, how? Have the NAFTA's trade rules or its environmental provisions hurt the NAFTA parties environmentally? What elements of the NAFTA environment program should be included in the FTAA? What elements should be omitted?

Inescapable Linkages

From a descriptive perspective, trade-environment links are real and cannot be avoided. International trade and investment inescapably affect the environment and environmental regulation. Where they promote economic growth and industrialization without adequate strategies to control pollution and manage natural resources, international trade and investment liberalization can lead to more pollution and resource consumption (Esty 2001; Nordstrom and Vaughan 1999). In this regard, the increasing number and scale of pollution-spewing maquiladoras[8] along the US-Mexico border provides a paradigmatic example of such "scale effects" and the risks that unsophisticated and unmanaged trade liberalization can pose (Weintraub 2000).

But freer trade can also be environmentally positive. Expanding trade opportunities can also be a useful mechanism for the transfer of pollution-control devices and other technologies that facilitate cleaner production processes and more efficient use of natural resources (OECD 1999 and 2000). Trade provisions limiting subsidies may also yield environmental gains. For example, where government subsidization results in overexploitation of energy, agriculture, forestry, fisheries, and water resources, trade agreements may commit governments to reducing the harmful subsidies. The empirical question is whether the efficiency gains from such technique and composition effects more than offset the degradation from scale effects (Esty 2001).

To the extent that trade liberalization spurs economic growth, it can help generate resources for governments to dedicate to enhanced environmental protection (Grossman and Krueger 1993). And, as anticipated by economic theory, trade-induced economic growth and structural changes produce distributive impacts—some industries and sectors gain, others lose—over both the short and the long run. Through its impact on growth, international trade can help to reduce poverty and the ecological degradation that almost always accompanies it.

Trade agreements also affect the environment insofar as countries accept market-access "disciplines" that constrain their regulatory freedom with respect to pollution control, management of natural resources, and other matters. Beyond the question of whether trade commitments will be invoked to "trump" environmental rules and programs through challenges brought under dispute-settlement procedures lies the question as to whether trade policies and objectives take precedence over environmental policies and objectives.[9] On the one hand, some economic integration initiatives, such as those advanced within the European Union, involve the harmonization of standards, including environmental laws and regulations. Such broad-based collaboration could strengthen environmental-protection efforts. On the other hand, many environmentalists fear that such policy coordination will translate into downward harmonization at the lowest common denominator or some other low level.

At the same time, environmental policy can significantly affect international trade flows. Public health rules, air and water emission limits, food safety standards, waste management regulations, labeling and recycling requirements, and other environmental policy demands channel—and may constrain—the flow of goods and services across national borders. Dispute-settlement cases challenging such regulatory interventions as trade barriers are already common in the World Trade Organization context and will likely arise in the FTAA context as well. The question is how to distinguish legitimate environmental standards from disguised barriers to trade (Esty 1994a).

Normative Necessity

Beyond the fact that the trade-environment link exists as a matter of descriptive reality, this relationship requires attention as a function of the demands of good economics and rational policy making. Absent attention to trans-boundary pollution and the management of environmental resources, the integrity and efficiency of international commerce cannot be maintained. Common resources, such as fisheries and the atmosphere, will be overexploited, and the result will be a "tragedy of the commons." Similarly, failures to "internalize" externalities will distort economic competition and generate air- and water-pollution spillovers that will be costly in economic terms as well as in political, environmental, and human terms. A lack of systematic attention to the environmental issues that arise in a trade context threatens to lower social welfare and lead to unnecessary environmental deterioration (Dua and Esty 1997).

From a "political economy" standpoint, recent experience suggests that governments will not be able to bring the FTAA negotiations to a successful conclusion without some focus on the environment (Destler and Balint 1999). In the United States and Canada, ignoring the trade-environment link threatens to derail political support for new trade agreements. In the United States and Canada, advances on the trade policy front cannot be achieved without attention to related environmental issues. The NAFTA won ratification by the US Congress in part because of substantial efforts to address environmental issues (Esty 1994b; Audley 1997).[10] Recent attempts to renew trade promotion authority (formerly called "fast-track negotiating authority") have failed, in part, for lack of an acceptable environmental dimension. Simply put, the political coalition backing trade liberalization in the United States is narrow and fragile (Schott 2001; Bergsten 1992; Aaronson 2001). Without the swing votes of pro-trade, pro-environment representatives and senators, US trade legislation fails (Esty 1998a; Destler and Balint 1999). Thus, from the perspective of the political economy of trade, finding ways to fold environmental sensitivity into the regional trade agenda is imperative.

The choice ultimately is not *whether* to address trade and environment issues but *how* to address them. Governments can develop policies overtly, transparently, thoughtfully, and systematically, or they can do it in an unstructured and ad hoc fashion. We do not mean to suggest that countries take up the relationship between trade and the environment for purely instrumental reasons. Our argument is broader. International trade will not achieve the hoped-for efficiency gains and economic growth—and the net welfare effects could even be negative—if core environmental issues are not addressed. Undeniably, the refined and sophisticated trade policy we recommend requires structural analysis, careful decision making, and active implementation. In some cases, an environmentally sensitive trade strategy will argue for rapid liberalization, perhaps to improve access to pollution-control technologies. In other circumstances, countries may decide to liberalize at a slower pace to avoid severe economic and social dislocations that could lead to environmental damage.

Trade policy making inevitably entails tradeoffs among goals and decision making under conditions of uncertainty and complexity.[11] Factoring environmental variables into the calculus complicates the process. But denying the existence of a trade-environment link cannot be sustained. There are real issues—political, economic, and environmental—to be sorted out. The public (including the business community, civil-society organizations, research institutions, and environmental groups) can assist governments in this task by generating information, sharing empirical data, generating options, and suggesting creative strategies to maximize both economic and environmental results.

Public participation in the process of making trade policy also increases the legitimacy of any eventual agreements that emerge (Esty 1998b). Public understanding and acceptance of the benefits, tradeoffs, risks, and consequences of different policy proposals is essential. Legitimacy thus requires transparency, which not only builds understanding but also creates public confidence that government policies reflect careful consideration of the issues, attention to non-economic values as well as economic growth goals, and a full-blown debate over hard choices (Esty 2002). Decisions must be seen as reflecting the broad public interest and not just providing benefits for various special interests.

At present, the FTAA process fails to meet these basic requirements. Governments have launched their negotiations on a narrow track that fails to take seriously the broader set of issues implicated by any push for freer trade. Within Latin America, numerous calls for greater public dialogue on potential trade and environment issues in the FTAA context have gone unheeded by many governments (Gitli and Murillo 2000; CEFIR 1998; Lucas 2000). The negotiators' response—establishment of a Committee of Government Representatives on Civil Society—isolates those wishing to debate serious environmental questions in a meaningless sideshow[12]. With no authority, no work plan, and no real mandate, the FTAA Civil Society Committee fails to provide a serious forum for debate or much of a mechanism to develop understanding of how best to handle trade and environment issues. It serves mainly as a "drop box" for comments and submissions from civil society (Lucas 2000).[13]

No formal process links the civil-society dialogue and any of the FTAA's nine negotiating groups.[14] In fact, no procedures even exist to guide the consideration of submissions from civil society, let alone analysis of them. To date, for example, there has been no substantive analysis of the more than 80 submissions received from various groups and organizations since the committee was established. Although the Civil Society Committee nominally reports directly to the FTAA trade ministers, it does so in terms that are far too general to be of any real use. Such lip service to critical issues—and to the process of public participation—promises to become a serious obstacle when it comes time to ratify the FTAA.

There are many contentious environmental proposals floating around. Some recommendations are ill-conceived and irresponsible. And some of the suggestions advanced in the name of protecting the environment are stalking horses for protectionist interests. But this does not diminish the force of the argument for building environmental sensitivity and assessments into any regional trade negotiations or agreement for the Americas. Sorting out what makes sense for a regional trade and environment agenda will take some work. Negotiators still lack the clear analysis and recommendations on hemispheric trade and environment issues they need to proceed. While a number of studies examine prospects for the FTAA and hemispheric relations in general, few specifically address concerns about trade and the environment from a hemispheric perspective, and none focus on extracting lessons from the NAFTA for the FTAA.[15] This volume begins to fill the breach.

Specifically, this volume explores the NAFTA's environmental record, evaluates the provisions advanced in that accord, identifies the lessons to be learned about how to manage tensions between trade and the environment constructively, and seeks to stimulate discussion on the trade and environment challenges and opportunities for the Americas and ways to build on the past.

Plan of the Book

The contributors to this book include some of the leading commentators on trade and environment issues in the Americas.[16] The mix of contributions from academics and researchers on the one hand and policy makers on the other reflects a conscious effort to present a range of perspectives, bridge the theory/practice divide and give depth and texture to a complicated subject. Some of the policy-style chapters, including those from a representative of the Mexican trade ministry and those from environmental non-governmental organizations (NGOs), convey strong advocacy messages about how the trade-environment interface should evolve.

Part I addresses the political economy of the environment in the NAFTA negotiations. Part II investigates selected economic and environmental impacts of the NAFTA. Part III reviews the NAFTA's environmental provisions and institutions. Part IV, which focuses on the FTAA, offers recommendations on integrating environmental considerations into the emerging hemispheric trade structure.

Major Themes

Negotiation Dynamics and Issues

Much has already been written about the history of the NAFTA negotiations, the influence of domestic US politics and interest groups on trade policy, and the role of business in the NAFTA negotiations (Johnson and Beaulieu 1996; Destler and Balint 1999). The political economy of the NAFTA's environment negotiations has also been explored extensively (Audley 1997; Barkin 1999; Hogenboom 1998; Mayer 1998).[17] Far less attention has been devoted to the Mexican government's position on trade and the environment. During the NAFTA negotiations, Mexico complained that the inclusion of environmental provisions in the NAFTA might be used by protectionists to hinder trade as well as foreign investment and economic growth in Mexico (Orme 1996; Garciadiego 1994). Concerns about the impact of the NAFTA's environmental provisions continue to be a major component of Mexico's ongoing hostility toward the trade and environment issue.

The chapters in part I explore the negotiation dynamics of the NAFTA with the specific goal of trying to understand *why* the Mexican government continues to lead efforts to exclude environmental issues from the FTAA negotiations and to avoid broader debate over how to integrate trade and environment issues in the context of economic integration.

Javier Mancera, an official with Mexico's Ministry of Economy, articulates the case against attempts to link trade policies and environmental policies. He expresses frustration with "unsubstantiated concerns" that trade agreements pose potential threats to domestic environmental protection, public health, and worker welfare. Separating what Mexican negotiators consider real trade issues from "non-trade" issues (such as the environment), Mancera argues that extraneous issues will weigh down the process of trade liberalization and may undermine developing countries' market-access goals. He cites the NAFTA's threat of trade sanctions for environmental transgressions as a case in point.

Gustavo Alanis Ortega and Ana Karina González-Lutzenkirchen of the Mexican Center for Environmental Law explore the origins of Mexico's official position on trade and the environment. They suggest that the dynamics of the negotiations for the NAFTA environmental side agreement played a critical role in consolidating Mexico's antipathy toward efforts to link trade policy and environmental policy. They also review Mexico's domestic trade and environment politics, examining the relative roles of different actors in the formation of Mexican trade policy. They show that, within Mexico, policy makers lack research elucidating important trade and environment issues and their implications for Mexico. A real debate on the issues is thus yet to emerge between government ministries or between government and civil society.

Mónica Araya, Director of the Sustainable Americas Program at the Yale Center for Environmental Law and Policy, further investigates the origins of the Mexican Economic Ministry's rejection of attempts to craft initiatives at the intersection of trade policy and environmental policy. Araya contends that the Mexican position is not based on a substantive evaluation of the NAFTA's impacts. Instead, she advances a range of other explanations for Mexican hostility to linking trade policy and environmental policy, including historical political tensions between Mexico and the United States, environmental conflicts along North-South lines (which fuel Mexico's skepticism about the depth of the United States' commitment to the environment), Mexico's frustration with the GATT tuna-dolphin dispute, the NAFTA negotiating dynamics (which forced Mexico to accept environmental elements to which it objected), and a strategic stance to maximize leverage in future trade negotiations (especially the FTAA). Araya concludes that a sense of hurt "national pride" from the loss of two important battles in the trade arena—the burden of an "unfair" tuna embargo and the imposition of an environmental side agreement as an "entrance fee" to the NAFTA—drives the Mexican Economic Ministry's stance on trade and the environment. The residual frustration translates into a desire to "even the trade score" with the United States. Araya urges Latin American and Caribbean negotiators to engage in a real debate on trade and the environment based on a thorough, substantive evaluation of existing approaches and their own needs. To avoid further trade and environment "trauma," Araya emphasizes the need for policy makers in the Americas to drop their focus on what they do not want and instead work to identify a trade and environment agenda that would work. Araya suggests that there is no need to start from scratch, since the NAFTA's environmental model could be refined and adapted to the FTAA context.

Following up on Araya's argument that Mexico's stance on trade and the environment has little substantive foundation, Eric Miller of the Inter-American Development Bank examines the NAFTA record in detail. He explores and rejects the proposition that the NAFTA's environmental provisions caused damage to Mexico's economic performance. He concludes that the economic impacts on Mexico of the NAFTA's environmental provisions have been negligible. More broadly, he argues that countries have little to fear from a proliferation of NAFTA-like environmental provisions. In fact, Miller notes that in most of Mexico's bilateral agreements subsequent to the NAFTA at least some of the provisions are repeated almost word for word. If these provisions were in any way damaging economically to Mexico, the Mexican Ministry of Economy is not likely to have permitted their inclusion in its other trade arrangements.

Frederick Mayer of the Duke University Center for North American Studies explores the NAFTA negotiating dynamic, where Mexican complaints seem to have a solid foundation. He explains why the Clinton administration decided to demand new environmental negotiations after the NAFTA talks themselves had concluded. He also examines why the United States made particular demands in the side negotiations and why the final agreement took the form that it did. Drawing on a wide range of documents and interviews with officials in the United States, Mexico, and Canada, Mayer argues that the requirement for and the outcome of the

NAFTA's environmental negotiations reflect not only national interests and relative power but also the domestic political processes operating within Canada, Mexico, and the United States. Looking forward, Mayer observes a "disconnect" between the political realities for the United States and the current trajectory of the FTAA talks. He argues that the United States cannot effectively negotiate in the FTAA, nor will other countries seriously negotiate with the United States, until Congress grants President Bush some form of trade promotion authority. Mayer also emphasizes that environmental groups have sufficient leverage within the United States to insist on a continued trade and environment agenda. Still, the precise form of the environment-trade link in the FTAA must, he advises, be open for negotiation. He observes that even in the NAFTA the United States found limits in what it could impose on its bargaining partners. Those limits are likely to be far greater when the US is negotiating with many more countries, each of which is less economically dependent on the United States than Mexico.

Environmental Performance of the NAFTA

The chapters in part II explore lessons that can be learned from the environmental impacts of trade under the NAFTA. Dozens of studies by governments, academics, NGOs, and community-based organizations have evaluated the NAFTA's impacts on environmental quality, law, and policy making, particularly in the United States and Canada, but also on the US-Mexico border (Moss 1993; Johnson and Beaulieu 1996; Hufbauer et al. 2000).[19] A significant theoretical literature on the relationship between trade and the environment has also emerged (Esty 1994a, 2001; Geradin 1997; Pearson 2000; Runge 1994).[20]

One of the great methodological challenges of the trade and environment issue is to sort out why the conclusions of different environmental assessments of the NAFTA diverge so greatly. The idea that the outcome of any assessment depends on the questions asked is hardly novel, but it is often forgotten in the context of trade policy. The scope of many of the existing assessments of the NAFTA's environmental performance is tightly focused on whether governments have fulfilled or complied with specific environmental provisions incorporated into the NAFTA package. However, several studies probe more deeply. Some analysts ask whether existing mechanisms were able to facilitate an adequate response to the most pressing trade and environment issues that have emerged. Other analysts question the value of the NAFTA itself, seeking to cast doubt both on the ability of the agreement to advance the social priorities of the countries involved and on the underlying economic assumptions and theories promoting free trade.

The existence of sharply contrasting appraisals also has much to do with original fears of what the NAFTA might portend for the environment, with expectations generated by governments during the environmental side negotiations, with different methodological approaches to evaluation, and with the distinct kinds of environmental issues under examination. Existing studies of the NAFTA examine a range of potential environmental effects, including impacts on environmental policy (such as dispute-settlement issues); impacts on environmental quality (e.g., the US-Mexico border, conservation and use of natural resources, pollution, biodiversity conservation, climate change, energy consumption, and transportation trends); environmental cooperation and the performance of the NAFTA's environmental institutions; governance issues such as transparency, accountability, and participation; and systemic issues, including the balancing of trade policies and environmental policies and the NAFTA as an economic model (Vaughan 2000).[21]

Worries about the impacts of the NAFTA on different aspects of environmental quality abound, yet only recently have there been concerted efforts to undertake rigorous analyses and develop real data.[22] These efforts have been closely tied to national efforts in Canada and in the United States to develop methodologies for, and to implement, retrospective and prospective environmental reviews of trade policy proposals and agreements. In recent years, the NACEC has worked to shift attention toward assessment of the impacts of trade and trade rules on natural resources and their management (from water to biodiversity) and systematic issues such as patterns of energy and transportation use. In each of these areas, research remains both preliminary and methodologically challenging.

Part II continues with three case studies that provide new evidence on the impact of the NAFTA on environmental quality and policy. They focus on three of the issues that have attracted the most vocal public concern: the "pollution" effects of the NAFTA, the environment and social impacts of the NAFTA on Mexico's agricultural sector, and the impact of the NAFTA's investment provisions on environmental policy.

Kevin Gallagher of the Global Development and Environment Study at Tufts University enters into the lively debate (Revesz 1992; Esty 1996; Esty and Geradin 2001) over the risk that trade liberalization will trigger a "race to the bottom" in environmental quality, policy, and regulation. The possibility that lower Mexican standards or lax enforcement would induce "industrial migration" to Mexico as a regional "pollution haven" generated a great deal of discussion during the NAFTA ratification process. Empirically exploring the relationship between trade liberalization and industrial pollution in Mexico, Gallagher finds that trade liberalization has had both positive and negative effects on industrial pollution in Mexico. Many of the industries deemed the dirtiest in the world economy are actually cleaner in Mexico than in the United States. At the same time, some of the industries usually considered clean are dirtier in Mexico. Gallagher's study suggests that liberalized trade and investment can enable countries to take advantage of newer technologies in sectors where plant vintage determines pollution levels. Such trade-induced cleaner production appears to have occurred in the Mexican steel industry. On the other hand, Gallagher observes, if pollution levels depend on the use of end-of-the-pipe pollution-control technologies (as in the paper industry), regulatory requirements and enforcement play a large role in determining pollution levels. In such industries, the combination of trade and investment liberalization with Mexico's weak record of enforcement and compliance has led to more pollution.

Alejandro Nadal of El Colegio de México compares the anticipated development impacts of the NAFTA-led liberalization of Mexico's corn sector with the actual environmental, social, and economic outcomes.[23] Much of the Mexican government's concern about the NAFTA's environmental provisions can be traced to the view that, because the primary goal of trade liberalization is to promote development, any environmental provisions that impede trade liberalization or affect competitiveness should be avoided. Though trade and trade liberalization can be important tools for development and environmental protection, there is no doubt that they can also have negative environmental impacts. Nadal offers a case study of the interrelated environmental and social problems that can result from a rapid liberalization process unaccompanied by well-planned or well-implemented adjustment measures. To properly understand and prevent potential economic and environmental impacts of trade liberalization, Nadal argues, governments must consider the broader macroeconomic policy mix when forecasting and evaluating different trade policy options.

In their chapter, Howard Mann (of the International Institute for Sustainable Development) and Mónica Araya build on earlier studies (Mann and von Moltke 1999; CIEL et al. 1999) of the NAFTA's chapter 11 investment provisions. They review chapter 11's impacts on environmental policy and law making and offer a series of recommendations for countries to consider in refining the chapter 11 model for the FTAA context. Mann and Araya explain why reproducing a NAFTA-like investment model in the Americas could weaken the ability of governments to achieve environmental goals and protect public health and safety. Failure to address chapter 11's shortcomings in the FTAA negotiations could, they predict, reduce the benefits of foreign investment in Latin America and the Caribbean, inhibit environmental improvement, and erode public support for economic integration.[24]

Performance of the NAFTA's Environmental Provisions

Assessments of how well the NAFTA's environmental elements have performed vary considerably. The issues that have attracted the most debate concern the record of the NAFTA's environmental institutions and the notion that the NAFTA's environmental provisions failed the environment by not providing sufficient scope to address instances of conflict between trade goals and environmental objectives. It is noteworthy that the environmental provisions contained within the NAFTA's main text (e.g., provisions referring to the relationship of the agreement to multilateral environmental agreements and to maintaining domestic environmental standards) have generated little criticism.[25]

Some commentators observe that the NAFTA's environmental side agreement has produced a useful range of cooperative initiatives and that Mexico's environmental performance is better than it would have been in the absence of the agreement and the collaboration it spurred (Hufbauer et al. 2000). Others argue that the NAFTA's environmental provisions have failed to respond adequately to the most critical environmental issues (Sierra Club 2001; Public Citizen 2001).

Richard Fisher, a former Deputy US Trade Representative, captures succinctly the perspective of a NAFTA optimist. He argues that the NAFTA represents an important step toward making trade policies and environmental policies more mutually reinforcing. Fisher rejects fears of trade-environment linkage and suggests that a strong economy and clean environment go hand in hand. He advances the case for an ongoing trade-environment link, contending that the NAFTA, its internal environmental provisions, and its environmental side agreement constitute environmental and economic success stories. Fisher highlights the importance of taking environmental considerations into account from start to finish in trade negotiations and the value of environmental reviews of trade policy options.[26] Fisher emphasizes that the political legitimacy and acceptance of trade agreements can be substantially improved when governments work closely with non-governmental organizations and other interested parties to shape negotiating objectives.

The next two chapters offer quite different Mexican perspectives on the NAFTA model. Blanca Torres, an economist at El Colegio de México, provides an optimistic if guarded endorsement of the NAFTA's environmental provisions and their contribution to regional environmental performance. Like Fisher, Torres argues that the NAFTA environmental institutions—the BECC, the NADBank, and the NACEC—were groundbreaking efforts to promote regional cooperation, even if they have had uneven records. In light of the range of challenges faced by the NAFTA parties in advancing a cooperative environmental agenda (particularly the political, historical, and economic asymmetries between the United States and Mexico), Torres argues that the complexity of environmental issues makes progress difficult. The NAFTA's environmental side agreement has had to contend with a series of challenges that frustrate cooperation, such as the legacy of an unbalanced NAFTA negotiation process, limited political will, and financial resources that have been inadequate to the task of improving Mexico's environmental performance. Nonetheless, in Mexico, Torres reports, the side agreement has helped to improve the design and performance of national environmental institutions, stimulate more accurate diagnoses of Mexico's environmental problems, increase public participation in environmental decision making, improve environmental enforcement, and increase the private sector's participation in environmental initiatives.

Laura Carlsen and Hilda Salazar, who work with a network of grassroots Mexican NGOs, focus on the intersection of international economic policies and environmental and social impacts and challenge the assessments offered by Fisher and Torres. While they accept the logic of complementing trade agreements with an environmental package, Carlsen and Salazar provide a more critical appraisal of the character and performance of the NAFTA, its environmental provisions, and the related environmental institutions.

Carlsen and Salazar suggest that, although the NAFTA's environmental institutions were launched with good intentions, inadequate political will, limited public participation, and limited financial resources have stifled their potential. In their view, these institutions were given a long list of useful and necessary environmental objectives to pursue (ranging from research, information exchange, and public education to the protection of endangered species), but they have not been able to intervene usefully in the range of specific environmental problems caused wholly or partially by economic trends advanced by the NAFTA (e.g., border pollution, air and water degradation). In particular, Carlsen and Salazar argue that the NAFTA's environmental provisions provide no mechanisms to address instances where trade or investment trends exacerbate unsustainable natural resource depletion. In conclusion, they urge that future trade agreements avoid subordinating environmental considerations to the perceived "real issues" of commerce and financial flows.

Linking Trade and Environment in the Americas: Ways Forward

Marianne Schaper, an economist with the UN's Economic Commission for Latin America and the Caribbean, finds that trade liberalization emerged as the policy that most influenced the restructuring of productive and export activities in South America during the 1990s and that the current South American export growth path appears to be environmentally vulnerable. As trade has intensified, she argues, pollution and environmental degradation have become more acute in all countries, but in different ways and with different levels of intensity across the continent. Schaper cautions that the continuing specialization of South American countries in pollution-intensive and natural-resource-intensive industries places the region in a vulnerable situation as exporters face the environmental requirements of exporting to developed countries. Looking forward, she urges Latin American policy makers to make stronger efforts to address the problem of weak environmental enforcement in their countries. In practical terms, Schaper advises countries that they have much to gain economically from efforts to address trade-environment links. Devising initiatives to increase investment in environmental services, in technology and equipment, and in improved environmental management practices will help countries ensure the long-run sustainability of their productive base. Schaper also proposes that businesses, by maintaining and augmenting voluntary efforts to improve environmental performance through schemes such as ISO 14000 certification, can both prepare themselves for the likelihood of environment and trade policy linkages and generate higher profits.

Mario Matus and Edda Rossi of Chile's Ministry of Foreign Affairs similarly counsel their Latin America government counterparts against ignoring the realities of the trade-environment link. In contrast to Mexican trade officials, they advise that there need be no contradiction between protecting the environment and expanding external trade. Matus and Rossi urge Chile's hemispheric partners to acknowledge the trade and environment challenge and to address the issues raised in practical and concrete ways. They offer Chile as an example of how countries can pursue both agendas simultaneously, pointing to the Canada-Chile Agreement on Environmental Cooperation (negotiated in parallel to their bilateral free trade agreement) as a model that Latin American and Caribbean countries might want to consider in the FTAA context. Drawing directly from the NAFTA model, the Canada-Chile environmental side agreement embodies a strong focus on environmental cooperation. Going beyond the NAFTA, the Canada-Chile Agreement makes greater use of each country's existing institutional framework for environmental protection and eliminates the NAAEC's provisions for trade sanctions.

The late Eduardo Gitli and Carlos Murrillo of Costa Rica's International Center for Economic Policy push forward the Matus-Rossi line, arguing that it is time for Latin American countries to develop constructive, independent proposals on trade and the environment. They summarize the environmental issues and opportunities that are before each of the FTAA's nine negotiating groups, and they present a Central American view on how best to integrate environmental considerations into the proposed hemispheric agreement. They recommend that the FTAA adopt NAFTA-style environmental provisions within its main text (such as provisions that discourage countries from lowering environmental standards to attract investment or trade) and parallel environmental negotiations with the ultimate aim of developing a Canada-Chile-style environmental side agreement that drops the possibility of trade sanctions. In addition, they present a series of proposals for integrating consideration of environmental issues and opportunities into each of the FTAA's nine negotiating groups. Gitli and Murillo argue that governments can prepare themselves for the task of addressing the trade and environment reality by generating and dissemi-

nating more information about the status and content of FTAA negotiations and facilitating greater public participation in the FTAA negotiation
process. In their view, a clear lesson from the NAFTA experience is that
civil society and domestic private-sector actors are essential partners for
governments faced with the challenge of advancing trade objectives and
environmental priorities simultaneously.

Taking the current state of play in the FTAA negotiations as a starting
point, Jake Caldwell of the National Wildlife Federation offers a comprehensive environmental advocate's set of proposals for incorporation of environmental considerations and public input into the FTAA negotiations. Like
others, Caldwell sees the NAFTA environmental approach as a model on
which to build. He recommends that the FTAA include provisions that
guarantee deference to national environmental standards and multilateral
environmental agreements. Moving beyond the NAFTA, Caldwell recommends that the FTAA include specific provisions to promote environmentally responsible investment and clarify the right of parties to make
distinctions concerning market access based partially on the environmental impacts of production, as long as there is no violation of the national
treatment principle.[27] In recognition of the special needs of developing countries, he calls for hemispheric environmental cooperation and capacity
building on trade and environment issues as integral components of the
FTAA process. He highlights opportunities for negotiators to use trade rules
to eliminate environmentally damaging and trade distorting subsidies in
natural resource sectors such as fisheries and forestry. To better understand
the critical pollution and resource-management issues, Caldwell recommends, FTAA parties should undertake environmental impact assessments
and create a grant mechanism to offset the associated costs. Finally, he calls
on governments to promote public participation, information dissemination, transparency, and accountability as cornerstones of the FTAA process
as necessary steps for ensuring public confidence in the negotiations.

Marie Claire Cordonier Segger of Canada's International Institute for
Sustainable Development and Nicola Borregaard of Chile's Recursos e
Investigación para el Desarrollo Sustentable highlight the importance of
democratic procedures and openness to progress on the trade and environment front in the Americas. Moving beyond the NAFTA model, they propose that countries of the western hemisphere look to four sub-regional trade
models (Mercosur, CARICOM, the Central America Common Market, and

the Andean Community) and to a series of bilateral agreement for useful precedents. In each instance, they explore the extent to which environmental issues are integrated, and they seek out innovative models to which Latin American and Caribbean governments have already committed.

In the concluding chapter, we summarize the volume's central themes, the lessons learned from the NAFTA, and the recommendations offered to the FTAA negotiators. We conclude that the Mexican government's attacks on the NAFTA's environmental provisions have little substantive basis. There is no empirical evidence to support the suggestion that Mexico has suffered economically or environmentally from the NAFTA's *environmental* elements. However, Mexico's complaints about US bullying during the final stages of the NAFTA negotiation do have a factual foundation. As the chapters in this volume clearly suggest, the damage done at that time continues to color trade and environment discussions both in the FTAA context and more generally. We argue that governments across the western hemisphere should use the NAFTA environmental "package" as a starting point for developing an FTAA environmental negotiating agenda. Although the NAFTA's environmental package warrants refinement, a number of the NAFTA's elements—public participation, environmental cooperation, creative financing of environmental infrastructure—are good starting points for an FTAA environmental initiative. Future trade agreements should, however, learn from the mistakes associated with the NAFTA. NAFTA's chapter 11 on investment needs to be rethought. More attention should be paid to expanding the resources available for environmental infrastructure investments and capacity building. And the emphasis on punitive trade sanctions—a central element of the NAFTA environmental side agreement—should be dropped. As governments across the region work toward regional integration, they can also look to the sub-regional arrangements in which they are already involved. Some of these sub-regional agreements (particularly Mercosur) already recognize the trade-environment link.[28] Based on these lessons, we set forth a series of recommendations for a trade and environment agenda for the Americas.

Developing a serious and systematic Latin American-driven environmental agenda is an essential step toward the realization of a politically viable hemispheric trade agreement that advances the original Miami Declaration goals of greater social justice, economic prosperity, and a healthy environment. This volume seeks to support such efforts to build understanding across the hemisphere and to ensure that trade priorities and environmental priorities are advanced together.

Notes

1. GATT: General Agreement on Tariffs and Trade. The tuna-dolphin case is often used to signal the origin of the trade and environment debate. There is little question that this was the controversy that brought trade and environment issues to the attention of the public. Multilateral trade negotiators and policy makers, however, recall other triggering events, such as the heated controversy about Austria's labeling and other requirements applied to wood and wood products derived from ASEAN countries; the European Free Trade Area's push to jump start an Environmental Measures in Trade group and, more systemically, the communication gaps between the United Nations Conference on Environment and Development negotiations, the UNCTAD VIII conference, and the Uruguay Round of the GATT negotiations. The tuna-dolphin case arose and was played out against this backdrop. For more information, see http://www.wto.org.

2. It is important to note that, over the past decade, the scope of trade policy has expanded to encompass issues such as government procurement, investment, services, non-tariff barriers to trade, and intellectual property. Trade agreements are not, therefore, confined to traditional trade issues, such as tariffs. Similarly, the term trade itself is commonly used as shorthand for both trade *and* investment. In this volume, we employ the common shorthand, but include specific reference to investment where it is specifically relevant. Issues related to investment and the environment emerge most significantly in chapters 7 and 8.

3. References to the NAFTA's "environmental provisions" in this volume include both environmental elements of the Trade Agreement itself and the parallel environmental agreement that created the NAAEC. In addition, a US–Mexico Border Environmental Cooperation Agreement (BECA) established the Border Environmental Cooperation Commission (BECC) and the North American Development Bank. The NAFTA explicitly addresses environmental issues in its preamble and in five of its 22 chapters. Other chapters relate to the environment indirectly. For more information, see appendix A below and the NACEC web site (http://www.cec.org).

4. For critical reviews of the environmental impact of the NAFTA and the performance of its environmental provisions, see the following: Public Citizen et al. 1997; Barkin 1997; Spalding and Audley 1997; Mann and von Moltke 1999; Sierra Club 2001. DiMento and Doughman (1998) assess the implementation of the NAAEC. The journal *Borderlines* also hosts ongoing dialogue and assessments of the NAFTA's environmental impacts and provisions. There is no doubt that the NAFTA's environmental record is mixed. Ongoing environmental problems on the US-Mexico border attract legitimate frustration (Simon 1997). Though it is important to note that the NAFTA's environmental provisions were never meant to "solve" border problems related to air and water pollution and the disposal of hazardous waste, it is also apparent that the NAFTA's trade and investment provisions have worked to intensify border pressures (Hufbauer et al. 2000).

5. Lichtinger's continued commitment to and interest in trade and environment questions was signaled swiftly after he took office with his attendance as opening speaker at a UNEP-sponsored meeting on the same issues in Mexico. See UNEP et al. 2001.

6. As of November 30, 2000, SECOFI (Mexico's Ministry of Commerce and Industrial Development) changed its name to Secretaría de Economía.

7. The Latin American stance appears to be rooted in fears of disguised protectionism and that higher environmental expectations could undermine competitiveness and the achievement of development objectives, in concerns about power politics and the potential for the US to impose unworkable environmental provisions on them, and in a strategic desire to use the environment as a bargaining chip in the negotiation process.

8. A maquiladora is a factory benefiting from favorable tax and tariff treatment.

9. While the precedence of trade over environmental policy objectives can be a direct result of negligently negotiated international rules and commitments, it is often the result of national governmental choice or discretion. This raises broader questions about the process of national trade policy making and the balance of objectives pursued through trade policy. Often government choices are either simply ill-informed or delinquent in that they reflect private, commercial interests over broader public policy interests.

10. For a history of the NAFTA negotiations and key arguments that animated debate, see the following: Audley 1993, 1997; Esty 1993, 1994b; CEC 1996; Mayer 1998; Hogenboom 1998; González 1999. On the role of the private sector in the NAFTA negotiations, see Macarthur 2001 and Rugman et al. 1999.

11. For a detailed overview of legal, economic, and policy debates surrounding trade and the environment, see Zaelke et al. 1993 and Rodrik 1999.

12. The 1998 Ministerial Declaration of the Fourth Trade Ministerial in San José states that ministers "recognize and welcome the interests and concerns that different sectors of society have expressed in relation to the FTAA. Business and other sectors of production, labor, environmental and academic groups have been particularly active in this matter. We encourage these and other sectors of civil societies to present their views on trade matters in a constructive manner." The declaration is available at http://www.alca-ftaa.org/ministerials/costa_e.asp.

13. The thoughts expressed in the following three paragraphs draw extensively on Lucas 2000.

14. For an overview of the FTAA negotiations and the negotiating structure, see http://www.ftaa-alca.org. Negotiations are being conducted by nine negotiating groups: (1) market access, (2) investment, (3) services, (4) government procurement, (5) dispute settlement, (6) agriculture, (7) intellectual property rights, (8) subsidies, anti-dumping, and countervailing duties, and (9) competition policy.

15. The following are important texts on trade and environment issues in the Americas: Runge et al. 1997; Segger et al. 2000; Konz 2000. We are aware of at least of four new complementary publications: Hufbauer et al. 2000, Rosenberg 2000, Hemispheric Social Alliance 2001, and a forthcoming collection (in Spanish) to be issued by the United Nations Economic Commission on Latin America and the

International Center for Trade and Sustainable Development. The fourth item consists of papers that explore concrete sectoral trade and environment issues in Latin America. The Hemispheric Social Alliance, a network of peoples' movements and NGOs across the Americas have worked collaboratively to prepare a document that articulates proposals for "Alternatives for the Americas." The United Nations Environment Programme's Regional Office on Latin America and the Caribbean has released proceedings and papers from a conference on Latin American and Caribbean perspectives on trade, environment and sustainable development held in February 2001. The Global Development and Environment Institute at Tufts University has compiled a number of targeted trade and environmental studies on the Americas.

16. Many of the chapters in this volume were originally presented at a conference hosted by the Yale Center for Environmental Law and Policy and the National Wildlife Federation in Washington in April 2000. The title of the conference was "The FTAA and the Environment: What Can We Learn from the NAFTA Model?"

17. Magraw 1995 contains a selection of the key documents, position statements, and letters that fed into the NAFTA environmental negotiations.

18. SEMARNAT was formerly known as SEMARNAP.

19. For an environmental overview of the legal aspects, see Rubin and Alexander 1996.

20. Several organizations have produced studies at regular intervals. The Office of the US Trade Representative, the Economic Policy Institute, the Center for Strategic International Studies, Public Citizen, the Institute for Policy Studies, and the Institute for International Economics have produced multiple studies of the NAFTA, each asking different questions and adopting quite different perspectives. The most recently of these are Hufbauer et al. 2000, Anderson 2001, and Public Citizen 2001.

21. Estimates of changes and trends in trade and investment flows tell us little in their own right about the environmental impacts of trade policy reform. They reveal nothing about the effect of trade policy on compositional changes at the economy-wide level and ensuing changes that might occur around the pollution per unit of output of GDP, changes in the allocation of resources within and between sectors, the relationship between trade-related rates of growth and indicators of environmental quality, or the link between open trade and access to different types of production technologies.

22. The NACEC has, however, begun to make considerable progress in this area. Following a multi-year initiative to develop sound methodological approaches to the task of environmental assessments of trade, the NACEC commissioned a series of case studies to test the methodology. The text of these case studies, and a summary of two experts meetings convened to discuss their results, is available at http://www.cec.org/symposium.

23. Nadal's article in this volume is an adapted version of a full report published jointly by WWF-International and Oxfam International.

24. For critical analyses of the NAFTA's chapter 11, see also Soloway 1999. For a neat summary of the environmental critique, see Friends of the Earth and Public Citizen 1999.

25. That said, many environmentalists would like to see several additional environmental provisions incorporated into the NAFTA's text to explicitly permit eco-labeling and trade-related restrictions based on considerations of production and processing methods, but also clear interpretations of particular legal terminology (e.g., terms such as 'necessary' and 'legitimate').

26. The US is working to apply this lesson to the FTAA negotiations with its efforts to promote discussion of environmental issues, conducting environmental reviews and taking advantage of positive environmental opportunities. In 1997, the US made a formal proposal to create a Study Group on the Environment (which was, however, rejected by other countries). Subsequent to the 1999 Executive Order 13141 on Environmental Reviews of Trade Agreements, the US government initiated an environmental review of the FTAA.

27. The NAFTA's national treatment principle forbids members from treating imported products less favorably (e.g., through higher taxes) than "like" domestic products.

28. Three recent efforts to build environmental sensitivities into trade agreements have been the Canada-Chile Agreement on Environmental Cooperation (an environmental side agreement to a bilateral Canada-Chile trade agreeement), the environmental provisions of the US-Jordan trade agreement, and the environmental working group of the Mercosur trade arrangement. See Blanco and Borregaard 1998, Matus and Rossi 2002, and White House 2000.

References

Aaronson, S. 2001. *Redefining the Terms of Trade Policymaking*. Washington: National Policy Association.

Alanis-Ortega, G., and Ana Karina González-Lutzenkirchen. 2002. "No Room for the Environment: The NAFTA Negotiations and the Mexican Perspective on Trade and the Environment." In this volume.

Anderson, S. 2001. *Seven Years under NAFTA*. Washington: Institute for Policy Studies.

Audley, J. 1993. "Why Environmentalists Are Angry about the North American Free Trade Agreement." In *Trade and Environment*, ed. D. Zaelke, P. Orbuch, and R. Housman. Island.

Audley, J. 1997. *Green Politics and Global Trade: NAFTA and the Future of Environmental Politics*. Georgetown University Press.

Barkin, D. 1999. "Free Trade and Environmental Policymaking in Mexico." *Borderlines* 7, no. 9: 14–15.

Bergsten, F. 1992. "The Primacy of Economics." *Foreign Policy* 87, summer: 3–24.

Blanco, H., and N. Borregaard. 1998. *Mercosur y Medio Ambiente*. Santiago: Ediciones LOM.

CEC. 1996. "NAFTA Effects: Potential NAFTA effects—Claims and Arguments 1991-1994." In *Environment and Trade Series 2*. Montreal: Commission for Environmental Cooperation.

CEFIR. 1998. *Participatión de la Sociedad Civil en los Procesos de Integración.* Impresora.

Centro de Formación para la Integración Regional, Asociación Latinoamericana de Organizaciones de Promoción y Centro Latinoamericano de Economía Humana. 1998. *Participación de la Sociedad Civil en los Procesos de Integración.* Uruguay: Impresora Editorial.

CIEL, Instituto del Tercer Mundo, Preamble Center. 1999. *Investment Agreement of the Americas: Environmental, Economic and Social Perspectives.* Washington: Center for International Environmental Law, El Instituto del Tercer Mundo, and Preamble Center.

de la Calle, L. 1999. "Linkages between Trade and Environment Policies." Presentation at WTO High Level Symposium on Trade and Environment. Available at http://www.wto.org.

Destler, I., and P. Balint. 1999. *The New Politics of American Trade: Labor and the Environment.* Washington: Institute for International Economics.

DiMenta, J., and P. Doughman. 1998. "Soft Teeth in the Back of the Mouth: The NAFTA Environmental Side Agreement Implemented." *Georgetown International Environmental Law Review* 10, no. 3: 651–752.

Dua, A., and D. Esty. 1997. *Sustaining the Asia Pacific Miracle: Environmental Protection and Economic Integration.* Washington: Institute for International Economics.

Esty, D. 1993. "Integrating Trade and Environment Policy Making: First Steps in the North American Free Trade Agreement." In *Trade and Environment*, ed. D. Zaelke et al. Washington: Center for International Environmental Law and Island Press.

Esty, D. 1994a. *Greening the GATT: Trade, Environment and the Future.* Washington: Institute for International Economics.

Esty, D. 1994b. "Making Trade and Environmental Policies Work Together: Lessons from NAFTA." *Aussenwirtschaft* 49: 59–79.

Esty, D. 1996. "Revitalizing Environmental Federalism." *Michigan Law Review* 95: 570–653.

Esty, D. 1998a. "Environmentalists and Trade Policymaking." In *Constituent Interests and US Trade Policies*, ed. A. Deardorff and R. Stern. University of Michigan Press.

Esty, D. 1998b. "Non-governmental Organizations at the World Trade Organization: Cooperation, Competition or Exclusion." *Journal of International Economic Law* 1, no. 1: 123–148.

Esty, D. 2001. "Bridging the Trade Environment Divide." *Journal of Economic Perspectives* 15, no. 3: 113–130.

Esty, D. 2002. "The World Trade Organization's Legitimacy Crisis." *World Trade Review* 1, no. 1.

Esty, D., and D. Geradin. 2001. "Regulatory Co-Opetition." In *Regulatory Competition and Economic Integration*, ed. D. Esty and D. Geradin. Oxford University Press.

Friends of the Earth and Public Citizen. 1999. *NAFTA's Corporate Lawsuits*. Washington: Friends of the Earth.

Gallagher, K. 2000. Structural Reform and Trade Patterns in Mexico: Does Trade Liberalization Matter? GDAE Discussion Paper, Tufts University.

Gallagher, K. 2002. "Industrial Pollution in Mexico: Did the NAFTA Matter?" In this volume.

Garciadiego, J. 1994. *El TLC día a día. Crónica de una negociación*. México: Miguel Angel Porrúa.

Geradin, D. 1997. *Trade and the Environment: A Comparative Study of EC and US Law*. Cambridge University Press.

Gitli, E., and C. Murillo. 2000. Factors Hindering the Inclusion of Environmental Issues into Trade Negotiations: FTAA and a Positive Agenda. Costa Rica: INCA Project, CINPE.

González, A. 1999. Visión preliminar sobre el estado del debate comercio y medio ambiente en México, primera parte: La postura gubernamental. México: Centro Mexicano de Derecho Ambiental.

Grossman, G., and A. Krueger. 1993. "Environmental Impacts of a North American Free Trade Agreement." In *The Mexico-US Free Trade Agreement*, ed. P. Garber. MIT Press.

Hemispheric Social Alliance. 2001. *Alternatives for the Americas: Building a Peoples' Hemispheric Agreement.* Washington: Development Gap.

Hogenboom, B. 1998. *Mexico and the NAFTA Environment Debate: The Transnational Politics of Economic Integration*. Utrecht: International Books.

Hufbauer, G., D. Esty, D. Orejas, L. Rubio, and J. Schott. 2000. *NAFTA and the Environment: Seven Years Later*. Washington: Institute for International Economics.

Independent Review Committee. 1998. Four-Year Review of the North American Agreement on Environmental Cooperation: Report of the Independent Review Committee. Montreal: Commission for Environmental Cooperation.

Johnson, P., and A. Beaulieu. 1996. *The Environment and NAFTA. Understanding and Implementing the New Continental Law*. Island.

Konz, P. 2000. Trade, Environment and Sustainable Development: Views from Sub-Saharan Africa and Latin America. Geneva and Tokyo: ICTSD and UNU-IAS.

Lucas, N. 2000. Some Issues for Consideration on Participation. Washington: National Wildlife Federation.

Macarthur, J. 2001. *The Selling of Free Trade: NAFTA, Washington and the Subversion of American Democracy*. University of California Press.

Magraw, D. 1995. *NAFTA and the Environment: Substance and Procedure*. Chicago: American Bar Association.

Mann, H., and M. Araya. 2000. "An Investment Regime for the Americas: Challenges and Opportunities for Environmental Sustainability." In this volume.

Mann, H., and K. von Moltke. 1999. *NAFTA's Chapter 11 and the Environment: Addressing the Impacts of the Investor-State Process on the Environment*. Winnipeg: International Institute for Sustainable Development.

Matus, M., and E. Rossi. 2002. "Trade and the Environment in the FTAA: A Chilean Perspective." In this volume.

Mayer, F. 1998. *Interpreting NAFTA: The Science and Art of Political Analysis*. Columbia University Press.

McFayden, J. 1998. NAFTA Supplemental Agreements: Four Year Review. Working paper 98-4, Institute for Environmental Economics.

Moss, A. 1993. "Global Trade as a Way to Integrate Environmental Protection and Sustainable Development." *Environmental Law* 23, no. 2: 711–713.

Nordstrom, H., and S. Vaughan. 1999. *Trade and Enbvironment. Special Studies 4*. Geneva: World Trade Organization.

OECD. 1999. *Foreign Direct Investment and the Environment*. Paris: Organization for Economic Cooperation and Development.

OECD. 2000. *Assessing the Environmental Effects of Trade Liberalisation Agreements*. Paris: Organization for Economic Cooperation and Development.

Orme, W. 1996. *Understanding NAFTA: Mexico, Free Trade and the New North America*. University of Texas Press.

Pearson, C. 2000. *Economics and the Global Environment*. Cambridge University Press.

Public Citizen. 2001. *Down on the Farm: NAFTA's Seven-Years War on Farmers and Ranchers in the US, Canada and Mexico: Dwindling Incomes For Small Farmers in the US, Mexico, and Canada, Lost Farms, and Rural Crisis is NAFTA's Legacy*. Washington: Public Citizen.

Public Citizen, Sierra Club, and US Business and Industry Council Education Foundation. 1997. *The Failed Experiment, NAFTA at Three Years*. Economic Policy Institute.

Red Mexicana de Acción Frente al Libre Comercio. 2000. *El Tratado de Libre Comercio de Norteamérica (TLCAN) y el medio ambiente en México: Tendencias de integración económica y sus impactos ambientales*. Mexico City: RMALC.

Revesz, R. 1992. "Rehabilitating Interstate Competition: Rethinking the 'Race to the Bottom' Rationale for Federal Environmental Regulation." *New York University Law Review* 67: 1210–1254.

Rodrik, D. 1999. *Making Openness Work: The New Global Economy and Developing Countries*. Washington: Overseas Development Council.

Rosenberg, R., ed. 2000. *Environmentally Sound Trade Expansion in the Americas: A Hemispheric Dialogue*. North-South Center Press.

Rubin, S., and D. Alexander. 1996. *NAFTA and the Environment*. Kluwer Law International.

Rugman, A., J. Soloway, and J. Kirton. 1999. *Environmental Regulations and Corporate Strategy: A NAFTA Perspective*. Oxford University Press.

Runge, C. 1994. *Freer Trade, Protected Environment*. New York: Council on Foreign Relations.

Runge, C., E. Cap, P. Faeth, P. McGinnis, D. Papageorgiou, J. Tobey, and R. Houseman. 1997. *Sustainable Trade Expansion in Latin America and the Caribbean.* Washington: World Resources Institute.

Schott, J. 2001. *Prospects for Free Trade in the Americas.* Washington: Institute for International Economics.

Segger, M., M. Munos, P. Meirles, J. Taurel, and V. Paul. 2000. *Trade Rules and Sustainability in the Americas.* Winnipeg: International Institute for Sustainable Development.

Sierra Club. 2001. Five Environmental Reasons to Oppose the FTAA: From Swordfish to Services. Prepared for People's Summit of the Americas, Quebec City. Ottawa: Sierra Club Canada.

Simon, J. 1997. *Endangered Mexico: An Environment on the Edge.* San Francisco: Sierra Club Books.

Soloway, J. 1999. "Environmental Trade Barriers Under NAFTA: The MMT Fuel Additives Controversy." *Minnesota Journal of Global Trade* 8, no. 1: 55–95.

Spalding, M., and J. Audley. 1997. *Promising Potential for the US-Mexico Border and For the Future: An Assessment of the BECC/NADBank Institutions.* Washington: National Wildlife Federation.

UNEP, IISD, UNAM, COMEDES, and UNCTAD. 2001. Conferencia Internaciónal sobre Comercio, Medio Ambiente y Desarrollo Sostenible: Perspectivas de América Latina y el Caribe. Mexico: UNEP ROLAC.

Vaughan, S. 2000. "NAFTA and the Environment: Assessing the Impacts." Paper presented at Washington conference on "The FTAA and the Environment: What Can We Learn from the NAFTA Model?"

Weintraub, S. 2000. "Economic Impacts of NAFTA's Environmental Provisions." Paper presented at Washington conference on "The FTAA and the Environment: What Can We Learn from the NAFTA Model?"

White House. 2000. Press Statement Regarding the US-Jordan Free Trade Agreement.

Zaelke, D., R. Housman, and P. Orbuch, eds. 1993. *Trade and the Environment: Law, Economics and Policy.* Island.

I
Lessons from NAFTA Environmental Negotiations

2

A Mexican View on Trade and Environment

Javier Mancera[1]

For the past few years, Mexico has followed the debate on "trade and non-trade issues" with attention. For Mexico, trade liberalization is fundamental to long-term development and is an effective instrument for economic growth. The relationship between trade and non-trade issues, such as the environment and labor, is complex.[2] International and domestic trade and environmental communities have been grappling with this complexity for a number of years. These issues will, undoubtedly, continue to challenge us in the future. If we want to achieve a better understanding of the linkages between trade and non-trade issues, we must start with a balanced debate.

From a Mexican perspective, the debate on trade and the environment has been unbalanced on the following five counts:

• The significance of trade policy for environmental outcomes is overstated.
• Developed country concerns dominate the discussion.
• Discussion builds on an assumed premise that trade hinders the protection of environmental and labor standards.
• There exists a misconception that domestic regulations are undermined by the international trading system.
• Developing countries receive mixed messages from developed countries regarding the benefits of trade and investment liberalization.

Trade Is Not Everything, but It Is Important

Some trade advocates imply that opening markets is sufficient for economic development (O'Grady 1999). It is not. Trade is not the solution to all economic problems (Rodrik 1999). But it can make a significant contribution to economic growth and to higher living standards (Bhagwati 1993; Guigale et al. 2001). It would be nice if countries could achieve economic growth and development by merely joining the World Trade Organization (WTO) or completing the Free Trade Area of the Americas (FTAA). But we know the process is more difficult.[3]

Conversely, others perceive international trade as the source of most of the problems we face today (Mander and Goldsmith 1996; Public Citizen 1996). Clearly, it is not. It would be nice if it were—by simply closing our borders we could cause our problems to vanish.

Trade must be placed in the correct context. It is neither a panacea nor the root of all evil. Trade should be evaluated against its contribution to economic development. The overwhelming evidence across the world is that open markets create economic opportunities and foster industrial restructuring and modernization (World Bank 2001). These elements in turn increase productivity, which lead to higher standards of living.

Discussion Is Dominated by the Concerns of Developed Countries

In the international media, the concerns of developed countries dominate, while those of developing nations are covered less prominently. This yields an unbalanced debate. Over the past century, developed countries have succeeded in, and benefited from, international trade. Today, the question is not whether it makes sense for developed countries to participate in an open international trading system, but whether and under what conditions it is sensible for developing countries to do so (Rodrik 1999). In the last decade, most developing countries have significantly transformed their economies, moving away from a closed, inward-looking development strategy towards an open market system (World Bank 2001). This shift was not cost free. In the short run, it has forced developing countries to adapt and restructure their economies to face increased competition.

Mexico's experience with the North American Free Trade Agreement clearly illustrates this point. Mexico has had to restructure its industrial and agricultural sectors to successfully compete in the North American region. While this transition created costs, it also produced great benefits. Notably, from 1993 to 2000, US-Mexico trade more than doubled, growing at an average annual rate of 17 percent (NAFTA Works 2001). Mexico now is the second largest export market for US goods, and it recently displaced Japan as the second largest US trading partner overall (Canada is the largest US trading partner) (NAFTA Works 2001).

In 2000, Mexico-US trade reached US$263 billion, more than three times the 1993 pre-NAFTA levels (NAFTA Works 2001). US purchases from Mexico now exceed combined US imports from Hong Kong, Singapore, South Korea, and Taiwan. Growth in Mexico's exports to the United States similarly has far outpaced the rest of the world. Mexico's share of total US imports rose from 6.8 percent in 1993 to 11.2 percent in June 2000.

The NAFTA has also fostered a new partnership between Mexico and Canada. Though our bilateral trade historically has been small, in 2000 it reached a record high of US$11 billion, nearly triple the level of the corresponding 1993 pre-NAFTA period. Mexico has become Canada's main trading partner in Latin America and its third largest supplier worldwide, while Canada has become the second leading market for Mexican goods.

Neither those for or against the NAFTA in 1993 foresaw such impressive growth. Mexican firms have proved that their products can compete in the United States, the world's most competitive market. In Mexico, the NAFTA has made a significant contribution to economic growth and job creation. Over half of the 4 million jobs created in Mexico between August 1995 and August 2000 were related to export activities. Beyond increases in the number of jobs, there have also been improvements in terms of salary, benefits, and training.

The NAFTA has also boosted foreign direct investment in Mexico, enabling many Mexican industries to expand, modernize and reduce emissions. Between 1994 and 2001, Mexico received more than US$77.9 billion in investment in plant and equipment. The agreement accelerated the pace of economic restructuring in Mexico and opened up many business opportunities (e.g., for natural gas distribution systems, independent power producers, airports, maritime ports, telecommunications, railroads, and environmental technologies and services). Not only are capital inflows to Mexico greater; they have brought technological advancements to Mexico, including state-of-the art environmental technologies and practices (Gallagher 2002).

Since the NAFTA, many regions of Mexico that had little hope of participating in the modern economy are becoming players in the arena of international trade. The most significant untold story of the NAFTA is that since 1994 most new exporting maquiladoras have been located not near the US border but in such places as Yucatan, Tlaxcala, and Puebla (Christman 2000). These plants provide tens of thousands of new jobs to workers in their own communities far from the US-Mexico border. This shift in industrial activity has eased environmental pressure along the border and positively affected business opportunities across Mexico.

But the NAFTA is only part of the story of Mexico's recent success in international markets. In addition to the NAFTA, Mexico has negotiated seven other free trade agreements (FTAs) with countries in Central and South America: one with Chile (1992), one with Bolivia (1995), one with Costa Rica (1995), one with Colombia and Venezuela (1995), one with Nicaragua (1998), one with Guatemala, Honduras, and El Salvador, and one with Uruguay. These FTAs are also showing strong results. Bilateral trade with Chile increased 571 percent since the agreement went into force. In 2001, Mexico was Chile's seventh largest supplier, up from sixteenth in 1992. Since 1995, Mexican trade has grown 18 percent with Bolivia, 35 percent with Colombia, 225 percent with Costa Rica, and 57 percent with Venezuela. Aside from participating in the negotiations for an FTAA, Mexico is also working to develop stronger relationships with Mercosur.[4] Elsewhere, Mexico has negotiated free trade agreements with Israel and the European Union, both of which entered into force in 2000.[5]

This network of trade agreements establishes Mexico as the only country in the world with preferential access to the largest and most dynamic markets in the world—North America and the European Union—as well as to Israel and its Latin America neighbors. The key point here is that Mexico's trade opening and its network of free trade and investment agreements have made an important contribution to its long-term economic development (Guigale et al 2001; Heath 1998). Ten years ago, few could imagine that Mexico would become one of the leading trading nations in the world[6] and the top trading economy in Latin America, with 47 percent of the region's exports and 45 percent of its imports. In 1999, Mexico's foreign trade totaled US$279 billion, up 185 percent since 1993. Mexico exported a total of US$154 billion to its NAFTA partners in 2000, 238 percent more than in 1993 (NAFTA Works 2001).

The ability of developing countries to increase volumes of international trade is in large measure a function of their incorporation into the international trading system. Whether the increasing number of developing countries that wish to be part of the global trade regime are successfully added will be the measure of our collective economic policy success over the next twenty years.

Starting with the Wrong Premise

The current debate on international trade and the environment is misleading. The premise of the discussion is that international trade and investment are damaging to the environment. Even when critics concede that trade and investment promote economic growth, they often argue that economic growth is a threat to environmental protection (Daly 1993). The empirical record not only fails to support this argument, it actually demonstrates the opposite (Runge 1994). Increased trade and investment between developed and less developed countries can actually strengthen environmental performance (OECD 1999; Esty and Gentry 1997; Frederiksson 1999).

Liberalized international investment has, for example, expanded opportunities for environmental investments in Mexico (Dominquez-Villalobos 2000; Gentry and Fernandez 1998). New plants that are being built in Mexico often use the same state-of-the art technology used in Japan, in the United States, or in Europe. Not only do these plants deliver the same quality products; they also have the same high environmental standards. In the case of the NAFTA, increased US and Canadian trade with, and investment in, Mexico is serving to advance environmental practices in Mexico (GAO 2001; Gallaghar 2002; Hufbauer et al. 2000).

The Mexican water sector, for example, attracts one of the highest rates of US investment—particularly in wastewater treatment projects (Frost and Sullivan 1994). In 1998, Mexico's Ministry of Environment and Natural Resources (SEMARNAT) estimated that the wastewater and potable water treatment in Mexico had a market value of US$1.2 billion.[7] The total value of investments in wastewater is even larger if some of the municipal projects being financially assisted by the North American Development Bank (NADBank) and the Border Environmental Cooperation Commission (BECC) are included.[8]

Through the NADBank and the BECC, the NAFTA's environmental institutions are also providing some economic benefits to Mexico. By the end of the year 2000, the BECC and NADBANK had, working together, assisted 36 projects, including two wastewater treatment plants. In total, the BECC has identified 133 current or future projects—not all of them in the water sector—that represent around US$900 million in investment.[9]

Unjustified Fears about the Impacts of International Trade Agreements on Domestic Regulation

There is also a common misconception that international trade agreements can overturn domestic regulations geared to enhance environmental protection or worker rights. This is not so. Both WTO rules and Mexico's FTAs permit measures to protect the environment as long as they are based on scientific evidence and are not discriminatory in their application to imports. In the case of the NAFTA, if a dispute panel finds that a domestic regulation is inconsistent with the NAFTA rules, it does not mean that the country must change the regulation. The principle is this: If the domestic regulation in question unjustifiably diminishes access for other countries, some compensation must be provided.

Widespread misunderstanding of this principle, particularly in developed countries, leaves developing countries vulnerable to developed countries' actions designed largely to respond to misplaced fears among their environmental communities. Developing countries worry that the access obtained in trade negotiations might later be reversed on environmental or labor grounds, depriving them of the benefits for which they bargained. This fear creates an awkward basis for negotiations, since there is no guarantee that the market access a developing country negotiates in exchange for a reciprocal opening will be maintained.

Negotiating Anxiety: Do As We Say, Not As We Do

Finally, the debate on trade and non-trade issues is unbalanced because when developing countries come to the negotiating table at WTO, FTAA, or other international trade forums they now find a puzzling increased hesitation and anxiety on the part of developed countries about the desirability and benefits of freer trade. One must remember that for decades developing countries were urged to join the international trading system based on the evident success of more open trade for developed nations and their intellectual drive. Currently, renewed demands from many developed countries for "managed trade" seem to echo development theories of the

1950s and the 1960s which proclaimed that developing countries should employ protectionist trade policies to build domestic industrial bases strong enough to face international competition. At that time and in the following decades, developed countries argued in favor of free trade and against these theories. Now that many developing countries have adopted market-oriented development strategies, some groups in developed countries seem to be resorting to earlier protectionist arguments to defend their markets from developing country competition.

In particular, demands for greater harmonization of environmental, labor, production, and labeling standards are recreating a North-South divide (Bhagwati and Audec 1996; Esty and Geradin 1998). Influential groups within developed countries (including NGOs and businesses) argue that products from developing countries should not be granted market access until they have achieved developed country industrial, environmental, and labor standards (Milner 1998; Mayer 1998). Such arguments violate the core economic principle of comparative advantage and destroy the potential for welfare gains from trade. Whether based on concerns about a sufficiently strong industrial base, on sufficiently effective environmental protections, or on sufficiently progressive labor laws, demands for harmonization undermine the trade liberalization process.

Developing countries are also puzzled that segments of civil society in developed nations are apparently opposed to trade liberalization, since the opening of markets has been a key element for fostering civil society in developing nations. The history of the twentieth century suggests that society's participation in social and economic policy was hindered in nations with protectionist or closed economies. In the nineteenth century, the liberal theorist John Stuart Mill argued that a free market of goods corresponds to a free market of ideas. This liberal principle is as true today as it was then. A free and open society with a vigorous and vocal citizenry—what is today called "civil society"—needs an open trading system. It requires the recognition of individual rights, be they of speech or property. The flow of goods from region to region and from country to country inevitably facilitates a flow of ideas.

The flourishing of an organized civil society depends on citizens and groups being independent from political and economic grants from government. Thus, a thriving market economy is one of the best means to support it. Open markets have created and will continue to create opportunities for the citizens of developing nations; these opportunities increasingly provide them with the means to rise above narrow self-interest and participate in institution building through an emerging and strengthening civil society. It is not by chance that market liberalization and Mexico's democratization happened concurrently. The cultural changes behind one stimulate the other.

Growth and development in the twenty-first century will only be achieved through increasingly participatory politics and open markets. The latter creates wealth, the former distributes it. A balanced debate on international trade between developed and developing nations is required for this to happen. Special-interest environmental groups cannot be allowed to dictate the terms of the dialogue.

Notes

1. The views expressed in this chapter are those of the author and do not necessarily reflect those of the Mexican government.

2. On the empirical challenges and the interplay of trade policy and environmental policy, see Bhagwati 1993, Fredriksson 1999, Esty 1994, and Grossman and Krueger 1993.

3. Developing countries face challenges both in terms of the implementation of international trade agreements and with ensuring that trade policies are complemented by appropriate domestic policy choices.

4. Mercosur (Mercado Comun del Sur, Southern Common Market) is a regional trade agreement among Argentina, Brazil, Paraguay, and Uruguay. The Mercosur countries have declared the goal of becoming a customs union with tariff-free internal trade and a common external tariff for all of South America by 2006. For more on Mercosur, see http://www.mercosur.org.

5. For the text of these two agreements, see http://www.economia.gob.mx.

6. According to the 1998 WTO Annual Report, Mexico is ranked 13th among the leading exporting and importing countries in the world, and 8th and 7th, respectively, when considering the European Union as one country (WTO 1998).

7. SEMARNAT was formerly known as SEMARNAP. For further information, see http://www.semarnat.gob.mx/.

8. These two institutions were created by a 1992 bilateral US-Mexico environmental agreement negotiated as part of the overall NAFTA process. For complete descriptions of these institutions, see http://www.becc.org and http://www.nadbank.org. These sites also contain extensive bibliographies.

9. Updated information about BECC and NADBank activities is available on their respective web sites.

References

Ayres, J. 1998. *Defying Conventional Wisdom: Political Movements and Popular Contention against North American Free Trade.* University of Toronto Press.

Barkin, D. 1999. "Free Trade and Environmental Policymaking in Mexico." *Borderlines* 7, no. 9: 14–15.

Bhagwati, J. 1993. "The Case for Free Trade." *Scientific American* 269, no. 5: 41–49.

Bhagwati, J., and R. Hudec. 1996. *Fair Trade and Harmonization: Prerequisites for Free Trade? Economic Analysis.* MIT Press.

Christman, J. 2000. "The Maquila Industry Outlook: 2000–2005." Presented XXXIX Maquiladora Industry Meeting, Mexicali.

Commission for Environmental Cooperation. 1996. NAFTA Effects: Claims and Arguments 1991–1994. Montreal: CEC.

Daly, H. 1993. "The Perils of Free Trade." *Scientific American* 269, no. 5: 50–55.

Delal Baer, M., and S. Weintraub. 1994. *The NAFTA Debate: Grappling with Unconventional Trade Issues.* Lynne Rienner.

Destler, I., and P. Balint. 2000. *The New Politics of American Trade: Trade, Labor, and the Environment.* Washington: Institute for International Economics.

Dominquez-Villalobos, L. 2000. "Environmental Performance in the Mexican Chemical Fibres Industry in the Context of an Open Market." In *Industry and the Environment in Latin America*, ed. R. Jenkins. Routledge.

Esty, D. 1994. *Greening the GATT.* Washington: Institute for International Economics.

Esty, D., and B. Gentry. 1997. "Foreign Investment, Globalization, and Environment." In *Globalization and the Environment,* ed. T. Jones. OECD.

Esty, D., and D. Geradin. 1998. "Environmental Protection and International Competitiveness: A Conceptual Framework." *Journal of World Trade* 32, no. 3: 5–46

Frederiksson, P. 1999. *Trade, Global Policy and the Environment.* Discussion Paper 402, World Bank, Washington.

Gallagher, K. 2002. "Industrial Pollution in Mexico: Did the NAFTA Matter?" In this volume.

GAO (General Accounting Office). 2001. United States General Accounting Office Reports to the Chairman, Subcommittee on Trade, Committee on Ways and Means, House of Representatives. US Experience with Environment, Labor and Investment Disputes. GAO-01-933.

Gentry, B., and L. Fernandez. 1998. "Mexican Steel." In *Private Capital Flows and the Environment*, ed. B. Gentry. Elgar.

Grossman, G., and A. Krueger. 1993. "Environmental Impacts of a North American Free Trade Agreement." In *The Mexico-US Free Trade Agreement*, ed. P. Garber. MIT Press.

Guigale, M., M. Lafourcade, and V. Nguyen. 2001. *A Comprehensive Development Agenda for the New Era*. Washington: World Bank.

Heath, J. 1998. "The Impact of Mexico's Trade Liberalization." In *The Post-NAFTA Political Economy*, ed. C. Wise. Penn State Press.

Hufbauer, G., and J. Schott. 1993. *NAFTA: An Assessment*. Washington: Institute for International Economics.

Hufbauer, G., D. Esty, D. Orejas, L. Rubio, and J. Schott. 2000. *NAFTA and the Environment: Seven Years Later*. Washington: Institute for International Economics.

Johnson, P., and A. Beaulieu. 1996. *The Environment and NAFTA*. Island.

Low, P., ed. 1992. International Trade and Environment. Discussion paper 159, World Bank, Washington.

Mander, J., and E. Goldsmith. 1996. *The Case Against the Global Economy and For a Turn Toward the Local*. San Francisco: Sierra Club Books.

Mayer, F. 1998. *Interpreting NAFTA: The Science and Art of Political Analysis*. Columbia University Press.

Milner, H. 1988. *Resisting Protectionism: Global Industries and the Politics of International Trade*. Princeton University Press.

NAFTA Works. 2001. "7 Years of US-Mexico Trade Under NAFTA." *NAFTA Works* 6, no. 3: 1–3.

OECD. 1999. *Trade Investment and Development: Policy Coherence Matters*, Part I. Paris: OECD.

O'Grady, M. 1999. "First, Open Markets." In *1999 Index of Economic Freedom.*, ed. B. Johnson, K. Holmes, and M. Kirkpatrick. Washington: Heritage Foundation and Dow Jones.

Public Citizen. 1996. *NAFTA's Broken Promises: The Border Betrayed*. Washington: Public Citizen.

Rodrik, D. 1999. *The New Global Economy and Developing Countries: Making Openness Work*. Washington: Overseas Development Council.

Runge, C. 1994. *Freer Trade, Protected Environment*. New York: Council on Foreign Relations.

Weintraub, S. 1997. *NAFTA at Three*. Washington: Center for Strategic and International Studies.

World Bank. 2001. *World Development Report: Attacking Poverty*. Washington: World Bank.

WTO. 1998. *Annual Report*. Geneva: World Trade Organization.

3

No Room for the Environment: The NAFTA Negotiations and the Mexican Perspective on Trade and the Environment

Gustavo Alanis-Ortega and Ana Karina González-Lutzenkirchen

In Mexico, understanding of trade and environment issues varies considerably. While some studies (primarily academic ones) have examined the environmental effects of North American Free Trade Agreement, there has been little serious analysis in Mexico of the intersection of trade and the environment and its implications for that country.[1]

The Mexican government advances a distinct position in international discussions. The position of the Mexican Ministry of Economy (formerly known as the Ministry of Trade and Industrial Promotion), expressed at both national and international forums, is that there is no convincing rationale for linking trade issues and environmental issues in Mexican public policy, and that environmental considerations are not relevant subject matter for international trade negotiations. In Latin America, Mexico is clearly leading the pack in opposing the incorporation of environmental considerations into negotiations for the Free Trade Area of the Americas (FTAA). There is no question that Mexico's attitude toward environment issues in the FTAA is informed by its NAFTA experience. To understand Mexico's current position and, more importantly, to devise future strategies, one must analyze the state of the debate on trade and the environment within Mexico, the negotiating dynamics of the NAFTA, Mexico's expectations of the NAFTA, and Mexico's perception of the NAFTA's final content and impacts.

This chapter explores the history of the debate on trade and the environment in Mexico to determine why the Ministry of Economy adopted its original position on trade and environment issues, and why it continues to maintain that position. In particular, we suggest that the Mexican government's official view on trade and the environment reflects an attitude developed during the bitter final stage of the NAFTA negotiations—and not a coherent policy position.[2]

The Debate on Trade and the Environment in Mexico

The following facts reveal how little salience trade and environment issues currently have for public or private agendas in Mexico:

• The Mexican President's Office has never explicitly addressed the question of trade and the environment.[3] The President's Office only infrequently addresses matters related to the environment; if they are addressed at all within the Federal Executive Branch, it is by either the Cabinet or the Social Development Coordination Mechanism (Roemer 1999).[4] The failure of the President's Office indicates the low priority assigned to it within Mexico's national political agenda.

• Many in the Mexican Senate perceive environmental issues as "subjects of ladies and eccentrics" (Aguilar 1999). A recommendation that the Senate assess the environmental effects of the NAFTA five years after it came into force immediately ran into problems.[5] Mexico's two largest industrial organizations, the Cámara Nacional de la Industria de la Transformación (CANACINTRA) and the Confederación de Cámaras Industriales de los Estados Unidos Mexicano (CONCAMIN), focus on fundamental problems in the Mexican economy (such as domestic over-regulation, a fluctuating regulatory environment, inadequate support for small and medium-size exporters, and lack of access to credit).[6] In this context, trade and environment issues are too far removed from their immediate interests and concerns to warrant attention, time, and resources (Tornel 1999).

• Some business groups do recognize that the debate on trade and the environment deserves attention. The Centro de Estudios del Sector Privado para el Desarrollo Sustentable (CESPEDES), for example, notes the potential for developed countries to apply trade barriers for environmental reasons.[7] According to a former leader of CESPEDES's activities, while there is some interest in the debate, it is unlikely that the government will become engaged in the discussion (Villanueva 1999). Indeed, given the government's position that there is no scope or need to link trade with environmental issues, it is difficult to imagine how discussion about "non-existent" problems would be handled.

Despite the absence of significant public debate in Mexico on trade and environment issues both during and after the NAFTA debate, the Mexican government—through its Ministry of Economy—has taken very firm positions on trade and the environment in international forums. The Mexican Economic Ministry determines Mexico's stance on trade and the environment; there is no formal inter-agency process for formulating, sharing, and taking into account the views of the range of relevant government agencies in determining Mexico's policy in this area. For Mexican trade officials, trade expansion is the highest-priority trade issue. The Mexican Ministry of Economy avoids discussing any subject that might interfere with trade or even deflect attention from trade liberalization. It perceives any effort to address trade-environment linkage as a dangerous precedent.

In the view of Mexico's Economic Ministry, any discussion of environmental matters in trade negotiations could result in protectionist barriers to trade. Therefore, their position that trade matters should not be linked to environmental matters should come as no surprise. In a speech to the World Trade Organization (WTO) High Level Symposium on Trade and the Environment in March 1999, Luis de la Calle, Mexico's Under Secretary for International Trade Negotiations, provided a clear statement of the Mexican government's perspective: ". . . it is a grave mistake to mix up environmental problems with international trade rules and market access concessions. Trade measures are not the best way to address environmental concerns. Moreover, they can be easily abused by protectionist forces. Environmental problems require environmental solutions, not trade measures or restrictions." (de la Calle 1999)

As was the case in the negotiations for the NAFTA and its side agreement, the Ministry of Economy's position remains Mexico's official position at international forums such as the World Trade Organization, the Organization for Economic Cooperation and Development (OECD), ongoing discussions of the appropriate work plan of the NAFTA's Commission for Environmental Cooperation (NACEC), and the negotiations for the creation of a Free Trade Area of the Americas.[8] During the selection process for a new WTO Director-General, former Mexican Trade Minister Herminio Blanco Mendoza, said in a radio interview that Mexico would support a candidate who focused on trade negotiations within the framework of a possible "Millennium Round" without being distracted by subjects that would obstruct or delay free trade agreements, "such as labor and environmental issues" (Blanco 1999).[9]

The Mexican Economic Ministry's strategy of denying the existence of a trade-environment link has several implications. It means that the Ministry refuses to discuss related subjects such as eco-labeling and the precautionary principle openly, refuses to recognize the implications of environment-trade connections for national development and environmental health, and has no goals or policies designed to improve its understanding of the implications of the debate on trade and the environment, and lacks studies analyzing the likelihood and potential impacts of threats of protectionism on various Mexican economic sectors.

Within Mexico, the Economic Ministry exerts extensive influence on environmental policy decision making, regulations and practices that could influence economic or trade activity. At the international level, Mexico's Economic Ministry actively negotiates for Mexico in multilateral environmental forums where trade questions arise. In April 1997, a Multilateral Environmental Affairs Section was created within the Economic Ministry alongside the Multilateral Trade Organizations Section. Representatives of the Multilateral Environmental Affairs Section attended, for example, the negotiation of multilateral environmental agreements, meetings of the WTO Committee on Trade and Environment, and meetings within the framework of Organization for Economic Cooperation and Development. However, their official position on trade and the environment was the same as the one promoted by the rest of Mexico's Ministry of Economy. The Multilateral Environmental Affairs Section was recently closed following a departmental reorganization.

Despite the Economic Ministry's dominant role in all trade-related matters, debate within Mexican government on trade and the environment has recently begun. The Ministry of Foreign Relations (SRE) and the Mexican Ministry of the Environment and Natural Resources (SEMARNAT) (formerly the Ministry of Environment, Natural Resources and Fisheries—SEMARNAP) have both shown some interest in the debate.[10] Through growing efforts at research and policy analysis, there is some hope among environmental groups that SEMARNAT will develop a stronger role on the trade and environment front, balancing the traditional leadership of the Economic Ministry. SEMARNAT distinguishes itself from the Ministry of Economy with its recognition of links between trade and the environment and its interest in further discussion of them.[11] Indeed, SEMARNAT has proposed exploring the variables that might explain how trade and the environment affect each other. In 1999, several newspapers published statements from the former Minister of Environment, Julia Carabias Lilló, in which she lamented that "the slow progress in the understanding that free trade and the environment must be coupled together, has caused delays and a lack of sufficient documents that could allow Mexico to be prepared to face embargoes–such as the tuna embargo" (Carabias 1999a).

When Carabias returned from a meeting of the North American Commission for Environmental Cooperation's Council of Ministers in June 1999, she said: "Now, the challenge is to begin to formally work on the trade and environment issues." (Carabias 1999b)

The NAFTA and the Origins of Mexico's Position on the Environment

To understand Mexico's position on the environment in the NAFTA debate, we need to appreciate the expectations that Mexican government and pro-free trade advocates had of the NAFTA. The completion of the NAFTA negotiations was an urgent priority for the Mexican government as the NAFTA formed a central plank in the Salinas government's economic development strategy (Mayer 1998). Both the Mexican government and Mexican pro free-trade advocates spent considerable resources to convince the Mexican public and the US Congress of the economic benefits of the NAFTA (ibid.).

Mexico's Political Commitment to the NAFTA

The Mexican government's approach to the NAFTA was influenced by the particular nature of Salinas's government—especially its focus on modernizing Mexico's economy, its close relationship with the country's business leaders, and its emphasis on realigning Mexico's institutional frameworks. The Mexican government's movement toward free trade began slowly under the de la Madrid administration in the mid 1980s. In 1982, a new Ministry of Trade and Industrial Promotion (SECOFI) was created (replacing the Ministry of Industry and Trade) and it proceeded to recruit free trade proponents (Pastor and Wise 1994, p. 478). The Mexican Ministry of Finance, the trade and development banks, and other financial institutions also emerged as active supporters of a more open trade policy and Mexico's participation in the NAFTA.

The Salinas administration intensified the emphasis on economic management and trade expansion as the government's highest priorities (Hogenboom 1998). With this focus came an increasingly important role for economic "technocrats." In Mexico, the driving force behind the NAFTA was not simply the government as a whole, but rather the small corps of US-trained economic advisors who dominated former president Salinas's Cabinet from 1988 to 1994 (ibid.). The Salinas technocrats forged better relationships with large private-sector interests than their predecessors. Over time, the business elite worked closely with the government economic team to advance their common interest in trade liberalization policy. By government request, business leaders organized into a private-sector coordinating body, the Coordinating Organization of Business Agencies of Foreign Trade (COECE).[12] COECE was responsible for communications between the private sector and the Mexican negotiators of the NAFTA. During the NAFTA negotiations, COECE supported the Ministry of Economy by providing it with information and drafting technical proposals; the two institutions also cooperated in a Mexican working group that commissioned assessments of the NAFTA's impact on various economic sectors. Through COECE, large private-sector interests were also able to exert influence on the scope of negotiations through the "cuarto de junto" (side-room negotiations), in which top business leaders worked out strategies and positions with top Salinas administration officials. This influence was, however, restricted to a small set of large companies; most Mexican companies, especially small businesses, lacked access to government negotiators. The Mexican government did convene an Advisory Council on the Free Trade Agreement that was charged with channeling participation by representatives from labor, agriculture, academic, and business. However, the Council failed to play a significant role in the NAFTA negotiations (Hogenboom 1998, p. 172).

Promotion of the NAFTA Project by Mexico in the United States

In response to considerable resistance from some groups within the United States to a free trade agreement with Mexico, the Mexican government joined with its business elite and began lobbying in Washington.

In January 1991, the Mexican Embassy in the United States established a NAFTA office to act as a direct liaison between the United States and Mexico on NAFTA issues. During negotiations, the NAFTA office researched the possible effects of the treaty and provided negotiators with specific, detailed information. They also tackled criticisms against the NAFTA in the United States. In total, from 1991 to 1993, the Mexican government spent over US$30 million on NAFTA-related activities; this included lobbying contracts for well-connected and former highly placed North American governmental officials (Mayer 1998, p. 236).

Through COECE, Mexico's private sector also started its own campaign in the United States. Among other initiatives, it hired a well-connected trade expert to lobby for the NAFTA in the US Congress, and arranged for US congressional staff to visit Mexico (Hogenboom 1998, pp. 173–174). As the Mexican pro-trade lobby strengthened, support for the NAFTA took on an increasingly trans-national character. Mexico's public and private sector worked with their counterparts in the United States. The governments met frequently throughout the preparations and the negotiations of the treaty, and afterwards during the negotiation of parallel agreements. In addition, the private sectors in Mexico and the United States worked together to build support for the treaty (Mayer 1998: 38–39).

Mexican Economic Interests in the NAFTA Project
According to free trade advocates, the NAFTA and its removal of trade barriers would benefit all three countries by: increasing the volume of trade; promoting competition and modernization; generating greater economies of scale for production; and freeing the flow of investment (Weintraub 1993; Dornbusch 1992; Aspe 1993; Edwards 1993; Hufbauer and Schott 1992).

From the outset, the Mexican government's case for the NAFTA focused heavily on the objective of increasing short and medium-term flows of foreign capital (Mayer 1998; Trigueros 1994). For Mexico, investment liberalization was expected to generate an even larger positive economic impact than increases in trade volumes. (While trade between Mexico and the United States was already significant prior to the NAFTA, the rate of investment growth remained below the government's targets.) To convince Mexico's various economic sectors of the NAFTA's importance, the Mexican government focused on the benefits—more jobs and higher salaries—that it hoped would result from investment liberalization. It sold the NAFTA to the Mexican private sector by arguing that foreign investment would promote economic expansion by supplementing, not replacing, the initiatives of Mexico's domestic interests. In turn, many Mexican business leaders perceived the NAFTA as a useful tool for permanently locking in the new model of pro-market economic policy in Mexico and guaranteeing permanent access to the US market (particularly in light of continued US efforts to protect against Mexican manufacturers in the 1980s).

The actual economic forecasts of the NAFTA's impact on Mexico, however, were mixed (Lustig 1994). The projections for most sectors of the

Mexican economy focused on the trade aspects of the NAFTA framework. Estimates suggested that the Mexican exporting sector would be strengthened. It was unclear, however, how other sectors would be affected by an expanding market and the possibility of access to cheaper loans, but also increased competition (Hogenboom 1998). For instance, while large manufacturing companies focused on external markets stood to gain from the NAFTA, textile companies that produced for internal markets were expected to lose (Lustig 1994).

In public, the Mexican government argued that the NAFTA would generate economic growth, development and more and better jobs in Mexico (Heath 1998). The official line was that Mexico could expect to gain jobs as US and Canadian companies transferred production to Mexico; over time, increases in Mexican exports and investment in Mexico would create even more employment and higher wages in the export industries. However, the potential effects of the NAFTA on employment in Mexico were not thoroughly investigated or discussed before or during the NAFTA negotiations (Heath 1998; Hoogenboom 1998).

A number of outstanding questions remained unanswered: What would be the impact of increased competition on employment, inequality and poverty? Would companies that modernized production be forced to lay off employees? How many small Mexican companies, unable to compete in the NAFTA's new economic climate, would disappear? From an environmental point of view, how would these economic changes affect Mexico's efforts to control pollution and manage its natural resources? It is not clear whether answers to such questions were available to the Mexican government during the NAFTA negotiations (Hoogenboom 1998).[13] It is not clear whether some of these questions were even asked.

The Position of the Promoters of the Treaty regarding Environmental Questions

Shortly after Mexico and the United States signed a declaration in June 1990 launching free trade negotiations, the link between free trade and environmental protection became an issue of trans-national debate. Environmentalists, particularly in the United States, pointed out that the economic, political, and policy changes brought about by the NAFTA might have damaging environmental side effects (Audley 1993, 1997; CEC 1996; Mayer 1998). US non-governmental organizations focused on US-Mexico environmental issues, particularly around the US-Mexico border: industrialization along the border was expected to exacerbate existing problems of pollution and inadequate waste management, water supply and treatment, and transportation (Audley 1993).[14] The spotlight was also placed on Mexico's lax environmental policy and inadequate enforcement of existing environmental legislation (Audley 1993, 1997).

As concern about the potential environmental impacts of the NAFTA gained increased media and political attention, governments in both the United States and Mexico reacted (Mayer 1998). Mexico's free trade advocates articulated very general responses (González 1999a,b). They argued that trade increases economic growth and economic growth increases environmental protection. Inadequate environmental protection, in their view, was a problem of a lack of resources which economic growth would resolve.[15] In the United States, the Mexican government devised a proactive approach of defending the NAFTA and Mexico against criticism. Through a series of image-building speeches, press releases, interviews and English-language publications, they focused on highlighting recent improvements in Mexican environmental policy and enforcement and on countering Mexico's poor image in the United States.

In Mexico, the Mexican government worked to downplay the importance of the issue. Given little public awareness of these issues and a relatively weak environmental NGO movement, the Mexican government's efforts to counter environmental arguments against the NAFTA were less extensive in Mexico than its efforts in the United States. The fact that pro-NAFTA propaganda was pervasive and solid in Mexico, and that criticism by Mexican environmentalists was not always publicly disclosed also help to explain the absence of a nation-wide debate on the NAFTA's environmental impacts in Mexico. However, a number of Mexican environmental NGOs were active around the NAFTA negotiations. In general, Mexican NGOs criticized the free trade initiative and the government for largely ignoring sustainable development and environmental protection. They questioned the compatibility of free trade and sustainability as presented by the NAFTA's promoters, referring to earlier experiences with the maquiladoras. Mexican groups or networks that actively opposed the official proposal for the NAFTA included the Mexican Action Network on Free Trade (RMALC), the Union of Environmental Groups (UGAM), the Grupo de los Cien, and the Mexican Ecologist Movement (MEM).[16]

In the course of the NAFTA debates, the Mexican government (depending on the speaker, circumstance, and the audience) represented the relationship between trade and the environment as either a *positive* causal relationship or as a more complex interaction. The Mexican government made some limited statements to the US media and released data attempting to show that expanded trade would reinforce Mexican environmental efforts. Interestingly, the domestically released data was substantially less informative than the data used in the United States. In an address to the Mexican Senate, for instance, former Mexican Trade Minister Jaime Serra Puche simply asserted that the greater the economic development, the better the environmental protection (Hogenboom 1998, p. 170). Luis Donaldo Colosio, then Mexico's Minister of Social Development (who was responsible for environmental policy before SEMARNAP was created in 1995), stated that "free trade and a better environment go hand in hand" and that "a North American free trade zone would increase Mexican sensitivities to the environment" (Hogenboom 1998, p. 179).[17] President Salinas de Gortari was more cautious, stating that free trade represented a chance to improve environmental protection that should be accompanied by other suitable policies (Hogenboom 1998).

In the United States, the Bush administration defended the NAFTA (and Mexico's participation) to the US Congress, and indirectly to the US public, against environmental challenges with a similar argument: increased resources and access to technology resulting from economic growth would benefit environmental protection. Bush (1992) emphasized that, "the major remaining obstacle to improved environmental quality in Mexico was not standards or enforcement but the availability of economic resources."

Business organizations largely refrained from public activity on the environmental issues as long as the Bush administration, assisted by the Mexican government, promoted the parallel-track approach. The World Bank also maintained distance from the public debate. However, during the later stages of the NAFTA debate, private companies and the World Bank reacted to the environmental criticism, and each developed lobbies with close relationships to the government efforts. The World Bank argued, for example, that non-ratification of the NAFTA could reduce the attractiveness of investment in Mexico. At the same time, the World Bank announced an unprecedented environmental loan package for Mexico (Hogenboom 1998). Through documents and speeches, pro-NAFTA business interests

nstitute of Latin American St.

31 Tavistock Square

London WC1H 9HA.

reacted actively to criticism regarding possible environmental damage stemming from trade and investment liberalization (Mayer 1998).

In short, it is clear that, from the outset, the focus of Mexican trade officials on the rapid completion of the NAFTA negotiations created a context
in which environmental considerations were bound to be perceived as an
unwelcome hindrance. Mexico's response to the questions and criticisms of
US, Mexican, and Canadian NGOs was to improve its environmental image
and downplay potential negative environmental implications, rather than
to acknowledge the importance of the linkage or proactively formulate a set
of policy responses. Curiously, the drumbeat of questions and criticisms from
US (and Canadian) environmentalists, journalists, and politicians never
served to firmly establish trade and environment issues as deserving government attention. Instead, Mexican officials saw the issue as an irritant, a
distraction and a non-issue flagged by NGOs as a way to advance their own
narrow agendas, and used by protectionists who sought to scuttle the
NAFTA. Mexican trade officials feared that linking the environment issue
to trade would expose Mexico to the risk of loss of trade opportunities if
the country failed to meet US environmental expectations and standards.

Mexico and the NAFTA's Environmental Side Agreement

Enter the Clinton Administration
While the Mexican, Canadian, and US governments improved environmental cooperation efforts during the period of the NAFTA negotiations,
differences of opinion remained. These differences arose most starkly in the
last stage of the NAFTA negotiations. During the 1992 US presidential campaign, candidate Bill Clinton announced that he would support the NAFTA
only if the partners to the agreement negotiated supplemental protection
for the environment and for workers in the form of a parallel agreement on
the environment. When Clinton was elected president, he fulfilled his campaign promise by demanding broader assurances of improvements in environmental policy and enforcement in Mexico than the Bush administration
had been willing to seek. Mexico's efforts to deflect environmental concerns
could not dissuade the Clinton administration, which continued to press
for a side agreement "with teeth." The character, substance, and context of
the ensuing 1993 supplemental negotiations proved harmful enough to
Mexican sensibilities to have resulted in steadfast Mexican government

opposition to environmental considerations linked to trade negotiations that persists to the present.

Negotiations for both the environmental and labor supplemental agreements were problematic for six reasons.

First, since the US Congress was not certain to approve the NAFTA, the negotiators were under enormous pressure to develop a package that could win US Congressional approval. Every issue was viewed through the lens of its US congressional impact. US Trade Representative Mickey Kantor never tired of reminding the Mexican negotiators that the only real issue was what the US Congress wanted.

Second, the negotiations were driven entirely by domestic US political considerations. Mexican views and needs were essentially ignored. The negotiating teams from Mexico and Canada had little interest in establishing parallel agreements on either labor or environmental issues. From the outset, both the Mexican and Canadian governments were reluctant to negotiate parallel agreements, and adamantly opposed renegotiating the NAFTA. Indeed, the call for a side agreement ultimately split the previous NAFTA alliance among the government agencies and the Presidents of Mexico and the United States. Insofar as it wanted the NAFTA to win approval in the US Congress, Mexico faced overwhelming pressure to mitigate the US public's concern about economic integration with Mexico.

Third, Mexico was under political and economic pressure to complete the NAFTA negotiations as swiftly as possible because the NAFTA was so central to its economic strategy. The Salinas administration saw defeat of the NAFTA as a disaster to be avoided at all costs. The US negotiators knew that or the Salinas team the NAFTA was more than a trade agreement; it represented the centerpiece of the Mexican government's strategy for economic modernization.

Fourth, improvements in cooperation between Mexico and the United States, developed during the period 1989–1993, were not enough to diminish Mexico's lack of trust in US intentions. Mexican negotiators saw little mutual gain in the Clinton administration's proposals, and took a cautious attitude toward the NAFTA side agreement.[18]

Fifth, there were no international precedents for parallel environmental accords linked to international trade agreements.

Sixth, the United States' supplemental environmental demands (which tended to reflect the concerns of a small set of US environmental NGOs) differed significantly from the Mexican and Canadian negotiating positions on environment issues. The US negotiating positions included preferences for an autonomous secretariat for the proposed Commission for Environmental Cooperation, for the opportunity for citizens to submit petitions to the Commission, for granting power to the commission to investigate such petitions, for the creation of a regional public advisory committee, for transparency, and for citizens' right to information. Mexico considered the US focus on dispute settlement and on Mexico's non-enforcement of environmental legislation prejudiced and motivated by US protectionist tendencies. Canada and Mexico shared concern about proposals that they considered to pose potential threats to national sovereignty. Mexico opposed the formation of a supranational environmental commission as well as proposals that would lend it considerable independence and the authority to investigate private complaints (Hogenboom 1998). They also opposed the US focus on punishment of environmental offenders through trade sanctions and preferred a supplemental agreement focused on stimulating cooperation. Canada joined Mexico in its opposition to trade sanctions; indeed, for the Mexican government, sanctions were the most sensitive issue in negotiating the parallel agreements. Mexico feared the US would abuse trade sanctions by masking economic protection measures under the guise of environmental protection (as it believed had happened in the case of the tuna embargo).[19] There were also concerns about the imposition of environmental taxation and the obligatory harmonization of environmental standards. Mexican negotiators worked to limit the side agreement provisions regarding transparency, public participation, and public access to information.

Ultimately, Mexico reluctantly conceded to the inclusion of three controversial provisions: trade sanctions as the maximum penalty for persistent failure of the countries to enforce, or their industries to comply with, environmental laws[20]; the creation of a North American Commission for Environmental Cooperation (articles 8–12); and a mechanism allowing civil society to register complaints about compliance with environmental laws (articles 14 and 15). US negotiators, especially Mickey Kantor, made almost no effort to convince the Mexicans or the Canadians of the wisdom or rationality of the US proposals. Instead, the proposals were presented as non-negotiable requirements. Thus, rather than bring Mexico around to the US position on environmental issues, the process and content of the side agreement negotiations reinforced Mexico's opposition to the linkage of environment with trade issues. The new demands and the way they were advanced reflected the asymmetry of power between the two countries. The abusive US negotiating style consolidated Mexico's initial discomfort with criticism of its enforcement of environmental policy, reinforced the belief within the Mexican government that the trade and environment issue was "political" rather than real, and hardened the Mexican desire to avoid protectionist barriers to trade (including those in an environmental guise). By compelling Mexico to accept concessions in the form of a parallel agreement on the environment, and in particular, provisions for trade sanctions, the NAFTA negotiations served to consolidate Mexico's opposition to future discussions of trade and the environment.

Reactions to the Parallel Agreements

In Mexico, many trade advocates and members of the public believed that the Mexican negotiators made unacceptable concessions in negotiations for the NAFTA's environmental side agreement. In particular, the possibility of trade sanctions against Mexico was perceived as an attack on Mexico's sovereignty.[21] In both governmental and non-governmental circles, the sanctions were a source of particular concern.

To soften the domestic impact of the side agreements, both President Salinas and Trade Minister Jaime Serra Puche affirmed that Mexico's sovereignty was guaranteed in the parallel agreements. Salinas and Puche emphasized the advantages the NAFTA offered in terms of improved environmental cooperation and assured Mexicans that given the highly complex procedures, it was highly unlikely that trade sanctions would be imposed on Mexico. Behind the scenes, however, it is widely acknowledged that Mexican negotiators were angered by the negotiation process, felt humiliated by the fact that they had few negotiating options if they wanted to see the NAFTA win US Congressional approval, and remain embittered about the process by which the environmental side agreement was initiated and evolved.

Most Mexican environmental NGOs opposed the inclusion of trade sanction provisions in the NAFTA environmental package. They argued that provisions for sanctions unfairly targeted Mexico and could infringe on Mexico's sovereignty.[22] However, the views of environmental groups on the range of other issues related to the side agreement varied. Some NGOs, like the Mexican Union of Environmental Groups (UGAM), considered the remaining aspects of the side agreement acceptable (Hogenboom 1998, p. 220). But in the eyes of the Mexican Action Network on Free Trade (RMALC), however, the final agreements did not come close to satisfying demands made by either Mexican, Canadian, or US civil-society organizations (RMALC 1993a,b). Mexican NGOs also raised concerns that the process for NGO participation in the NACEC remained vague, and that actual funding for environmental activities was restricted to the border region (Hogenboom 1998, p. 220).

Conclusion

We have demonstrated that Mexico's hostility to discussion of trade and environment issues can be traced to the deeply unsatisfactory NAFTA environmental side agreement experience. The scars from the 1993 negotiations remain, and they largely explain the government's continued opposition to incorporating environmental considerations into trade arrangements. We have also noted that the relative absence of debate within Mexican society or government agencies on trade and environment questions has enabled the Economic Ministry to retain policy leadership on this issue, ignoring other voices more favorable to an environmental agenda in the trade context. We have argued that the Salinas administration's economic fears and sense of urgency about passage of the NAFTA heightened their frustration with environmental negotiations. The legacy of these negotiations remains clear.

The way in which environmental considerations were handled in the NAFTA negotiations (by forcing Mexico to accept a parallel agreement) consolidated Mexican opposition to the "environment question." Mexican negotiators were irritated by political pressures to concede, particularly on the issue of trade sanctions. They were particularly frustrated that it was domestic US political considerations (such as the power of US environmental NGOs) that put them in this position. It is not hard to understand why Herminio Blanco, the former Mexican Trade Minister who has set the tone for Mexican trade policy in recent years, having experienced the pressure and strain of the NAFTA negotiations, felt strongly that injecting environmental issues into future trade negotiations (such as the Free Trade Area of the Americas negotiations) would distort, complicate, and perhaps endanger trade negotiating processes and goals.

We share the Mexican government's view that Mexico's environmental challenges are related to the ineffective enforcement of environmental laws, regulations, and technical norms. That said, there is no escaping the fact that the NAFTA, or any other trade agreement, exists in the context of both market and government failures on the environment front in Mexico. Thus, both negative and positive lessons from the NAFTA, and other trade agreements, must be appropriately evaluated.

The point here is that the Mexican government must acknowledge that the trade-environment linkage is real and warrants systematic attention. It will remain a permanent agenda in international trade forums. The time has come for the Mexican government to move beyond its view of environmental issues as screens for protectionism. Mexico needs to adopt a proactive attitude to trade and environment issues, starting with a commitment of human and material resources to better understand them.

Notes

1. Notable Mexican studies on trade and the environment include Alfie 1995, Barkin 1992, RMALC 1993a, Sanchéz 1991, and Nadal 2000.

2. Our understanding of this process has benefited considerably from Hogenboom 1998.

3. The President's Office usually assists the president with his internal and international communication, internally and internationally. It also provides advice on political, economic, and social matters.

4. The Social Development Coordination Mechanism within the President's Office looks and advises on issues such as agricultural promotion, health, social policy investment needs, and, eventually, environmental issues.

5. Ultimately, a 5-year evaluation of the NAFTA was undertaken. It consisted of interviews with representatives of different sectors. The report is being held by the Mexican senate as an internal document.

6. Mexico's small and medium-size industries—which represent more than 80% of national industry—face a situation of domestic overregulation that constrains their access to international markets. Furthermore, the decision-making environment is affected by constantly changing regulations. They also suffer from inadequate access to credit from private banks and inadequate incentives or support from the government, especially the Ministry of Economy. Raúl Tornel, president of the Ecology Commission of the National Confederation of Industrial Chambers of Mexico (CONCAMIN), argues that "it is not the same to protect the Mexican industry as to protect industry in Mexico . . . the actions of SECOFI have tried to protect industry in Mexico, notwithstanding where the capital comes from, and not Mexican industry" (Tornel 1999).

7. The Centro de Estudios del Sector Privado para el Desarrollo Sustentable (CES-PEDES) was launched by a group of businessmen concerned about the inadequate attention the private sector had dedicated to issues of sustainable development (Villanueva 1999).

8. The fact that Mexico has denied a link between trade and the environment at the previously mentioned forums has been frequently confirmed to CEMDA by officers working at such forums, as well as from the minutes of official meetings and reports in various publications.

9. This comment was made before it was determined that Mike Moore of New Zealand would succeed Renato Ruggiero as Director-General of the WTO for the subsequent two years.

10. The relative weakness of Environment Ministries compared to more "traditional" Economic Ministries of the Environment is a reality in all countries—not just Mexico.

11. SEMARNAT officials have been able to follow the debate by attending, with representatives from the Ministry of Economy, meetings of the WTO's Committee on Trade and the Environment, and other meetings carried out in the context of the Organization for Economic Cooperation and Development.

12. The Coordinadora de Empresas para el Comercio Exterior (COECE), which is headed by Mexico's Coordinating Council of Entrepreneurs (Consejo Coordinador Empresarial), was created in 1990. It represented eleven major business organizations. Financial capital played a leading role in the COECE, which, like the CCE, overrepresented the interests of large companies (Pastor and Wise 1994; Hogenboom 1998, p. 171). The COECE carried out the same function in the framework of the negotiation of the Interim Agreement (Trade) between Mexico and the European Union.

13. Some did evaluate the employment effects of the NAFTA in Mexico before or immediately after the NAFTA was signed. See, e.g., Leamer 1992 and Lustig 1994.

14. NGOs working on US-Mexico border issues include the Border Ecology Project, the Environmental Health Coalition, and the Texas Center for Policy Studies. Environment and development issues along the U.S-Mexico border are covered in the monthly *Borderlines*, published by the Interhemispheric Resource Center. See http://www.us-mex.org/borderlines.

15. For a definitive articulation of this perspective, see Bhagwati 1993.

16. For details on the positions presented by each of these groups, see Hogenboom 1998, pp. 142–146.

17. Colosio's speech can be found in the records of the Mexican Senate at 25/VII/92.

18. The change of US presidents also changed the role of the US private sector in the NAFTA environment debate (Hogenboom 1998). The private sector, which had concurred with Bush's parallel-track approach, opposed Clinton's idea of supplemental agreements. In March 1993 members of the business community submitted a proposal supporting the creation of a regional environmental commission that would encourage cooperation. They opposed a commission that would hear

complaints from citizens or NGOs, investigate those complaints, or impose sanctions. Clinton´s team was not very accessible to them. On the other hand, the US business community was in constant contact with Mexican trade representatives in Washington and with Mexican business organizations (Hogenboom 1998).

19. In 1991 the US government imposed a second tuna embargo on Mexico. It differed from the embargo imposed in 1980 in that it had apparent environmental foundations. Based upon the 1988 modifications to the US Marine Mammal Protection Act (MPPA), an embargo was imposed on Mexican tuna exports to the US on the grounds that the Mexican tuna fleet had surpassed the limits of dolphin incidental killings according to the MMPA. The embargo has lasted almost a decade and is having serious economic effects on the Mexican tuna industry (González 1999a; SEMARNAP 1998).

20. Articles 22–36 of the NAAEC established a mechanism for Consultations and Conflict Resolution. As a last step in the process, if the Action plan determined by a panel has not been implemented, "NAFTA benefits may be suspended in an amount no greater than that sufficient to collect the monetary enforcement assessment." See NAAEC, article 36.

21. Mayer (1998) offers a discussion of the importance of sovereignty issues to the trade debate in Mexico. He highlights that since the early days of NAFTA negotiations, Mexican policy makers worked to ensure that the agreement was not seen as a "sellout" to the US.

22. Interview with Regina Barba, former president of the Mexican Union of Environmental Groups (Union de Grupos Ambientalistas de México-UGAM), March 1999.

References

Aguilar, Z. 1999. Interview with author.

Alfie, M. 1995. "Ecología: a un año del TLC." *El Cotidiano* 67.

Aspe, P. 1993. *El Camino mexicano de la transformación económica.* Mexico City: Fondo de Cultura Económica.

Audley, J. 1997. *Green Politics and Global Trade: NAFTA and the Future of Environmental Politics.* Georgetown University Press.

Audley, J. 1993. "Why Environmentalists Are Angry about the North American Free Trade Agreement." In *Trade and Environment,* ed. D. Zaelke, P. Orbuch, and R. Housman. Island.

Barkin, D. 1992. "El Impacto ambiental del Tratado de Libre Comercio en el campo mexicano." In *El Sector Agropecuario Mexicano al Tratado de Libre Comercio,* ed. C. González Pachecho. Mexico, D.F.: Instituto de Investigacions Económicas, UNAM.

Bhagwati, J. 1993. "Trade and Environment: A False Conflict?" In *Trade and Environment,* ed. D. Zaelke, P. Orbuch, and R. Housman. Island.

Blanco, H. 1999. Radio program "Para empezar," June 16.

Bush, G. 1992. Report of the administration on the North American Free Trade Agreement and Actions Taken in Fulfillment of the May 1 1991 Commitments.

Carabias, J. 1999a. Quoted in *La Jornada*, July 5.

Carabias, J. 1999b. Quoted in *El Universal*, July 5.

CEC. 1996. *NAFTA Effects. Potential NAFTA Effects: Claims and Arguments 1991–1994*. Montreal: Commission for Environmental Cooperation.

de la Calle, L. 1999. "Linkages Between Trade and Environment Policies." Presented at WTO High Level Symposium on Trade and Environment.

Dornbush, R. 1992. "The Case for Trade Liberalization in the Developing Countries." *Journal of Economic Perspectives* 6, no. 1: 69–85.

Edwards, S. 1993. "Openness, Trade Liberalization, and Growth in Developing Countries." *Journal of Economic Literature* 31, no. 3: 358–393.

González, A. 1999a. Visión Preliminar sobre el Estado del debate Comercio y Medio Ambiente en México, primera parte: La postura gubernamental. México D.F.: Centro Mexicano de Derecho Ambiental (CEMDA).

González, A. 1999b. "Comercio y medio ambiente en el TLCAN: Crónica de una negociación." *Puentes* 2, no. 1: 5–6.

Heath, J. 1998. "The Impact of Mexico's Trade Liberalization." In *The Post NAFTA Political Economy*, ed. C. Wise. Pennsylvania State University Press.

Hogenboom, B. 1998. *Mexico and the NAFTA Environment Debate, the Transnational Politics of Enconomic Integration*. Utrecht: International Books.

Hufbauer, G., and Schott, J. 1992. *North American Free Trade. Issues and Recommendations*. Washington: Institute for International Economics.

Leamer, E. 1992. "Wage Effects of a US-Mexico Free Trade Agreement." Working paper 3391, National Bureau of Economic Research.

Lustig, N. 1994. "NAFTA: Potential Impact on NAFTA's Economy and Beyond." In *Economic Integration in the Western Hemisphere*, ed. R. Bouzas and J. Ros. University of Notre Dame Press.

Mayer, F. 1998. *Interpreting NAFTA: The Science and Art of Political Analysis*. Columbia University Press.

Nadal, A. 2000. *The Environmental and Social Impacts of Economic Liberalization on Corn Production in Mexico*. Gland and Oxford: Oxfam GB and World Wide Fund for Nature.

Pastor. M., and C. Wise. 1994. "The Origins and Sustainability of Mexico's Free Trade Policy." *International Organization* 48, no. 3: 459–489.

RMALC. 1993a. Propuesta de RMALC para la creación de una comisión ambiental de América del Norte (CAAN), 7/VI/93.

RMALC. 1993b. Declaración de prensa sobre la terminación de las negociaciones de los acuerdos paralelos al TLC, 14/VIII/93.

RMALC. 2002. *El TLCAN: Objectivos y Resultados 7 años después,* ed. A. Arroyo Picard.

Roemer, A. 1999. Technical Secretary of the Social Development Cabinet. March. Telephone interview with author.

Sanchéz, R. 1991. "El Tratado de Libre Comercio en América del Norte y el Medio Ambiente de la Frontera Norte." *Frontera Norte* 3, no. 6.

SEMARNAP. 1998. "Pesca del Atún y Protección del delfín." Cuadernos de la SEMARNAP.

Tornel, I. 1999. Personal interview. Ecology Commission of the National Chamber of the Transformation Industry.

Trigueros, I. 1994. "The Mexican Financial System and NAFTA." In *Mexico and the North American Free Trade Agreement*, ed. V. Bulmer-Thomas, N. Craske, and M. Serrano. St. Martin's Press.

Villanueva, L. 1999. Telephone interview with author.

Weintraub, S. 1993. "US-Mexico Trade: Implications for the United States." In *Assessments of the North American Free Trade Agreement*, ed. A. Moss. Miami: North-South Center.

4

Mexico's NAFTA Trauma: Myth and Reality

Mónica Araya

The trade community in Latin America and the Caribbean has steadfastly opposed any attempt to include an environmental dimension in the negotiations for a Free Trade Area of the Americas (FTAA).[1] The lack of interest in replicating the North American Free Trade Agreement's environmental model in the FTAA, or at least in taking it into consideration, reflects widespread opposition to linking environment goals to trade agreements.[2] Yet the NAFTA model has delivered significant environmental benefits.[3] This chapter explores why neither the elements of the NAFTA model nor the lessons from the US-Canada-Mexico experience are being seriously explored in the context of the FTAA negotiations.

The Latin American trade community remains skeptical about the legitimacy and appropriateness of efforts to link trade policy and environmental policy, in general, and about the NAFTA environmental model in particular. This hostility to a trade and environment agenda emerges most forcefully from Mexico.

As a member of both the Latin American community and the NAFTA, Mexico is in a unique position in the FTAA process. It could act as an intermediary between the US and Canadian "pro-trade-and-environment" position and the "no trade-environment linkage" position supported by most other countries in the Americas. Instead, Mexico has widened the division between trade and environment advocates and skeptics by actively encouraging opposition to efforts to include environmental considerations in the trade agenda. Mexico's message to the rest of Latin America has been clear and consistent: Signing the NAFTA environmental side agreement (NAAEC) was a "big mistake," which should not be repeated. As a result, it is surprise that a discussion of lessons learned from the NAFTA's environmental model has not yet been welcomed by the Latin American and Caribbean trade community.

This chapter examines Mexico's NAFTA experience in more depth, seeking to understand Mexico's hostility to the trade and environment issue.

The Mexican Critique of the Trade and Environment Agenda

Environment in a Trade Agenda = Protectionism
Mexican trade officials follow the debate on trade and the environment with considerable concern. Their prevailing position has been to deny the existence of a link between trade and the environment (González 1999a; Hogenboom 1998). During the 1999 World Trade Organization (WTO) High Level Symposium on Trade and Environment, Mexico was one of the most vocal opponents to the existing—let alone an expanding—environment agenda in the trading system. Throughout the symposium, the Mexican Ministry of Economy (then known as SECOFI) advanced its view that folding environmental concerns in any way into the trade agenda represented a mistake (de la Calle 1999).[4] The Mexican Ministry of Economy's position on trade and the environment appears to derive from an assumption that environmental considerations are inevitably a guise for protectionism rather than legitimate attempts to achieve environmental goals (Mancera 2002; González 1999a).

First, Mexican trade officials deny the legitimacy of addressing environmental concerns within the trade agenda (Mancera 2002). They argue that environmental advocates, while promoting noble goals, have found "strange bedfellows," meaning protectionist groups. Their concern is that environmentalists have fallen prey to rent-seeking economic interests focused on inhibiting free trade. The Mexican Ministry of Economy also argues that Mexican businesses and workers have directly suffered from such alliances. Mexican trade officials cite the US tuna embargo as an example of environmentally driven protectionism that is harmful to Mexico but beneficial to certain US economic interests (de la Calle 1999).

Second, the Mexican Ministry of Economy argues against any efforts to promote the harmonization of environmental standards through trade negotiations. In its view, harmonization efforts threaten developing countries' access to markets in developed countries. The Mexican Ministry of

Economy has also expressed concern that uniform environmental standards set by developed countries undermine developing countries' competitive advantages. They believe that protectionists in the developed countries, unhappy about the increasing trade participation of the developing countries, are always looking for new ways to create trade barriers, and thus that environmental standards are likely to be manipulated into obstacles to developing countries' exports.

Third, officials from the Mexican Ministry of Economy reject what they characterize as misleading assumptions that trade harms the environment (Mancera 2002). They argue, consistent with traditional economic logic, that the environment's worst enemy is poverty, not trade liberalization (Esty 2001). Their conclusion is that inserting environmental considerations into a trade agenda threatens decades of progress in market liberalization—a process they consider critical not only to development but also to reducing pressures on the environment (González 1999a).

Mexico is not alone in its opposition to a trade and environment policy agenda. The vast majority of developing countries share the Mexican Ministry of Economy's worries, especially in the WTO context. The mainstream position among developing countries is that pressures to adopt higher environmental standards could negatively affect market access for otherwise competitive developing country products, thereby benefiting protectionist interests (Araya 2001a). Before the WTO Ministerial Meeting in Seattle, a group of fifteen developing countries (the "G-15") clearly set forth this position.[5] The G-15, which includes Mexico, has defended a pro-status-quo approach, according to which the environment "is ab initio a non trade issue . . . and all legitimate environmental concerns can be accommodated within the existing WTO provisions, including article XX of GATT 1994" (*BRIDGES* 1999).

The Imposition of US Environmental Values

When the NAFTA side issues (environment and labor), handled on a parallel track under the G. H. W. Bush administration, moved to center stage under the Clinton administration (Esty 1993), the Mexican government felt threatened and mistreated (Mayer 1998; Audley 1997).[6] The Mexican negotiators accepted the fact that domestic US political dynamics required them to pay some degree of attention to environmental concerns raised in the NAFTA context. However, they also thought that the 1992 mix of direct environmental provisions in the NAFTA and cooperation commitments were an appropriate compromise.

Mexican negotiators interpreted the re-opening of the NAFTA negotiations to establish an environmental side agreement as an attempt by the United States to foist its environmental values on Mexico (and Canada) (Hogenboom 1998). Mexico felt it was forced to accept the new environmental agreement to avoid risking the entire North American trade agreement (González 1999b). The fact that these the supplemental negotiations were unexpected increased Mexico's discomfort.

The Mexican government perceived the expansion of the scope of negotiations as politically expedient for the Clinton administration rather than analytically sound. Mexican trade officials understood that US politics would inescapably affect the NAFTA but resented that Mexico had to pay the price for candidate Clifton's ill-considered campaign promises pandering to environmental interests.[7] Despite their discomfort with the environmental side agreement, Mexican negotiators ultimately accepted a softened side agreement because they regarded the NAFTA as the centerpiece of the Salinas administration's economic strategy (González 1999a).

Threats to the Mexican Economy

According to Mexican trade officials, environmental protection (together with labor) is an issue that can easily be abused in the trade context (Mancera 2002). Mexico's former Minister of Trade, Herminio Blanco, for example, argued that any focus on a trade and environment agenda was likely to obstruct, delay, and hurt the process of trade liberalization (González 1999a). But we must ask: Has this been the case in the NAFTA context? Have the NAFTA environmental provisions hurt the Mexican economy? Is there empirical evidence that these environmental provisions have slowed Mexico's trade liberalization process?[8] The answer is No (Hufbauer et al. 2000; Miller 2002).

The assumption that all environmental efforts are tantamount to protectionism is misplaced. While some protectionists may exist within the environmental community, many environmentalists support freer trade. More importantly, there is no evidence that the environmental provisions of the NAFTA have ever been used in a protectionist manner. The lack of any actual evidence in support of the Mexican Ministry of Economy's claims that protectionism is the chief driving force behind the NAFTA trade and environment agenda is notable.[9]

Another argument that might be advanced to bolster Mexico's case against trade-environment linkages is that the NAFTA's environmental provisions and/or environmental side agreement hurt Mexico environmentally. For example, was Mexico forced to shift its environmental priorities away from its own pressing environmental challenges? A misplaced environmental agenda could have retarded rather than advanced Mexican environmental progress by distracting the country from its own priorities. Again, there exists no evidence to suggest that the NAFTA's environmental provisions have hurt Mexico.

Additional Explanations for Mexico's Hostility to a Trade and Environment Policy Agenda

A number of additional reasons for Mexico's rejection of a trade and environment policy agenda can be identified. But we must consider further whether any of these explanations are relevant to the possible application of the NAFTA's environment model in the FTAA context.

• Historical-political tensions. Mexico and the United States share a border of 3,200 kilometers and deep historic ties and tensions. Their bilateral relationship has been complex: "It is intense, in view of proximity and interdependence; it is asymmetric, in view of the preponderance of the US power; and it has been fraught with misunderstanding, in view of cultural and developmental differences." (Smith 1999, p. 225) Moreover, Smith suggests that the US-Mexico relationship is likely to become more difficult to manage, more resistant to control, and more unpredictable. US-Mexico political tensions are largely not trade-related. Smith identifies three contentious US-Mexico issues: illicit drugs, illegal immigration,[10] and pollution spillovers. Given the historical political antagonisms, reaching agreement on any issue between the two countries is difficult. The NAFTA negotiations, particularly on the environmental front, were bound to be made more difficult by these deeper historical and political antagonisms.

• Conflicts on North-South lines. The differences between the interests of developed and developing countries have also played a role in Mexico's approach to trade and the environment. The undeniable economic might of the United States in Latin America and the Caribbean creates suspicion that US positions—including its environmental agenda—are shaped by narrow economic interests. For this reason, even environmental nongovernmental organizations (NGOs) in Mexico condemned the NAFTA's provisions allowing for the use of trade sanctions to punish environmental non-compliance (Gonzalez 1999a,b). Along with many other developing countries, Mexico remains skeptical of the depth of the US commitment to environmental issues. Doubts about US sincerity on the environmental front are fueled by US failure to ratify important international environmental agreements (such as the Kyoto Protocol and the Convention on Biological Diversity), by minimal US financial support for the implementation of the agenda hammered out at the 1992 Earth Summit, and by high US consumption rates.

• Frustration with the tuna-dolphin case.[11] The Mexican government cites the tuna-dolphin case as evidence of both the problems caused by using trade measures to address environmental concerns and the threat of protectionism under the guise of environmental protection.[12] For the Mexican government, its first trade and environment experience—in fact one of the first cases ever at the trade-environment interface—remains traumatic. According to the Mexican Ministry of the Environment and Natural Resources (SEMARNAT, formerly SEMARNAP), the economic impacts of the tuna embargo have been severe.[13] In Mexico, and across Latin America, the tuna embargo is widely perceived as a measure to protect US tuna-fishing interests, and thus as misguided and unfair. This case has also badly damaged the credibility in Mexico of US environmental groups, whose support for the US embargo on Mexican tuna has seemed insensitive to the Mexican interests at stake and unappreciative of the significant Mexican efforts to reduce dolphin deaths.[14] In many ways, this case has framed the trade and environment debate in Latin America and the rest of the developing countries. This decade-long unresolved controversy has given the trade and environment skeptics powerful and not totally groundless reasons to argue that linking the trade and the environment leads to protectionist mischief and to economic harm for developing countries.[15]

• The negotiating dynamic of the NAFTA. The NAFTA's supplemental negotiations on the environment represented a deviation from the original terms of negotiation agreed upon in September of 1992. Moreover, the Clinton administration's demand for an environmental side agreement was an eleventh-hour surprise and placed Mexican trade officials at a serious negotiating disadvantage. Mexico badly needed the NAFTA but it had already expended all its negotiating "chips." The re-negotiation that started in March of 1993 quickly emerged as Mexico's second "traumatic" trade and environment experience. Not only was Mexico subjected to strong US pressure to agree upon environmental and labor issues; the lack of precedents to follow impeded Mexico's prospects for limiting US demands (Hogenboom 1998). The Mexicans were under pressure to submit to an agreement to which they did not fully subscribe and about which they felt inadequately informed. The chief Mexican NAFTA negotiator (Herminio Blanco) later became the Minister of Trade in Mexico's Zedillo administration, taking with him the bitter experiences of the NAFTA negotiation and a firm opposition to environmental linkages of any sort (González 1999a). Clearly, signing the environmental side agreement did not mean that the Mexican officials were convinced it was appropriate to address environmental issues explicitly in a trade context. Rather, they felt that the negotiating dynamics left them no other choice. Mexico's official position on trade and the environment is driven and sustained, at least in part, by a continuing grudge against the unbalanced dynamic of the negotiations for the NAFTA environmental side agreement.

• Future leverage. Another possible explanation for Mexico's opposition to environment linkages is that it provides the Mexican government future opportunities to use "the environment" as a bargaining chip in negotiations with the United States. The logic might be "Why give up early on an item of considerable importance to the United States or Canada?" In this scenario, Mexico is strategically rejecting any inclusion of environmental concerns in the early stages of the FTAA negotiation in order to extract concessions at a later stage (i.e., on agriculture or more flexible intellectual property rules). Although civil society's pressure to negotiate greener trade agreements has only increased since the NAFTA days, it still remains uncertain what the bargaining dynamic in FTAA context will be.[16]

• A hurt "national spirit." It may also be that the Mexican Ministry of Economy's anti-environment stance in the FTAA on the environment is driven by a desire to "even the score." For Mexico, the persistence of the trade and environment debate may reinforce a sense of bitterness about a lost battle with the United States. Every mention of trade-environment linkage reminds Mexican trade officials of what they perceive as an unfair tuna embargo and the imposition of an environmental side agreement as the entrance fee to the NAFTA.

It is not hard to understand why imposed agendas and threats of trade sanctions make countries uncomfortable. The Mexican government stands in the shadow of a giant neighbor. As the only developing country that shares a border with the most powerful nation in the world, Mexico remains constantly vulnerable to bullying. In short, Mexico's pride renders it unable to entertain losing another environmental battle in the trade context.

Lessons for the FTAA

What lessons can the NAFTA environmental experience provide for the FTAA context? The lessons fall under the four following categories.

Negotiation-Related Aspects: No Last-Minute Surprises
The United States extracted an environmental side agreement from Mexico by using the threat of not having a final trade agreement. This negotiating dynamic has generated negative long-term repercussions. Not only did the US negotiating strategy serve to entrench Mexican political opposition to trade-environment linkage; it also generated the perception that inclusion of environmental considerations is the price to be paid for a trade agreement, rather than a legitimate goal in its own right.

If the goal is a sound trade and environment arrangement in the FTAA context, the agenda should be taken up early, not as a last-minute add-on. The United States and Canada, as the two leading countries advocating trade-environment linkages in the FTAA, should focus on building trust among their partners in the Americas and assuming more effective leadership roles. They must explain the logic of their positions and convince their interlocutors of the wisdom of a systematic trade and environment agenda with reasoned arguments, not simply pronouncements about the demands of US politics. The United States cannot succeed by bullying. Even if threats work in the short run, they do not generate real commitment to the environment.

Rather than leave the field open for the United States to advance a US-defined package as a sine qua non for any hemispheric agreement, countries across the American should propose and negotiate a range of environmental options.[17] In the FTAA context, any efforts to impose an environmental package, particularly in the later stages of the FTAA negotiation, will likely undermine prospects for a constructive long-term approach to trade and environment issues.

Emphasize Compliance and Flexibility
By addressing the environment primarily (although not exclusively) through a "parallel approach," the NAFTA offered countries a considerable degree of flexibility. The NAFTA model acknowledged the political importance of recognizing differences between countries in terms of their trade and environment goals and left countries room to define and accomplish their own level of environmental protection standards. The FTAA should use the same approach. Given widespread mistrust of US environmental intentions in trade negotiations and the fear of unilateral imposition of environmental standards, the only viable option in the FTAA context is likely to be envi-

ronmental provisions that replicate the NAFTA's focus on compliance with domestic regulations.

Address Environmental Issues Up Front

Several benefits of the NAFTA model are clear. It addressed environmental concerns before the agreement entered into force, rather than develop responses in an ad hoc fashion as problems arose. Pre-defined institutional responses to potential and unexpected environmental consequences of trade liberalization could also add legitimacy to the FTAA negotiation process. Moreover, through the NACEC, the NAFTA has also spurred efforts to develop methodologies for environmental assessment of trade initiatives that could stimulate a more empirically based trade and environment debate in the Americas.[18] Specifically, these assessment methodologies could be employed by governments to review the environmental implications of different trade policy options before agreements are finalized, to pre-emptively identify critical areas of concern that require environmental measures parallel to the trade liberalization process, and to help address specific environmental harms before they occur, without necessarily harming trade flows. An ex ante approach could also increase the efficiency of environmental policy, since prevention tends to be cheaper than ex post environmental cleanups and remedies.

Pursue a Cooperative Approach

The need for cooperative approaches to global and regional environmental problems is increasingly clear. In the Americas, many of the most pressing environmental problems (e.g., deforestation and air pollution) spill across territorial boundaries.[19] The NAFTA's environmental side agreement created a North American Commission for Environmental Cooperation (NACEC), which provides institutional support to ensure ongoing attention to issues arising at the trade-environment interface.

This cooperative approach has enhanced communication flows among the three members, increased citizen input, and engaged civil society in setting environmental priorities. The NACEC's development of methodologies to assess trade-related environmental problems and related research have benefited all three countries by providing an analytical foundation for trade and environment issues.

For the FTAA, the main goal should be to find a formula that provides institutional support for environmental cooperation and explicitly addresses the potential environmental problems created by expanded trade.[20]

Increase Transparency and Openness

Negotiations for the NAFTA's side agreements were stimulated by the need to solve domestic political problems in the United States. The US presidential election of 1992 empowered labor and environmental interests, weakened business interests, and compelled an environmental commitment from candidate Clinton that endured beyond the elections into his presidency. The NAFTA negotiations offer a number of political economy lessons that will be useful to FTAA negotiators.[21]

First, the NAFTA experience teaches us that US domestic politics matter to US trade policy. Despite divisions in the US environmental community, NGOs clearly influenced the US position in the negotiations. Although the Republicans now hold the presidency and a majority in the House of Representatives, pro-environment senators and representatives are powerful swing voters in US trade debates (Destler and Balint 1999; Esty 1998).

Second, the NAFTA negotiations suggest that we must not underestimate the power of civil society. The 1999 Seattle protests illustrate that failure to consider and to respond to civil-society concerns about trade can erode long-term support for the trade liberalization process.[22] Civil-society actors in all countries of the Americas want assurances that their values inform economic policy making. Similarly, the NAFTA negotiating story shows us that inadequate support from US environmental constituencies has the potential to threaten the entire negotiation of an agreement.[23] Wisely, the governments chose to respond by opening processes of consultation and dialogue.

Third, we know that the NAFTA's "parallel agreement" lent considerable legitimacy to the trade agreement (especially in the United States and in Canada) by including the concerns of important constituencies such as environmental and labor groups. This is a lesson that FTAA negotiators have yet to appreciate. The FTAA's only mechanism for input from civil society in the FTAA process is a rather weak Committee of Government Representatives whose terms of reference fail to provide effective opportunities for influence on trade decision making. Despite the differences and the lags in the "culture of participation" among Latin America, Canada, and the United States, the FTAA will increasingly require a more encompassing model for openness and transparency.[24]

Looking Forward by Looking Backward to the NAFTA

The debate on trade and the environment is not going to fade away. It is continuously evolving, and the number of issues, actors, and proposals is expanding.[25] A smart strategy for Latin American countries would be to build constructive approaches that go beyond the classic opposition to whatever is proposed by the developed countries in either the WTO or the FTAA context (Araya 2001a).

The NAFTA environment model, despite its limitations, offers several advantages and opportunities, including lessons that remain unexplored in the Americas. Mexico's concerns about the negotiating dynamic of the environmental side agreement are understandable. However, it is important to separate the evaluation of the procedure from the evaluation of the substantive approach to the environment. Using environmental protection as an excuse for protectionism and unilateralism is neither legitimate nor fair. But it is neither rational, legitimate, nor fair to use the perceived threats of protectionism and unilateralism as excuses for ignoring environmental considerations. Developing country trade officials, and Mexican Ministry of Economy officials in particular, must stop using the tuna-dolphin case as a defensive tool against all discussions about the environment in the trade context. The trade and environment agenda is real (Araya 2001b; Esty 2001). It cannot be boiled down to merely a "Northern plot" to discriminate against Southern exporters.

The time has arrived for a debate on trade and the environment in Latin America and the Caribbean. It is time for Southern leadership to establish a constructive trade and environment agenda for the FTAA. With civil society's growing awareness of trade issues throughout the Americas, the legitimacy of the FTAA will depend on transparent and serious efforts to address trade, the environment, and other social issues. Debate must go beyond the trade negotiations themselves. It must address how to make trade liberalization and environmental protection mutually supportive. This exercise might require discussion of possible coordination mechanisms between trade policy makers and environmental policy makers, training of environmental negotiators on trade rules and principles (and vice-versa), addressing of concerns from the private sector about competitiveness, and assessment of environmental effects of trade initiatives in order to mitigate them *before* they happen.

In any case, the question is less *whether* this debate should take place than *what* kind of debate is needed. The best way for Latin American countries to avoid further trade and environment "trauma" is to identify approaches that address the linkages in a way that works and makes sense for Latin Americans. Latin America needs to stop focusing on what it does not want and focus on what it does want.

Notes

1. During the governmental discussions about the content of the 1998 Joint Declaration of the Fourth Trade Ministerial of the Summit of the Americas in San José, Costa Rica, the US proposed the creation of an FTAA Study Group on Trade and Environment. This proposal was supported only by the government of Canada. Civil-society organizations have also made a number of proposals; see FOCAL et al. 1999, WWF 2001; Sierra Club 2001; Sierra Club of Canada et al. 2001, and Hemispheric Social Alliance 2001.

2. A detailed analysis of the reasons for this rejection in the context of the FTAA is beyond the scope of this chapter. However, it is clear that among the reasons are: a) widespread distrust of US environmental intentions in the trade negotiations; and b) fears that environmental safeguards could serve as non-tariff barriers to hemispheric trade (regardless of whether or not this is their intent).

3. Full discussion of its benefits is beyond the scope of this chapter. For detailed discussion of the environmental provisions and outcomes, see the chapter in this volume by Torres and that by Carlsen and Salazar.

4. Luis de la Calle's presentation on behalf of the Mexican Economic Ministry at the WTO High Level Symposium WTO on trade and the environment is available at http:www.wto.org/english/tratop_e/envir_e/hlmenv_e.htm. The chapter in this volume by Mancera, a trade official with the Mexican Ministry of Economy, provides another example. It is, however, difficult to find specific statements, policy papers, or official perspectives from the Mexican government that elaborate any further on this stance. Jagdish Bhagwati is one international trade expert who concurs with the Mexican Ministry of Economy's view that environmental considerations should be addressed outside the trade process; see Bhagwati 1993.

5. The "G-15" actually had seventeen members: Algeria, Argentina, Brazil, Chile, Egypt, India, Indonesia, Kenya, Jamaica, Malaysia, Mexico, Nigeria, Peru, Senegal, Sri Lanka, Venezuela, and Zimbabwe (*BRIDGES* 1999). The G-15 Ministerial Meeting was held in India in August 1999. The G-15 position was a reaction to other countries' attempts to insert more environmental concerns in the multilateral trade agenda. For an official "developed country" perspective at the WTO, see the

following documents at http://www.wto.org: WT/GC/W/265 (Switzerland); WT/GC/W/304 (US); WT/GC/W/176 (Norway), WT/GC/W/194 (European Union).

6. For a history of efforts to incorporate environmental considerations into the NAFTA, see Audley 1997. For a general political analysis and history of the NAFTA negotiations, see Mayer 1998.

7. For an overview of the concerns about NAFTA expressed by the US environmental community, see Audley 1993.

8. The overall conclusion reached by Weintraub (1997) was that the NAFTA is performing as expected. Weintraub argues that two-way trade between Mexico and the US has increased and is continuing to rise, and that the type of industrial specialization that is required to augment the competitiveness of companies is in fact taking place. Based on Weintraub's data (ibid.), one can venture a preliminary assertion that the NAFTA's parallel environmental agreement to the NAFTA has not impeded the trade objectives of the trade agreement. Yet one would have to assess the impact of the provisions of the NAFTA's environmental side agreement on Mexico's trade *explicitly* in order to arrive at definitive and empirically supported conclusions. For an analysis of the economic effects of the NAFTA's environmental provisions, see the chapter by Miller in this volume.

9. Trade-environment tensions and synergies are difficult to assess and explain (NACEC 1996). There is a lack of empirical case studies on the effects of environmental regulations on trade flows, let alone the environmental provisions of trade agreements on economic indicators. Even pure trade effects of the NAFTA on the Mexican economy are difficult to isolate and assess. Several efforts are underway to develop and test methodologies for assessing the relationships between trade and the environment (Vaughan 2000; WWF 1998; NACEC 1999). Mexican authorities will not be able to escape the need for case studies to examine the environment-trade linkage, particularly given the increasing number of Mexican non-governmental actors demanding more effective environmental protection.

10. Illegal immigration from Mexico accounts for 55–60% of unauthorized residents in the US (Smith 1999).

11. The tuna-dolphin dispute began in 1988, when the California-based Earth Island Institute claimed that Mexican commercial fishermen were violating the US Marine Mammal Protection Act (PL 92-522, 86 Stat. 1027). Earth Island Institute sued the federal government for failure to enforce a congressional mandate to curtail the incidental killing of marine mammals, in this case dolphins, by commercial fishermen. The US subsequently imposed a ban on tuna imported from Mexico, among other places, effectively shutting "non-dolphin-safe" tuna out of the US market (Esty 1994a and 1994b). In 1991, a GATT panel ruled that the US tuna embargo violated its GATT obligations. This report was never adopted (WTO 1999).

12. For an extensive and critical analysis of this controversial case and the "green" protectionism hypothesis, see Körber 2000. For an analysis of the legal issues surrounding the tuna-dolphin dispute, see Hampton 1998.

13. Impacts include a decline of the national fishing fleet, lower rates of tuna fishing, reduced tuna exports (tuna exports declined from 80,000 metric tons in 1988 to 20,000 metric tons in 1993). The loss of $300 million in exports to the US, and

the loss of jobs due to the reduction of the tuna fishing fleet (González 1999a; SEMARNAP 1998).

14. The tuna embargo was lifted in April 2000, but the controversy continues. In August 1999, the Earth Island Institute and a coalition of environmental and animal rights groups sued the US Department of Commerce, which had with the support of environmental groups such as the World Wildlife Fund, the Center for Marine Conservation, and Environmental Defense, sought to authorize a redefinition of "dolphin-safe" labeling standard for tuna. A US District judge ruled against the Clinton administration's claim that the decision to loosen the label standard was not justified by the scientific evidence (*Brower v. Daley* 1999). Without a change in the "dolphin-safe" label standard, the tuna dolphins remain, in practice, banned from the US market. Despite efforts by the US Department of Commerce to lift the ban and the opposition of several environmental NGOs to the tuna embargo, the US has not demonstrated sufficient political will to devise policy arrangements that could promote both dolphin protection and fair market access for Mexican tuna exports. See also Körber 2000.

15. Mexican frustrations have been sufficiently high that in August 2000 the Mexican government suggested it might work to solve the case through the WTO's dispute-settlement mechanism (Inter-American Tropical Tuna Commission 2000). See also BOL-00-91 at http://www.embassyofmexico.org.

16. In the NAFTA negotiations, it was clear that neither the US nor the Canadian negotiators could return to their capital with a trade agreement that lacked explicit provisions on the environment. For an evaluation of the new politics of American trade policy, particularly the role and political power of environmental constituencies, see Destler and Balint 1999.

17. Canada could play an important role here. The Canadian government has openly expressed interest in using the side agreement approach of the NAFTA and its bilateral free trade agreement with Chile as models for future trade agreements. Canada could help promote a pro-trade-and-environment position among Latin American countries by adopting a position independent from that of the US. Depending on the approach Canada chooses, its role could also help reduce North-South polarization in the FTAA context.

18. The NACEC, for example, held the first North American Symposium on Understanding the Linkages between Trade and Environment in 2000. The NACEC commissioned a series of research papers to test its methodologies for environmental assessments of trade policy. The studies and the proceedings of the meeting are available at http://www.cec.org/symposium/.

19. A cooperative hemispheric environmental initiative could be designed independent of an FTAA. However, the FTAA negotiations offer a rare political moment for hemispheric environmental action.

20. It depends on whether the problem can be classified as national (e.g. pollution of a watershed), regional (e.g. the pollution of a common river), or as a global externality (e.g. carbon dioxide emissions that contribute to the greenhouse effect).

21. For further discussion of political lessons from the NAFTA negotiations for the FTAA, see the chapter by Mayer in this volume.

22. Another example of the influence of NGOs in trade processes is being observed at the WTO. The WTO is sending concrete messages about its desire to open more space for dialogue with environmental NGOs. The WTO High Level Symposium on Trade and Environment (March 1999) is a good example of this new approach to non-state actors. For an analysis of the need for environmental NGO participation in the WTO, see Esty 1998. For a summary of the High Level Symposium on Trade and Environment, see http://www.iisd.ca/linkages/sd/wtohls.html.

23. The problems that arose in the negotiation among the countries of the Organization for Economic Cooperation and Development regarding a Multilateral Agreement on Investment, closed in December 1998, were related partially to the opposition led by an international coalition of NGOs. (See http://www.citizen.org/pctrade/mai/Sign-ons/maisign.htm.) Many of the opponents were environmental organizations. See also Alliance for Democracy et al. 1997.

24. For a comprehensive analysis of the increasing role of non-governmental actors in the international arena, see Charnovitz 1997. For a compendium of participation in integration processes in a Latin American context, see CEFIR et al. 1998.

25. In Seattle, for example, the WTO officially acknowledged that trade liberalization reinforces the need for environmental protection. It noted that there are some areas where environmental degradation can occur as the result of trade expansion where it is not complemented by good environmental policy (WTO 1999).

References

Alliance for Democracy et al. 1997. *The MAI: Democracy for Sale?* New York: Apex.

Araya, M. 2001a. "El Rechazo del Tema Ambiental en el ALCA: Una Taxonomía," *Foreign Affairs en Español*, May–August: 1–16.

Araya, M. 2001b. "Implications of the Canada-Costa Rica Free Trade Agreement." *BRIDGES Monthly 5, no. 4: 17 18.*

Audley, J. 1993. "Why Environmentalists Are Angry about the North American Free Trade Agreement." In *Trade and Environment*, ed. D. Zaelke, P. Orbuch, and R. Housman. Island.

Audley, J. 1997. *Green Politics and Global Trade: NAFTA and the Future of Environmental Politics*. Georgetown University Press.

Audley, J. 2001. "Return to the Bad Old NAFTA Days." Paper presented at Washington conference on "The FTAA and The Environment: What Can We Learn from the NAFTA Model?" (revised April 2001).

Bhagwati, J. 1993. "Trade and Environment: A False Conflict?" In *Trade and Environment*, ed. D. Zaelke, P. Orbuch, and R. Housman. Island.

BRIDGES. 1999. "G-15 Summit on Seattle: Mixed Signals on Content of WTO Forthcoming Negotiations." *BRIDGES Weekly Trade News Digest* 3, no. 33.

Brower v. Daley. 1999. United States District Court for the Northern District of California, no. C99-3892 THE, 28–29.

Carlsen, L., and H. Salazar. 2002. "Limits to Cooperation: A Mexican Perspective on the NAFTA's Environmental Side Agreement and Institutions." In this volume.

CEFIR (Centro de Formación para la Integración Regional), Asociación Latino-americana de Organizaciones de Promoción, and Centro Latinoamericano de Economía Humana. 1998. *Participación de la Sociedad Civil en los Procesos de Integración*. Uruguay: Impresora Editorial.

Charnovitz, S. 1997. "Two centuries of participation: NGOs and international conferences." *Michigan Journal of International Law* 18, no. 2: 183–286.

de la Calle, L. 1999. Presentation at WTO High Level Symposium on Trade and Environment.

Department of Commerce. 2000. Press release from US Department of Commerce, April 12 (http:www.nmfs.gov/prot_res/PDF_docs/bower_ruling_prsrls.pdf).

Department of State. 2000. Release Draft Guidelines on Implementation of Executive Order regarding Environmental Reviews of Trade Agreement. Press release, July 10, 2000.

Destler, I., and P. Balint. 1999. *The New Politics of American Trade: Trade, Labor, and the Environment*. Washington: Institute for International Economics.

Esty, D. 1993. " Integrating Trade and Environment Policy Making: First Steps in the North American Free Trade Agreement." In *Trade and Environment*, ed. D. Zaelke, P. Orbuch, and R. Housman. Island.

Esty, D. 1994a. *Greening the GATT*. Washington: Institute for International Economics.

Esty, D. 1994b. "Making Trade and Environmental Policies Work Together: Lessons from NAFTA." *Aussenwirtschaft* 49: 59–79.

Esty, D. 1996. "Revitalizing Environmental Federalism." *Michigan Law Review* 95: 570–653.

Esty, D. 1998. "Non-governmental Organizations at the World Trade Organization: Cooperation, Competition or Exclusion." *Journal of International Economic Law* 1, no. 1: 123–148.

Esty, D. 2001. "Bridging the Trade-Environment Divide." *Journal of Economic Perspectives* 15, no. 3: 113–130.

FOCAL, ECOS, and IISD. 1999. *Trade, Sustainable Development and Civil Society in the Free Trade Area of the Americas: How to Make the Link*. Ottawa: Canadian Foundation for the Americas.

Gitli, E., and C. Murillo. 1999. "Factores que desalientan la inclusión del tema ambiental: El ALCA y una posible agenda positiva." In *Comercio y Ambiente: Temas para Promover el Debate*, ed. M. Arayam. Washington: Organización de Estados Americanos.

González, A. 1999a. *Visión Preliminar sobre el Estado del debate Comercio y Medio Ambiente en México, primera parte: La postura gubernamental*. México D.F.: Centro Mexicano de Derecho Ambiental (www.inca.cr.or).

González, A. 1999b. "Comercio y medio ambiente en el TLCAN: Crónica de una negociación." *Puentes* 2., no. 1: 1–2.

Hampton, J. 1998. A Summary of Current Information on the Biology, Fisheries and Stock Assessment of Bigeye Tuna (*Thunnus obesus*) in the Pacific Ocean, with Recommendations for Data Requirements and Future Research. Noumea, New Caledonia: Secretariat of the Pacific Community, Oceanic Fisheries Programme.

Hemispheric Social Alliance. 2001. Alternatives for the Americas: Building a Peoples' Hemispheric Agreement. Available at http://www.asc-hsa.org.

Hogenboom, B. 1998. *Mexico and the NAFTA Environment Debate: The Transnational Politics of Enconomic Integration*. Utrecht: International Books.

Hufbauer, G., D. Esty, D. Orejas, L. Rubio, and J. Schott. 2000. *NAFTA and the Environment: Seven Years Later*. Washington: Institute for International Economics.

Inter-American Tropical Tuna Commission. 2000. Mexico Calls on the US Government for Urgent Political Consultations Regarding US Commitments on the Conservation of Dolphins and the Inter-American Tropical Tuna Commission. Press release, August 7, 2000.

Körber, A. 2000. *The Political Economy of Environmental Protectionism*. Elgar.

Mancera, J. 2002. "A Mexican View on Trade and Environment." In this volume.

Mayer, F. 2002. "Negotiating the NAFTA: Political Lessons for the FTAA." In this volume.

Mayer, F. 1998. *Interpreting NAFTA: The Science and Art of Political Analysis*. Columbia University Press.

Miller, E. 2002. "Did Mexico Suffer Economically from the NAFTA's Environmental Provisions?" In this volume.

NACEC. 1996. *NAFTA Effects. Potential NAFTA Effects: Claims and Arguments 1991–1994*. Montreal: North American Commission for Environmental Cooperation.

NACEC. 1999. *Assessing Environmental Effects of the North American Free Trade Agreement (NAFTA). An Analytic Framework (Phase II) and Issue Studies*. Environment and Trade Series, no. 6. Montreal: North American Commission for Environmental Cooperation.

Orme, W. 1996. *Understanding NAFTA: Mexico, Free Trade, and the New North America*. University of Texas Press.

SEMARNAP. 1998. "Pesca del Atún y Protección del delfín." *Cuadernos de la SEMARNAP*.

Sierra Club. 2001. Sierra Club Position on the FTAA: Statement by Robert Cox. Washington: Sierra Club.

Sierra Club of Canada, Council of Canadians, Canadian Labour Congress, Polaris Institute, and Canadian Alliance on Trade and Environment. 2001. *Five Reasons to Oppose the FTAA: From Swordfish to Services*. Ottawa: Sierra Club of Canada.

Smith, P. 1999. "Mexico." In *Pivotal States*, ed. R. Chase, E. Hill and P. Kennedy. Norton

Torres, B. 2002. "The North American Agreement on Environmental Cooperation: Rowing Upstream." In this volume.

Vaughan, S. 2000. "Understanding the Environmental Effects of NAFTA: What Can We Say?" Paper presented at Washington conference on "The FTAA and the Environment: What Can We Learn from the NAFTA Model?"

Weintraub, S. 1997. *NAFTA at Three*. Washington: Center for Strategic and International Studies.

World Trade Organization. 1999. *Trade and Environment: Special Study 4*. Geneva: World Trade Organization.

WWF. 1998. *Developing a Methodology for the Environmental Assessment of Trade Liberalization Agreements*. Gland: World Wildlife Fund.

WWF. 2001. *Achieving Sustainable Commerce in the Americas: WWF's Statement on the FTAA*. Washington: World Wildlife Fund.

5

Did Mexico Suffer Economically from the NAFTA's Environmental Provisions?

Eric Miller

During the negotiation of the North American Free Trade Agreement, Mexican proponents of free trade worried that the inclusion of environmental provisions as part of the NAFTA "package" could offer Canada and the United States an effective protectionist instrument for hindering increases in Mexico's exports, ability to attract investment and economic growth (Gilbreath 2001; Mayer 1998; Orme 1996).[1] Put simply, Mexican negotiators feared that including environmental provisions in the NAFTA would blunt the economic benefits accruing from liberalization (Gilbreath and Tonra 1994; von Bertrab 1997; Audley 1993). The Mexicans were not alone in their skepticism; the trade community in general feared that a focus on environmental issues would hinder progress toward trade liberalization.

Domestic political pressures forced Canadian and US trade negotiators to incorporate environmental considerations into their negotiating objectives (Mayer 2002). Mexican negotiators remained—and remain to this day—reluctant to consider any trade-environment linkage. They appear to perceive a grave risk of protectionism in the "green" agenda. The infamous US-Mexico tuna-dolphin dispute has clearly been a major factor driving Mexico's concerns about the economic impacts of environmental provisions (WTO 2000; Orme 1996).[2] From the perspective of Mexican trade officials, the tuna-dolphin case crystallized concerns about disguised protectionism, the intrusion of environmental provisions on national sovereignty, and the impure motives of environmental groups (Garciadiego 1994; Araya 2002).

In addition to concerns about hidden "green" protectionism, Mexican officials argued against environmental provisions for fear that they could provide an excuse for discrimination against Mexican imports because of the "dirty" nature of the natural resource extraction and production processes on which many Mexican (and other developing country) exports strongly depend, the difficulties faced by small and medium-size enterprises (which make up a sizable percentage of developing country economies) in complying with world-class environmental standards, and the impacts that compliance with world-class environmental standards might have on the competitiveness of Mexican goods (Vossenaar 2000).

Eight years later, one may fairly ask whether either NAFTA's environmental side agreement—the North American Agreement on Environmental Cooperation (NAAEC)—or the environmental provisions incorporated into the NAFTA have caused the sort of damage to Mexico's economic performance that critics feared.[3] In this chapter I seek to bring empirical evidence to bear on this question. First, I analyze Mexico's trade and economic performance in the aggregate since the entry into force of the NAFTA to determine whether or not the negative impacts that Mexican officials anticipated have materialized in Mexico under the NAFTA. Second, I examine developments in specific areas that the NAFTA's environmental provisions were most likely to have been affected. (Appendix A of this volume provides details on the environmental provisions contained within the NFTA and its environmental side agreement.) A lack of detailed econometric studies of the economic impact of the NAFTA's environmental measures complicates efforts to prove what the impacts have been. Nonetheless, a preliminary exploration of whether any significant trends have emerged seems useful. I focus on the narrow question of the economic effects of the NAFTA's environmental provisions. I do not attempt to grapple with a broader assessment of the costs and benefits to Mexico of the NAFTA or of the distribution of benefits within the Mexican economy, nor do I tackle the question of the extent and scale of the environmental impacts of the NAFTA's trade provisions on Mexico's economy or any economic implications arising from them.[4]

The Evidence: Impacts on Mexico's Overall Economic Performance

The NAFTA environmental provisions that caused Mexico so much distress fall into two broad categories: those contained in the text of the trade agreement itself and those set forth in the environmental side agreement. The environmental provisions of the NAFTA proper spread across a number of its chapters and have a variety of purposes, including thwarting the use of the NAFTA as a lever for reducing environmental protection levels, providing guidance for instances where the rights and obligations of the NAFTA might conflict with those contained in bilateral or international environmental agreements, discouraging the use of a "pollution haven" strategy for development, and clarifying the treatment of the environment in dispute-settlement processes.[5]

The environmental side agreement commits the NAFTA governments to report on the state of the environment, to strive for the improvement of environmental laws and regulations, to effectively enforce environmental laws, and to publish and promote the dissemination of environmental information. The parties established the North American Commission for Environmental Cooperation (NACEC) to turn these commitments into reality and create a dispute-settlement mechanism for the purpose of correcting problems of non-enforcement of environmental laws.[6]

The defining methodology of the NAFTA's environmental provisions and the NAAEC is one of "negative prescription" rather than "positive rule making" (Dymond and Hart 2000; Geradin 1993). Almost all of the NAFTA's environmental provisions are based on the notion of "thou shall not" (lower environmental protection to attract investment, downwardly harmonize environmental standards), rather than on the notion of "thou must" (toughen national environmental laws, adopt provision x). There is a requirement under the NAAEC that countries enforce their environmental laws, but the definition and the scope of those laws remain up to the individual country. The rest of the NAFTA's environmental provisions are either descriptive or procedural.

Agreements based on a "negative prescription" methodology tend to be much looser in punitive enforcement than those constructed on a "positive rule-making" basis. The effects of the NAFTA's environmental provisions and of the NAAEC are thus likely to be modest and their ability to inflict economic suffering on Mexico negligible. To prove this "negligible effect" hypothesis, one must piece together a myriad of disaggregated information.

The Big Picture

On aggregate, the NAFTA has been a tremendous boost to Mexican trade and investment. Table 1 illustrates the evolution of Mexico's exports to the United States and Canada from 1989 through 1999, the end of the NAFTA's fifth year. The average annual growth rate of Mexico's exports

Table 1
Mexico's exports to its NAFTA partners, 1988–2000, in millions of US dollars.
Source: IMF. Real AAG (average annual growth) calculated by author using 1995
constant dollars based on US CPI from IMF International Financial Statistics.

	Exports to Canada	Exports to US
1988	273	13,454
1989	272	16,163
1990	226	18,837
1991	1,139	33,953
1992	1,000	37,468
1993	1,541	43,117
1994	1,470	51,943
1995	1,979	66,475
1996	2,170	80,673
1997	2,157	94,531
1998	1,521	103,306
1999	2,391	120,393
2000	3,353	147,686
Real AAG	19.4%	18.3%

to both countries was over 20 percent. In dollars, Mexico's exports to the
United States have more than doubled since the entry into force of the
NAFTA, while exports to Canada have more than tripled. Mexico over-
took Japan to become the second largest trading partner of the United
States (after Canada) in 1999, and it remains in that position (US Census
Bureau 2002).

The story of foreign direct investment (FDI) in Mexico mirrors this history of robust growth. The NAFTA years coincide with unprecedented FDI
flows into Mexico. The 1980–1989 yearly average was US$357 million,
vs. US$3,324 million for 1990–1999 (see figure 1). Between 1989 and
1999, the total stock of US FDI in Mexico grew from a little over US$5.5
billion to over US$34 billion (This translates into an average annual
increase of 17.7 percent—see table 2.) During the same years, the stock of
Canadian FDI in Mexico grew by an impressive 25 percent per year. One
cannot know what these patterns would have been in the absence of the
NAFTA environmental provisions, but there is no obvious evidence of

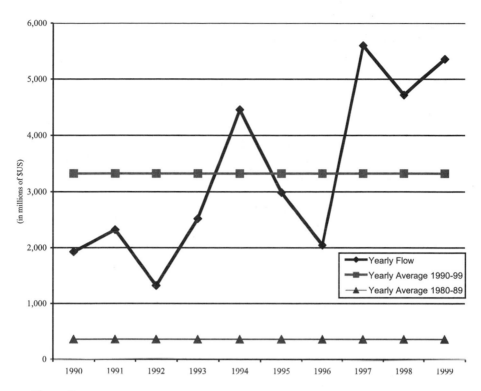

Figure 1
Flows of foreign direct investment from United States to Mexico.

harm to Mexico's economic growth. Indeed, there is no evidence of any harm whatsoever.

Specific Economic Impacts

The macroeconomic data offer no evidence that, overall, Mexico's growth, exports, or incoming FDI have suffered from the NAFTA. We can, however, ask a narrower question: Are there specific areas where the agreement's environmental provisions have hindered Mexico's economic progress or the attainment of the maximum possible benefit from NAFTA?

Resource-Extraction Industries
As was noted above, a principal reservation about environmental measures in trade agreements—heard from the governments of many developing

Table 2
Stocks of US and Canadian FDI in Mexico, 1988–2000, in millions of US dollars.
Sources: US FDI from OECD 1997 and Bureau of Economic Analysis 2002;
Canadian FDI from Statistics Canada 1998 and 2002. Real AAG (average annual
growth) calculated by author using 1995 constant dollars based on US CPI from
IMF International Financial Statistics.

	US FDI	Canadian FDI
1988	5,712	163
1989	8,264	200
1990	10,313	210
1991	12,501	174
1992	13,730	373
1993	15,221	411
1994	16,968	786
1995	16,873	691
1996	19,351	1,402
1997	24,050	1,396
1998	26,657	1,514
1999	32,292	2,158
2000	35,414	2,186
Real AAG	12.8%	20.3%

countries—centers on their potential to stifle exports of primary products
with "dirty" resource-extraction processes. Mexican officials expressed
some fear about such a dynamic in the NAFTA context. But the export
performance of Mexico in these dirty sectors provides evidence to the contrary. Table 3 presents Mexico's combined exports of mineral products,
wood products, and base metals and articles thereof to the United States
in the period 1992–1999.[7] Exports in these sectors in 1999 were almost
double the levels of 1992. Again, we cannot know what the results might
have been, but it is hard to imagine the growth could have been much
higher. The data thus suggest that the NAFTA's environmental provisions
were not a serious drag on the trade performance of these sectors. These
results should come as no surprise to anyone who has actually examined
the NAFTA's environmental provisions. (Indeed, US environmentalists cite
the fact that the NAFTA may have stimulated the expansion of exports in

sectors characterized by inadequate natural resource management strategies as one of the treaty's greatest weaknesses.)

NAFTA's Sanitary and Phytosanitary Rules

One of the most important places that the NAFTA incorporates environmental provisions is in chapter 7(B), pertaining to Sanitary and Phytosanitary (SPS) Measures. (See appendix A.) If the NAFTA's SPS measures, including its environmental components, had negative economic effects on Mexico, one could expect there to be a drop in trade over the pre-NAFTA period or at least a slower rate of growth. At the same time, however, reductions in tariff levels brought about by the agreement would stimulate trade. Establishing precise causality, therefore, is a difficult task.

Table 4 presents the value of exports of agricultural products from Mexico to the United States between 1992 and 1999.[8] The table also highlights exports that are specifically vulnerable to SPS measures (such as meat, fish, and vegetables). Exports have at least doubled between 1992 and 1999 in all cases except preparations of meat or fish, which increased by one-third.

Mexico's strong export performance in the agricultural sector as a whole tends to suggest that the NAFTA's SPS provisions, including environmental measures, had no discernible negative economic effect on Mexico. Undoubtedly, by combing through the export data, one could find particular subsectors that might have been hurt by the SPS rules. But such results are part of every trade agreement. And if such effects did occur, an argument can be made that the tougher SPS rules have actually increased Mexico's social welfare by raising health and safety standards. Not to mention that consumers in the United States and Canada now have greater confidence in Mexican products.

Investment

During the NAFTA negotiations, environmentalists worked hard to ensure that the NAFTA incorporated safeguards to prevent the parties from becoming "pollution havens," attracting investment with promises of low environmental standards or lax enforcement (Esty 1994b; Audley 1997). One might have expected that these safeguards could work to restrict investment that Mexico might otherwise have attracted. Article 1114.1 of the NAFTA explicitly permits adoption and enforcement of measures to

Table 3
Mexico's exports of selected natural resource products to United States, in millions of US dollars. Source: DataINTAL 2.0.

1992	1993	1994	1995	1996	1997	1998	1999
7,328,940	7,608,233	8,599,343	10,765,828	13,601,154	14,405,463	11,476,916	13,371,422

Table 4
Mexico's exports to United States of products vulnerable to sanitary and phytosanitary measures, 1992–1999, in millions of US dollars. Source: DataINTAL 2.0.

	1992	1993	1994	1995	1996	1997	1998	1999
Chapters 1–24	2,947,499	3,511,769	3,828,869	5,283,038	5,077,039	5,528,315	5,880,915	6,473,204
Chapter 2 (Meat Products)	20,087	24,962	33,880	53,967	89,454	116,946	98,278	125,246
Chapter 3 (Fish, Crustaceans)	268,230	352,807	395,947	567,835	565,278	598,567	497,445	575,871
Chapter 4 (Dairy Products, Eggs)	2,527	2,756	4,276	5,020	19,324	14,448	11,537	11,667
Chapter 7 (Vegetables)	904,605	1,200,998	1,252,231	1,675,527	1,571,406	1,635,084	1,882,108	2,039,670
Chapter 8 (Fruit, Nuts)	384,188	389,381	433,128	522,081	546,102	542,742	623,823	777,840
Chapter 16 (Preparations of Meat or Fish)	44,934	41,667	43,109	51,614	85,790	92,584	58,032	65,358
Chapter 20 (Preparations of Vegetables, Fruits, or Nuts)	132,617	156,604	187,940	209,461	227,906	240,249	286,629	302,457
Chapter 21 (Miscellaneous Edible Products)	58,157	72,615	79,731	98,714	109,384	132,232	143,940	166,969

ensure that investment activity proceeds in a manner sensitive to environmental concerns; article 1114.2 stipulates that parties should not waive or otherwise derogate from environmental measures in order to attract and/or retain investors. However, since the NAFTA's entry into force, article 1114 has seldom been mentioned. Indeed, no public record exists of any complaints against Mexico, which might have led to some subsequent reductions of investment or even of any consultations among the parties on such issues. Instead, it is the investor-state dispute-settlement mechanism set forth in section B of chapter 11 that has emerged as a centerpiece of the trade and environment stress in the NAFTA.[9]

Dispute Settlement

Is there evidence of negative economic impacts on Mexico from the NAFTA's environment-related dispute-settlement mechanisms? Once again, the answer is No. The environment-related provisions of the NAFTA's chapter 20 cannot in and of themselves generate economic suffering. (See appendix A for details of the provisions.) And to date, no cases involving a conflict between the NAFTA rules and International Environmental Agreements have been heard.

In terms of the dispute-settlement procedures under the NAFTA's environmental side agreement, by September 2001 no party had yet suffered from provisions regarding either denial of trade benefits (e.g., through trade sanctions) or monetary penalties. The only active area of dispute settlement under the NAAEC has been article 14, which permits any NGO or citizen to make submissions on cases of alleged non enforcement of environmental laws. As of September 2001, Mexico had been the subject of thirteen submissions, Canada has been the subject of ten, and the United States has been the subject of eight (CEC 2001). While important for transparency purposes, this mechanism is unlikely to directly inflict economic harm on the parties, because the procedures under which it operates only permit the creation of a Factual Record as an end result.[10]

The Cost of Environmental Cleanup

Throughout the NAFTA negotiation process, some environmental activists alleged that the agreement would exacerbate damage to the environment, especially along the US-Mexico border (Audley 1993, 1997). To the extent that the NAFTA has expanded economic activities along the US-Mexico border, it continues to generate social welfare burdens associated with environmental abuse (air and water pollution, worker safety concerns, and public health risks such as those associated with inadequate water supplies in unplanned development) (Hufbauer et al. 2000). In both the short and the long term, Mexican authorities and companies will have to confront the inevitable economic costs of cleaning up and restoring the environment of border communities. In the agricultural sector as well, NAFTA-driven shifts in production away from peasant agriculture and toward horticultural activities are generating significant short-term and long-term economic costs through overuse of marginal lands, soil erosion, and loss of biological diversity (Nadal 2000). Thus, freer trade can magnify existing environmental stresses (Dua and Esty 1997)—and the NAFTA has proved to be no exception.

This raises another question: Was Mexico forced to spend money on environmental problems that it would better have devoted to other issues? Neither the text of the NAFTA nor the text of the NAAEC obliges Mexico to spend money on environmental protection or cleanup. However, Mexico faced considerable political pressure from environmental groups in all three countries (and also from the governments of Canada and the United States, which understood the political importance of Mexico's environmental image and performance to the passage of the NAFTA through their legislatures) to improve its environmental performance as a precondition for its participation in the NAFTA. In this light, the NAFTA negotiations can be said to have imposed economic costs on Mexico inasmuch as they pushed Mexico to make considerable environment investments in the five years preceding the launch of the NAFTA that it might not otherwise have pursued (Gilbreath and Tonra 1994; Hogenboom 1998; Torres 2002).[11]

But the real question is whether these expenditures were suboptimal. A sweeping review of the post-NAFTA academic and policy literature provides not even the slightest hint that they were. Nevertheless, the question of how the NAFTA's pressures for stronger environmental performance have influenced Mexico's expenditures on environmental cleanup and protection warrants further exploration. Mexico has certainly initiated several schemes in the past few years to improve environmental performance (Torres 2002). And a number of joint US-Mexico and Canada-Mexico programs have been launched. However, it remains unclear whether these initiatives are responses to international pressure in the NAFTA context, or to domestic environmental pressures within Mexico, or to a growing recognition among Mexican policy makers of the economic, social, and environmental importance of more effective environmental policy, or to a combination thereof.

Did the NAFTA make it cheaper for Mexico to clean up and improve its environment, or more expensive? The evidence shows that the NAFTA has made the cost of importing and employing environmental equipment substantially cheaper, at least with respect to products coming from Canada and the United States. Before the agreement entered into force, tariffs on such equipment were relatively high. However, the NAFTA mandates the elimination of these border charges, in most cases, either immediately or within five years. Table 5 presents Mexico's pre-NAFTA tariff levels and years to elimination for a variety of types of centrifuges and filtering and purifying machinery for liquids and gases. Before the NAFTA era, the 10–20 percent tariff at the border might have influenced a firm's decision about employing this type of environmental technology. At this point in the NAFTA implementation process, this sort of disincentive has disappeared for the vast majority of such equipment.

The Border

The environmental situation on the US-Mexico border has long been in need of serious attention. In October 1993, (less than a month before the NAFTA vote in the US legislature), the United States and Mexico signed a "Border Environment Cooperation Agreement" (Tiemann 2000). This accord established the North American Development Bank (NADBank) and the Border Environment Cooperation Commission (BECC) and is generally considered part of the NAFTA "package." The NADBank has some US$3 billion in capital for financing environmental infrastructure projects designed to improve the health and welfare of the communities along the border. As of the year 2000, the NADBank had approved 42 projects (NADBank 2000). The role of the BECC is to assist communities in developing projects for financing by the NADBank. Both institutions have faced difficulties in implementing their mandates. In particular, there is considerable disappointment about the slow rate and limited effectiveness of loans issued by the NADBank.[12]

But these mechanisms, which almost certainly would not have been created without the NAFTA, have not hurt Mexico economically. To the

Table 5
Market access effects of NAFTA on specific environmental equipment entering Mexico. Source: NAFTA Schedule of Mexico.

HS number[a]	Pre-NAFTA tariff level	Years to elimination
8421.29.01	15%	5
8421.29.02	15%	10
8421.29.03	20%	5
8421.29.04	15%	5
8421.29.05	10%	0
8421.29.99	10%	5
8421.31.01	15%	0
8421.31.02	10%	0
8421.31.99	15%	5
8421.39.01	20%	5
8421.39.02	15%	5
8421.39.03	10%	5
8421.39.04	20%	5
8421.39.05	15%	5
8421.39.06	15%	5
8421.39.07	15%	5
8421.39.08	10%	5
8421.39.09	10%	5
8421.39.99	10%	5
8421.91.01	10%	0
8421.91.02	10%	0
8421.91.03	10%	0
8421.91.99	10%	0
8421.99.01	15%	5
8421.99.02	10%	0
8421.99.03	15%	5
8421.99.99	10%	5

a. HS: Harmonized Commodity Description and Coding System.

contrary, the NADBank and the BECC have helped to finance desperately needed border environmental projects. Granted, Mexico had to commit financial resources to these institutions, but these infrastructure investments have high rates of return in terms of improved social welfare. Moreover, investing in infrastructure in the fast-growing border region makes economic sense for Mexico. The NADBank needs institutional support and improvement; the projects that it finances pay significant economic dividends from which Mexico gains enormously.

Conclusions

This case study has a simple conclusion: There is no evidence that Mexico has suffered economically in any substantive way from the NAFTA's environmental provisions. While generalizations are always a bit problematic, the NAFTA record suggests that Latin American countries have little to fear from the adoption of NAFTA-like environmental provisions in the FTAA. While none of Mexico's post-NAFTA accords include an environmental side agreement, most of Mexico's bilateral agreements subsequent to the NAFTA repeat at least some of the NAFTA's environmental provisions. For example, the 1999 Mexico-Chile Trade Agreement mirrors the NAFTA's language on the relationship to international environmental agreements; the 1994 Group of Three accord (Mexico, Colombia, Venezuela) includes virtually word for word the NAFTA language on sanitary and phytosanitary measures and standard-related measures; and the 1998 Mexico-Nicaragua Trade Agreement includes language on investment that, like the NAFTA, permits adoption and enforcement of measures to ensure that investment activity is carried out in an environmentally sensitive manner.[13] If these provisions were in any way damaging economically, it is unlikely that Mexico would have sought their inclusion in its other free trade agreements.

Cleaning up and protecting the environment makes good economic sense. The NAFTA demonstrates that trade agreements can have an important stimulating effect on environmental cooperation. And with growing economic interdependence comes an increased sense of ecological interdependence (Esty 1996).

The central question in the debate on trade and the environment in the years ahead will be how to reconcile the impulse of certain developed countries for a greater use of the "positive rule-making" methodology with the resistance of certain developing countries for anything more than some "negative prescription" in this area. The developing countries are deeply suspicious, and not without reason, of disguised protectionism, and of giving their richer neighbors a new stick in the form of high environmental standards that all must meet; but this need not be the approach (Esty and Geradin 1998). In any case, the answer is not to reject all "trade and environment" discussions, but rather to craft carefully considered and sensible provisions. When trade negotiators sit down to define the linkages of the future, they would do well to remember this.

Notes

1. For a history of the environmental aspects of the NAFTA negotiations, see Audley 1997, Von Bertrab 1997, and Orme 1996.

2. In the late 1980s, a US court imposed a ban on imports of Mexican tuna claiming that this product was not harvested in a "dolphin friendly" manner. Mexico viewed the US ban as a protectionist non-tariff barrier to trade and requested a GATT dispute-settlement panel on the matter in February 1991 just weeks before the NAFTA negotiations were launched. The GATT panel ruled that the US ban did indeed constitute discriminatory behavior (WTO 2000). While the GATT ruling was in Mexico's favor, Mexico was by that stage in the midst of the NAFTA negotiations. Backlash against the ruling from the US environmental community meant that Mexican officials were politically unable to push the US to respond fully to the decision (Vogel 1999; Mayer 1998). Instead, the Salinas administration implemented new measures to reduce the dolphin mortality rate, Salinas smilingly declaring that he "likes dolphins too" (Orme 1996).

3. For an overview of the NAFTA's environmental provisions and its side agreement, see appendix A of this volume. For the full text of the NAFTA and the NAAEC, see http://www.nafta-sec-alena.org.

4. A range of different assessments of the environmental impacts of the NAFTA on environmental law and quality are available. Assessments date from the NAFTA's negotiation phase (Magraw 1995) and have been produced at various intervals by governments, academics, and civil-society organizations since its implementation (Moss 1993; Johnson and Beaulieu 1996; Hufbauer et al. 2000). For an overview of the legal aspects, see Rubin and Alexander 1996. Many of these assessments draw on earlier theoretical explorations of the relationship between trade and the environment (Dean 1992) and the debate around the General Agreement on Tariffs and Trade (Esty 1994a; Cameron et al. 1994).

5. The relevant provisions are listed in detail in appendix A of this volume.

6. For a summary of the commitments that governments made in signing the NAAEC, see http://www.cec.org. This web site also provides detailed descriptions of the NACEC. Two important component of the NACEC's cooperative functions

are that it may be called upon to provide expert assistance for chapter 11 disputes related to the waiving of environmental measures for the purpose of attracting investment and to help parties avoid environment-related disputes before they arise (including by identifying experts capable of providing needed technical advice).

7. These export products are classified by the Harmonized Commodity Description and Coding System (HS) as follows: Mineral Products (HS chapters 25–27), Wood Products (chapters 44–46), and Base Metals and articles thereof (chapters 72–83).

8. These figures are obtained by summing the trade for chapters 1–24 of the Harmonized Commodity Description and Coding System (HS).

9. As of December 2000, some 15 investor-state cases had been filed, 8 of which were environment-related. For a list of these cases, and further analysis, see Mann and von Moltke 1999 or Mann and Araya in this volume.

10. Mexico has, however, raised concerns that a series of Factual Records might establish a record of non-compliance that would be sufficient to stimulate a process leading up to trade sanctions. However, several analyses have concluded that the prospect of any complaint ever reaching the stage of dispute settlement, let alone the imposition of sanctions, is extremely unlike given the lengthy and cumbersome process that negotiators designed (Audley 2001). Indeed, several analysts highlight that the unwieldiness of the NAFTA's environmental dispute settlement provisions was an intentional effort (Audley 2001; Rosenberg 2000). Politically, negotiators needed to devise a set of "teeth" for trade agreements to satisfy public concerns (particularly within the environmental community) in the US. The question of whether they ever could or would be implemented dwindled in importance relative to other concerns (Audley 1997, 2001).

11. Gilbreath and Tonra (1994) argue that in the years preceding the NAFTA, Mexico "undertook one of the world's most rapid transformations of environmental policy." This included introducing tougher standards, overhauling institutional structures, and increasing the number of environmental inspections.

12. For further exploration of this issue, see Kelly et al. 2001 and the chapter by Carlsen and Salazar in this volume.

13. See article 1114 of the NAFTA.

References

Araya, M. 2002. "Mexico's NAFTA Trauma: Myth and Reality." In this volume.

Audley, J. 1993. "The 'Greening' of Trade Agreements: Environmental 'Window Dressing' and NAFTA." In *North American Free Trade Agreement: Opportunities and Challenges*, ed. K. Fatemi. St. Martin's Press.

Audley, J. 1997. *Green Politics and Global Trade: NAFTA and the Future of Environmental Politics*. Georgetown University Press.

Audley, J. 2001. "Return to the Bad Old NAFTA Days." Paper presented at 2000 Washington conference on "Environment in the FTAA: What Can We Learn from the NAFTA Model?" (revised April 2001).

Cameron J., P. Demaret and D. Geradin, eds. 1994. *Trade and the Environment: The Search for Balance,* London: Cameron May.

CEC. 2001. *Citizen Submissions on Enforcement Matters.* Montreal: Commission for Environmental Cooperation.

Dean, J. 1992. "Trade and the Environment: A Survey of the Literature." In International Trade and the Environment, discussion paper 159, World Bank.

Dua, A., and D. Esty 1997. *Sustaining the Asia Pacific Miracle.* Washington: Institute for International Economics.

Dymond, W., and M. Hart. 2000. "Post-Modern Trade Policy: Reflections on the Challenges to Multilateral Trade Negotiations After Seattle." *Journal of World Trade* 34, no. 3: 21–38.

Esty, D. 1994a. *Greening the GATT: Trade, Environment and the Future.* Washington: Institute for International Economics.

Esty, D. 1994b. "Making Trade and Environmental Policies Work Together: Lessons from NAFTA." *Aussenwirtschaft* 49: 59–79.

Esty, D. 1996. "Revitalizing Environmental Federalism." *Michigan Law Review* 95: 570–653.

Esty, D., and D. Geradin. 1998. "Environmental Protection and International Competitiveness: A Conceptual Framework." *Journal of World Trade* 32, no. 3: 5–46.

Garciadiego, J. 1994. *El TLC Día a Día: Crónica de una negociación.* Mexico City: Grupo Editorial Miguel Angel Porrua.

Geradin, D. 1993. "Trade and Environmental Protection: Community Harmonization and National Environmental Standards." *Yearbook of European Law* 13.

Gilbreath, J. 2001. *Environment and Trade: Predicting a Course for the Western Hemipshere Using the North American Experience.* Washington: Center for Strategic and International Studies.

Gilbreath, J., and J. Tonra. 1994. "The Environment: Unwelcome Guest at the Free Trade Party." In *The NAFTA Debate*, ed. M. Delal Baer and S. Weintraub. Boulder: Lynne Rienner.

Hogenboom, B. 1998. *Mexico and the NAFTA Environment Debate: The Transnational Politics of Economic Integration.* Utrecht: International Books.

Hufbauer, G., D. Esty, D. Orejas, L. Rubio, and J. Schott. 2000. *NAFTA and the Environment: Seven Years Later.* Washington: Institute for International Economics.

Johnson, P., and A. Beaulieu. 1996. *The Environment and NAFTA: Understanding and Implementing the New Continental Law.* Island.

Kelly, M., C. Reed, and L. Taylor. 2001. *The BECC and NADBank: Achieving Their Environmental Mandate.* Austin: Texas Center for Policy Studies.

Magraw, D., ed. 1995. *NAFTA and the Environment: Substance and Process.* American Bar Association.

Mann, H., and K. von Moltke. 1999. *NAFTA's Chapter 11 and the Environment: Addressing the Impacts of the Investor-State Process on the Environment.* Winnipeg: International Institute for Sustainable Development.

Mann, H., and M. Araya. 2002. "An Investment Regime for the Americas: Challenges and Opportunities for Environmental Sustainability." In this volume.

Mayer, F. 2002. "Negotiating the NAFTA: Political Lessons for the FTAA." In this volume.

Mayer, F. 1998. *Interpreting NAFTA: The Science and Art of Political Analysis.* Columbia University Press.

McCarthy, S. 1998. "Gas War: The Fall and Rise of MMT." *Globe and Mail* (Toronto), July 24.

Moss, A., ed. 1993. *Assessments of the NAFTA.* Transaction.

Nadal, A. 2000. *The Environmental and Social impacts of Economic Liberalization on Corn Production in Mexico.* Gland and Oxford: Oxfam GB and World Wide Fund for Nature.

NADBank. 2000. Summary of Project Development and Loan Activities. North American Development Bank, San Antonio, Texas.

Orme, W. 1996. *Understanding NAFTA: Mexico, Free Trade, and the New North America.* University of Texas Press.

Rosenberg, R. 2000. "Can NAFTA's Approach to Environmental Cooperation Instruct the FTAA Negotiating Process? Emerging Institutionality and the Struggle for Political Will." Paper presented at Washington conference on "Environment in the FTAA: What Can We Learn from the NAFTA Model?"

Rubin, S., and D. Alexander. 1996. *NAFTA and the Environment.* Boston: Kluwer Law International.

Tiemann, M. 2000. NAFTA: Related Environmental Issues and Initiatives. Washington: Congressional Research Service.

Torres, B. 2002. "The North American Agreement on Environmental Cooperation: Rowing Upstream." In this volume.

Vogel, D. 1999. The Politics of Trade and Environment in the United States. Working paper 94, Berkeley Roundtable on the International Economy.

Von Bertrab, H. 1997. *Negotiating NAFTA: A Mexican Envoy's Account.* Praeger/CSIS.

Vossenaar, R. 2000. "Medidas Ambientales con Efectos Comerciales Sustanciales: La Creación de Barreras y Oportunidades de Mercado." In *Comercio y Ambiente*, ed. M. Araya. Washington: Organization of American States.

WTO. 2000. Mexico etc versus US: "Tuna-Dolphin." Geneva: World Trade Organization.

Statistical Sources

Statistics Canada. 2002, 1999. Canadian Direct Investment Abroad by Country and Geographical Area. Statistics Canada, Ottawa.

DataINTAL. 2000. DataINTAL 2.0. INTAL, Buenos Aires.

IMF. Direction of Trade Statistics (2000, 1996, 1992). IMF, Washington.

OECD. International Direct Investment Statistics Yearbook (1997, 1998). Organization for Economic Cooperation and Development, Paris.

Bureau of Economic Analysis. 2000. Top 50 Partners in Total US Trade: 1991–1999. Bureau of Economic Analysis, Washington.

Bureau of Economic Analysis. US Direction Investment Abroad (1994–1999). Bureau of Economic Analysis, Washington.

US Census Bureau. 2002, 2000. US Trade with Mexico and with Japan in 2000. Foreign Trade Division, US Census Bureau, Washington.

6

Negotiating the NAFTA: Political Lessons for the FTAA

Frederick W. Mayer

In 1990, when George Bush of the United States, Carlos Salinas of Mexico, and Brian Mulroney of Canada agreed to negotiate a North American Free Trade Agreement, none of the three leaders imagined that environmental issues would be on the negotiating table. By 1993, these "side issues" had become central to the effort to conclude the NAFTA. During intense negotiations, which rivaled those of the commercial talks concluded the year before, Canada, Mexico, and the United States agreed to a set of common rules, procedures, and institutions on the environment (and labor) along with those governing trade and investment in North America.

It is commonplace to regard the outcome of international negotiations as the product of national interests and relative power. Certainly the United States got much of what it demanded in the environmental and labor negotiations. Mexico made significant concessions, and Canada made somewhat fewer. These basic facts are consistent with an interpretation that the more powerful United States forced a weaker Mexico and Canada to accept an agreement which favored US interests more than Mexican or Canadian interests. But an analysis that looks only at the international dynamic ignores the most interesting questions. If side agreements were in the United States' "interest," why did the US decide to demand them in 1993, not in 1991 or 1992 during the trade negotiations? Why did US demands at the bargaining table take the form that they did? In particular, why did the United States insist on the largely symbolic inclusion of trade sanctions as a possible method of enforcement? And why did the final side agreements take the particular form that they did? To answer these questions, one must consider the political processes operating within each of the NAFTA countries, particularly the United States.

In this chapter, I will argue that the eleventh-hour demand by the United States for labor and environment side agreements in the NAFTA, the particular demands made in the negotiations, and the final form of the side agreements all reflected a new domestic politics of trade that differed considerably from the historical norm. Much to the surprise of the trade policy community, the NAFTA negotiations mobilized societal forces (including many environmental groups) that had never previously involved themselves in policy debates on trade. The nature of the US political system, particularly the need for the Congress to approve trade agreements and the Congress's sensitivity to external pressures (i.e., lobby groups), created a strategic opportunity for environmental groups to exercise their newfound leverage.

Although much has changed in American politics since 1993 (most notably, a Republican now sits in the White House, and Republicans have a majority in the House of Representatives), the political forces that first mobilized around the NAFTA continue to define the US political landscape. As in 1993, no agreement will get through the Congress without some provisions to address environmental and labor issues. Indeed, Congress will likely not even grant negotiating authority without promises to include such provisions. Yet the FTAA talks are proceeding without such provisions, reflecting the determination of Latin American countries not to include them. There is, therefore, a disconnect between the negotiation process and the realities of US politics that, if not addressed, will prove a formidable obstacle to the creation of a Free Trade Area of the Americas.

A Brief History of the NAFTA's Environmental Negotiations[1]

To understand why the NAFTA was ultimately accompanied by an environmental side agreement requires a close look at the history of negotiations and the sequence of choices made by the many actors engaged in the process. The history can be divided into three distinct stages: the period leading up to the congressional vote in May 1991 to grant negotiating authority, which put environment on the agenda for the first time; the negotiation of the NAFTA itself by the Bush administration culminating in August 1992 (which included a "parallel track" environmental negotiation); and the "side" negotiations on environment and labor conducted by the Clinton administration in 1993.

Putting the Environment on the Agenda

Perhaps the first public clue that environmental issues might be important to the NAFTA's fate was a one-day conference held on Capitol Hill in January 1991, four months before Congress would vote on whether to give the president "fast track" authority to negotiate the NAFTA (and to conclude the Uruguay Round of the General Agreement on Tariffs and Trade (GATT)). Historically, obtaining fast track authority had been a rather pro forma exercise. In the fall of 1990, however, a small group of labor, environmental, and citizen activists had begun planning a campaign to make the fast track vote a referendum on the NAFTA and to insist that their concerns be added to the agenda. Organizers expected only a modest turnout for the January meeting. Instead, the hall was packed, revealing for the first time the beginnings of a coalition of environmental groups and labor unions, as well as human rights, religious, and other interests, that would eventually threaten the NAFTA.

Environmental activists voiced three concerns about the NAFTA: that it would encourage US businesses to move to Mexico to take advantage of lower environmental standards or lax enforcement, that it would exacerbate the already terrible environmental problems of the border region, and that it could be used to challenge US environmental regulations (Mayer 1998). These concerns found sympathy among Democrats in Congress (then the majority party), who were generally more responsive than their Republican counterparts to environmental causes and who had no particular reason to make things easy for President Bush.

From the outset, the labor and environmental issues were politically connected. Neither issue alone would have been sufficient to force the hand of Congress or the Bush administration. For environmental groups, a labor connection greatly increased the clout of environmental groups on Capitol Hill, where union support was often crucial to Democrats. The environmental connection helped labor unions broaden their appeal to the media and to the general public, which often appeared more interested in dramatic images of environmental degradation than in the plight of workers. "Environment became a means of drawing attention to poor company practices in the border," recalled the AFL-CIO's Mark Anderson. "Nobody cared about a worker losing his job in Illinois. They were much more sensitive to toxic dumping in Mexico."[2]

In early March, the two most important figures in Congress for trade leg-islation—Dan Rostenkowski, chairman of the House Ways and Means Committee, and Lloyd Bentsen, chairman of the Senate Finance Committee—sent a letter to the president with a clear message: without some promise to address environmental and labor issues, they could not muster enough Democratic support to approve fast track negotiating authority (Bentsen and Rostenkowski 1991). An even stronger message was sent by Representative Richard Gephardt, then House majority leader. "I request that you not limit the talks to what used to be traditionally known as 'trade issues'—tariffs, trade-related investment restrictions, dispute res-olution and the like—but rather that we address North American Free Trade systematically," Gephardt wrote (1991). "To do so will require dis-cussing issues like transition measures, wage disparity, environmental pro-tection and worker rights."

The Bush administration, after some internal debate, grudgingly recog-nized the political necessity of responding to the positions established by Bentsen, Rostenkowski, and Gephardt. Without their support, fast track was doomed. Bush's response was an "Action Plan" in which he promised to address the issues raised by the congressional leaders. On the environ-mental front, the plan included "an expanded program of environmental cooperation in parallel with the free trade talks" as a promise to produce a comprehensive plan for cleaning up the US-Mexico border; it also included the appointment of environmentalists to trade advisory committees (Bush 1991). On the labor front, the plan promised to provide "adequate adjust-ment assistance and effective retraining for dislocated workers" and to "expand US-Mexico labor cooperation" (ibid.).

The promise was sufficient to win the backing of several major environ-mental organizations, including the World Wildlife Fund (WWF), the National Resources Defense Council (NRDC), and the National Wildlife Federation (NWF). In fact, the NWF's president published an editorial let-ter of support in the *New York Times* (Mayer 1998). In the judgment of these groups, the US president's actions were an important step toward putting the environment on the international trade agenda. After represen-tatives of the WWF, the NRDC, and the NWF met with the US president on May 8, WWF President Kathryn Fuller said: "For the first time, envi-ronmental issues are being intertwined throughout the negotiations and dis-cussions on an international trade agreement." (Greenwire 1991) The plan also won Gephardt's grudging support, although he made clear that he intended to hold the administration to its commitments.

Many in the environmental community, however, were not won over by the president's promises. The Sierra Club, Friends of the Earth, Greenpeace, and allies such as the increasingly active Public Citizen viewed the endorsement by their environmental colleagues as a sellout. "At the very least," said a spokesperson for Friends of the Earth, "we expected the big environmental groups to stay on the fence" (Lee 1991).

The split verdict by environmental groups reflected a deep divide in the environmental community. Those groups endorsing fast track were generally Washington-based organizations accustomed to the give and take of Washington politics. Those groups opposed to fast track were generally grassroots organizations, less willing or able (because of their internal structure) to compromise on causes they supported. For supporters, the NAFTA was as much an opportunity as a threat; it represented a chance to address the existing environmental problems on the border and to set a new precedent for inclusion of environmental concerns in trade negotiations. For opponents, however, the NAFTA was perceived primarily as a threat that would exacerbate environmental problems on the border and could weaken environmental regulation at home.[3]

Labor remained united in opposition. The labor activist Pharis Harvey, a leader of the opposition coalition, denounced Bush's promises as an "inaction plan" and charged that it failed to deal with concerns about "environmental destruction, suppression of labor rights and standards, growing wage disparities, increased immigration, dangers of pesticide use, human rights abuses, and the need for adjustment assistance for affected workers" (Federal News Service 1991). Labor unions were quietly furious with Gephardt for his decision to support fast track.

The vote in May was conducted under a provision in the 1988 Trade Act that gave the president an automatic extension unless either the House or the Senate voted to deny it. This unusual circumstance made it considerably easier to obtain fast track authority in 1991 than it has been subsequently, since opponents could not use the blocking tactics that commonly prevent controversial legislation. In the end, Congress, with the support of a sizable minority of Democrats, approved fast track authority on a narrow vote. For those Democrats, the support of several large environmental groups was crucial. It neutralized the environment as an issue and helped overcome the continued opposition of labor. It was now clear that it would be difficult, if not impossible, to pass the NAFTA without support from some elements of the environmental community, a lesson not lost on environmental leaders. Now the promises made in order to obtain fast track created a benchmark against which progress on the environment could be evaluated. The environment was on the agenda.

The NAFTA Negotiations and the Border Plan

For the next 15 months, as the Bush administration negotiated the NAFTA, a growing number of people in the environmental community monitored the negotiations and plans for cleaning up the US-Mexico border. They became increasingly unhappy. When the Bush administration released a draft border plan in February 1992, fast track supporters such as the NRDC joined opponents of fast track in virtual unanimous condemnation. "They are gambling that they can turn the environmental issues into a high-profile side show and that nobody's really going to look at the content of this plan and whether the money's really there," said Mary Ellen Kelly, a leader in the border environmental coalition (BNA 1992a). When a draft of the NAFTA negotiating text leaked in March, there was a firestorm of protest from across the environmentalist spectrum.

In the NAFTA talks themselves, perhaps the central issue was what standard to use when environmental standards came into conflict with principles of free trade. Environmentalists feared that trade agreements would be used to gut environmental standards. The draft text, borrowing language from GATT, put the burden of proof on environmental regulations to show that they were the "least trade restrictive" option. The environmental community insisted that these terms be renegotiated to give primacy to national environmental laws.

To maximize their leverage, the environmental groups enlisted the support of important legislators on Capitol Hill. Senator Max Baucus (D-Montana), chair of the Trade Subcommittee of the Senate Finance Committee, wrote letters to US Trade Representative Carla Hills and EPA Administrator William Reilly insisting that these issues be addressed. In the House, Representative Bill Richardson (D-New Mexico) warned: "What will decide the passage of the free-trade agreement in the Congress probably next year will be the issue of the environment." (BNA 1992b) The Bush administration got the message. Although many changes requested by the environmental community were not adopted, the language on environmental, health, and safety standards was renegotiated. The new language shifted the burden of proof so that standards could not be challenged if they were supported by legitimate scientific reasoning. In addition, negotiators added hortatory language stating that standards should not be lowered to attract investment.

When the NAFTA negotiations concluded in August, however, environmental groups were far from satisfied. In their view, the border plan was still underfunded, and they were concerned that parallel discussions to address environmental cooperation had languished. Sensing that the NAFTA was in trouble, the Bush administration promised more money for the border region and announced its support for the creation of a North American Commission on the Environment that would facilitate environmental cooperation (although it would have no enforcement powers). In hearings on Capitol Hill, the WWF signaled its likely willingness to support the agreement. As November and the presidential election approached, it became apparent that the NAFTA would be voted upon in a Clinton presidency. Environmental organizations anticipated that this development might create new possibilities for the environment. The environmental community upped the ante.

Candidate Bill Clinton was under considerable pressure from labor unions to insist that "Bush's NAFTA" be renegotiated or even scrapped. President Bush's promises to beef up worker retraining and to improve cooperation with Mexico on labor issues had never addressed the unions' concerns about wage disparities, worker rights, and job loss. From their perspective, Bush's negotiators had not even lived up to their minimal promises. Some environmental groups shared a similarly negative assessment of the NAFTA. However, the groups that had supported fast track in 1991 were less inclined to insist on scrapping the NAFTA than on adding stronger environmental provisions to it. They pressed Clinton to insist on negotiating side agreements to "fix" the NAFTA.

Clinton was wary of backing away from the NAFTA entirely, both because he generally supported free trade and because he was concerned that he would be seen as caving in to pressure from special interests. On the other hand, Clinton wanted to be responsive to certain groups because he needed their political support. After an intense internal debate within the campaign, Clinton opted for a middle stance. He would support the

NAFTA, but before sending it to Congress for a vote he would insist on negotiating supplemental environmental and labor agreements with Canada and Mexico.

The Environmental Side Negotiations

It would take the Clinton administration until the middle of August 1993 to translate a vague campaign promise into an international environmental agreement with Canada and Mexico, neither of which had any interest in negotiating it. Two contentious issues emerged: (1) the degree of independence that the international institutions this agreement created would have, including the important question of whether private citizens would have standing to initiate investigations, and (2) the nature of the dispute settlement and enforcement provisions, including whether trade sanctions could be used for enforcement.

The Mexican government was unhappy that it had been compelled to negotiate at all. (The Canadian negotiators were also unhappy, but they recognized that the push for these side agreements had to do with perceptions about Mexico, not Canada.) From the perspective of Mexico's negotiators, the United States was trying to change a deal already fully negotiated. When President Salinas of Mexico met with US President-Elect Clinton, he insisted on two conditions for talking: that there be no reopening of the NAFTA text itself and that there be no attempt to negotiate common North American standards. Clinton agreed: the talks would be supplemental negotiations focused on finding means to ensure "national enforcement of national laws." This was a crucial concession, as it meant that there would be no change in the traditional assumption that a nation can adopt whatever level of protection it chooses for environmental and natural resources.

As *demandeur* in the environmental negotiations, the United States had first to decide what it wanted. Mexico and Canada waited. During this stage, environmental groups played a crucial role in helping to define the US bargaining position. In part this was because they had more expertise on these matters than the trade negotiators at the Office of the US Trade Representative (USTR), but in large part environmental groups were thought to hold the key to passing the NAFTA. Labor leaders enjoyed considerable access to the Clinton administration, and they quietly used it to push for a labor agreement that would address their concerns about wage rates and labor rights in Mexico. They could not credibly promise to support the NAFTA if the administration negotiated a strong side agreement. As a consequence, the administration initially focused more on formulating a position for the environmental negotiation and then, not wanting to take a weak initial position in the labor negotiations, used the environmental position as a template for the labor talks.

Almost from the outset, the ideological split in the environmental community became apparent to the Clinton administration. Like the labor unions, the Sierra Club, Friends of the Earth, Greenpeace, and their ally Public Citizen remained so critical of the NAFTA that there was little likelihood that their support could be won. The Sierra Club, for example, called for renegotiating part of the NAFTA text, for an independent international agency "with the power to investigate claims of environmental wrongdoing," for a right of citizens to "seek damages for lax environmental protection within courts of law of any party," and for a provision requiring US corporations operating outside the United States to adhere to US laws (Sierra Club 1993). None of these demands was judged to be negotiable with the Mexicans.

The administration, therefore, quickly focused on those environmental organizations which might eventually support it. The NWF, the WWF, the EDF, and the NRDC perceived the NAFTA as an opportunity to advance environmental causes of concern to them. Stewart Hudson of the National Wildlife Federation later put it this way: "We felt at the end of the day we would get more if there is a quid pro quo."[4] These environmental groups now pushed for the creation of a strong and independent international institution with powers to investigate allegations by citizens of failures to enforce national environmental, health, and safety laws and to impose penalties (including trade sanctions) for violations.

Clinton administration officials also consulted with the business community, since its support of the NAFTA would be essential when Congress considered it later in the year. Business was very wary of the new side negotiations, and worried that they would create trade barriers in the guise of environmental protection. A March letter from the US Chamber of Commerce to new USTR Mickey Kantor warned that the "use of trade sanctions as a remedy for environmental violations could evolve into a new set of non-tariff barriers" (Workman 1993). In March, the pro-trade faction of the Clinton administration, headed by Treasury Secretary Lloyd Bentsen, prevailed on Kantor not to include a demand for trade sanctions in the negotiating position of the United States.

The internal battle on the US side was not over, however. Early in his tenure as USTR, Mickey Kantor had said that the NAFTA agreement was flawed and had promised, as a remedy, side agreements with "teeth"—a metaphor that defined much of the subsequent public debate in the United States. In testimony before the Senate Environment and Public Works Committee in March, Kantor told the senators: "The process must result in decisions that have real teeth and meaningful results."[5] Environmental (and labor) advocates worked with their allies on Capitol Hill to equate teeth with trade sanctions.

In May, the environmental organizations with which the administration had been consulting stated explicitly what would be required to gain their support. Seven groups—the WWF, the EDF, the NWF, the Nature Conservancy, the Audubon Society, the NRDC, and Defenders of Wildlife—sent a public letter to Mickey Kantor detailing their demands. That self-styled "Group of Seven" called for creation of a strong and independent environmental commission with the power to initiate investigations and make enforcement recommendations. In this letter, the seven groups abandoned the call for a commission that could itself impose sanctions. Instead, they proposed that if the commission found a pattern of failure to comply with its recommendations, governments could initiate dispute settlement proceedings, which might culminate in trade sanctions.

The "Group of Seven" was now firmly and publicly committed to trade sanctions. Anything less would be a symbolic defeat. And given the increasing strength of opposition to the NAFTA (former independent presidential candidate Ross Perot had by then made opposition to the NAFTA a crusade for his United We Stand America organization; failed Republican presidential candidate Pat Buchanan was railing against the NAFTA as a threat to US sovereignty; on the left, a coalition of labor, environmental, citizen, and other groups continued to grow), the support of these mainstream environmental groups was crucial to the NAFTA's prospects. At a full cabinet meeting shortly thereafter, a decision was made to press for the inclusion of trade sanctions (Mayer 1998). Mickey Kantor later recalled: "We couldn't have credible labor and environmental side agreements without sanctions. It just wouldn't work. How far we could go, how far the Mexican government would allow us to go, we didn't know, but that was a part of the negotiation. We had to have something at the end of the day that both Congress and the American people and the press would see had some teeth to it."[6]

The focus now shifted from the internal US negotiation to the international arena. On May 17, in Hull, Canada, the United States finally tabled its full bargaining position. It included a demand for an independent commission with the power to investigate claims made by citizens and for a dispute process that included the possibility of trade sanctions. The position drew an immediate negative response from both Canada and Mexico. "The Mexicans opposed trade sanctions and opposed independent secretariats on both labor and environment," recalled Chip Roh, the assistant USTR playing a lead role in the negotiations (Mayer 1998). Canada was only slightly less perturbed. John Weekes, Canada's chief trade negotiator, went public with his opposition. "We . . . have encountered some serious difficulties, particularly regarding the manner in which the secretariat might operate," he said. "We have difficulties as well in the area of dispute settlement, including the use of trade sanctions."[7] The US business community was furious. A letter to Kantor signed by every major business organization stated: "[The proposal] threatens to create a new, politically unaccountable bureaucracy. . . ." It characterized trade sanctions as "unnecessary" and "counterproductive" (Business Roundtable et al. 1993).

All sides were now becoming entrenched in their incompatible positions. The side negotiations were at an impasse. The NAFTA's backers in the business community, on Capitol Hill, and in the administration searched for a way to reach a resolution. The United States modified its bargaining position on the issue of trade sanctions; in cases of persistent failure to enforce national laws, panels could impose fines of up to US$20 million rather than trade sanctions that could be used only if the offending party failed to pay the fines. The scheme was slightly more palatable to the Mexicans, who could credibly claim that, since they would pay any fines, trade sanctions would never be invoked. The Canadians remained dead set against the possible use of trade sanctions.

As summer passed, the NAFTA's prospects in the US Congress became ever more tenuous. Unable to promote the NAFTA until the side agreements were completed, the Clinton administration had left the field to the opponents, who were clearly making inroads in public opinion and in Congress. Alarmed, the Clinton administration increased its efforts to woo uncommitted representatives—most notably Richard Gephardt, who was still considered crucial to the NAFTA's prospects in the House of Representatives. For its part, the Mexican government was persuaded that the situation was dire, and grudgingly moved toward the US position. Mexico was much less willing to compromise in the labor negotiation than in the environmental talks, with the result that what had initially been parallel discussions took quite different paths. In the labor talks, Mexico initially refused to even consider the possibility of fines or sanctions, and never accepted the idea of an international institution not directly controlled by national governments. The United States eventually dropped its insistence on identical environment and labor institutions and began to back away from insistence on sanctions as a possible remedy for all types of labor disputes.

Finally, in marathon negotiating sessions in Washington during the first week of August, Mexico largely acceded to US demands in the environmental side negotiation. The environmental side agreement would create a North American Commission for Environmental Cooperation (NACEC) with power to hear complaints brought by citizens or governments, and with modest capacity to investigate them. And Mexico accepted that in extreme cases governments could initiate a dispute settlement process, which could lead to fines and even (at the very end) to trade sanctions. The acceptance of trade sanctions was a bitter pill for Mexico to swallow; Mexico viewed the idea of international sanctions for failure to enforce its domestic laws as a violation of sovereignty, but the Mexican government recognized that without this concession the NAFTA would never get through Congress.

Canada, however, much to the surprise of the American negotiators, refused to budge on the sanctions issue. Even though it was clear that the dispute settlement process was so laborious that it was highly unlikely that they would ever be used (to date, there have been no trade sanctions imposed in environmental dispute cases), the symbolism of trade sanctions was too much for the new government of Prime Minister Kim Campbell (who had replaced a beleaguered Brian Mulroney that summer). In Washington, it seemed that the whole deal might collapse, but the political reality was that the NAFTA's opponents were much more focused on Mexico than on Canada. The United States quietly agreed to a different arrangement with Canada that did not include the possibility of trade sanctions.

In the United States, the side agreements did not sway many NAFTA critics. The labor agreement fell far short of the demands made by unions and their allies. There were no provisions for dealing with wage disparities (although President Salinas promised to raise Mexico's minimum wage), and no enforcement provisions for dealing with violations of worker rights. Richard Gephardt, whose support had been sought by all three countries, announced his opposition. The Citizens Trade Campaign, the opposition coalition that included many environmental groups, blasted both side agreements as wholly insufficient. The environmental groups with which the Clinton administration had been working announced that they were satisfied with the environmental side agreement and would now support the NAFTA. In their judgment, the agreement was strong enough, and they were pleased that it was accompanied by several US-Mexico bilateral measures intended to address the environmental problems of the border region. Although their support led directly to few converts in the US Congress, their stance largely neutralized the environment as an issue for the NAFTA, and gave members of Congress political cover on this front. Legislation to implement the NAFTA and the side agreements were passed by Congress in late November, with more than 100 House Democrats voting in favor.

Lessons Learned

What are the lessons of this history? How were a relatively small number of environmental groups able to push the environment onto the agenda of an international trade negotiation when none of the three countries intended initially to include it? How were these groups able to achieve as much as they did in that negotiation, given the opposition of US business and the Mexican and Canadian governments? And what accounts for the limits on what they could achieve?

An analysis of these questions must begin with the observation that this was not simply an international negotiation, but rather a "two-level" bargain, in which the nature of the internal, domestic-level negotiation determined bargaining positions in the international negotiation.[8] It is not particularly helpful to talk in terms of the "US interest" in an environmental side agreement, as that raises the question of where that interest came from and why it appeared to change so markedly from 1990 to 1993. Nor can one make sense in these terms of why the United States took particular stands in the side negotiations, risking the whole agreement for a largely symbolic capacity to impose trade sanctions. Rather, the determination of the "national interest," as revealed by particular bargaining positions and commitments in the side negotiations, reflects the outcome of a domestic political process, a process in which environmental groups enjoyed considerable leverage.[9]

The question then becomes: What gave environmental groups leverage in the US domestic-level negotiation?

One part of the answer has to do with the nature of the US Congress. The Mexican Congress would accede to whatever President Salinas asked of it (in 1993, Mexico's Institutional Revolutionary Party was still clearly predominant). In Canada, the parliamentary system ensured that Mulroney (and later Campbell) could count on a solid Conservative majority in Parliament. The US Congress was much less predictable. The side negotiations were always conducted in the shadow of an uncertain vote in Congress. Moreover, the relative weakness of political parties and the relative strength of interest groups in the United States make the US Congress particularly responsive to outside political pressure.

A second part of the answer concerns the breadth of the political opposition mobilized in the NAFTA context. Without the virtually unanimous opposition of labor, the strenuous opposition of many other environmental groups, and the vociferous opposition by Ross Perot and Pat Buchanan, trade politics as usual could have ensured passage of the NAFTA without any support from environmental groups. Historically trade politics has been a rather specialized affair, conducted without public fanfare by trade lobbyists and trade policy makers in Congress and in the executive branch. The historical challenge for trade policy has been to fend off narrow protectionist interests, typically domestic industries seeking competitive advantage over international competitors.[10] The politics of the NAFTA were qualitatively different. Business interests were almost unanimously for free trade, and most of the important opposition came from societal forces that had not previously engaged in trade issues.[11] One of the ironies of the NAFTA fight was that, in tactical, political terms, it forced groups with completely different ideological outlooks into collusion, even though they never actually formally cooperated. Ross Perot, who made opposition to the NAFTA his rallying cause in 1993, never set foot in the offices of Democratic politicians on Capitol Hill or communicated directly with the unions or with the Citizens Trade Campaign. Pat Buchanan, a right-wing populist who spoke for a collection of conservative nationalist groups, and Ralph Nader, a left-wing populist allied with the Citizens Trade Campaign, both invoked concerns about sovereignty. However, Buchanan was concerned that international agreements might compel regulation, and Nader was concerned that they might undermine it. The two never spoke. Nevertheless, the breadth and depth of opposition to the NAFTA, from many quarters, gave leverage to environmental organizations still in play.

Third, and just as important, the Democrats' majority in both houses made obtaining at least some Democratic votes for the NAFTA a necessity, further empowering environmental interests. Most Republican members of Congress, responsive as they tended to be to business interests, were likely to support the NAFTA without any side agreements. Only a distinct group on the far right opposed the NAFTA. Democrats, on the other hand, were under tremendous pressure to oppose the agreement from many of their traditional constituencies, most importantly the labor unions. By the time the side negotiations concluded in August 1993, it was clear that the majority of Democrats would oppose the NAFTA, and that the vote would be extremely close in the House of Representatives. The Republican leadership insisted that 100 Democrats would have to be in favor of the NAFTA before they could deliver enough Republican votes to pass it, making centrist Democrats in the House the swing votes. These members generally favored free trade, but few could not afford to vote for any agreement that was unanimously condemned by the environmental community. The closer the vote, the more evident it was that these votes were needed, and the more leverage environmental groups willing to endorse the NAFTA enjoyed.

Last, environmental organizations were strategically sophisticated. From the outset, a number of leaders in the US environmental community saw an opportunity in the NAFTA. Most of them did not subscribe to the more alarmist views about its negative effects. They had real concerns about how to balance trade and the environment, as well as about environmental problems in the US-Mexico border, but they were determined to extract as much as possible on these issues from the side negotiations. What seemed possible changed as political circumstances shifted. With Clinton's election, in particular, and with the rising tide of general opposition to the NAFTA, leaders in the environmental movement recognized their political importance. The May 1993 letter from the Group of Seven[12] represented a sophisticated reading of what was possible in the political environment of the moment and a clear commitment strategy to capitalize on it. Once the

"Group of Seven" publicly committed to particular priorities, the Clinton administration was virtually forced to negotiate an agreement that included the group's calls for both an independent commission with standing for citizens to bring complaints and for at least some possibility of trade sanctions as an enforcement mechanism.[13]

Of course, there were also domestic politics at work in Mexico and Canada, which were reflected in their stances in the side negotiations. Although the Salinas administration had staked so much on the NAFTA that it had relatively little bargaining power in the side negotiations, there were limits beyond which the Mexican government would not go. These were most evident in the labor talks, which threatened business interests in Mexico close to the Salinas administration. At the margin, therefore, Mexico was willing to concede more on environment than on labor, particularly because of the lack of strong independent unions in Mexico.[14] To the surprise of US negotiators, Canadian domestic politics made it impossible for Canada to accept any form of trade sanctions in the side agreements. Sanctions had long raised sovereignty issues for Canadians, and the weak position of Prime Minister Campbell and her Conservative Party after Mulroney's resignation made it doubly difficult for Canada to accept them.

Implications for the FTAA

Domestic political circumstances in each of the NAFTA parties at particular moments during the NAFTA negotiations determined the requirement for, and many of the characteristics of, the environmental side agreement. Turning to the present, and the ongoing Free Trade Area of the Americas (FTAA) negotiations, the analytic challenge is to assess and to anticipate the political landscape through which this agreement must navigate.

In the United States, some features of the political landscape look reasonably certain.

First, all roads lead through Congress. Although much preliminary work can and is being done, the United States cannot truly make commitments, nor will other countries make tough concessions, until Congress grants the president some form of "fast track" negotiating authority (which the Bush administration now seeks to call "trade promotion authority"). This is a very tall order. As of April 2002, Congress had not voted for new fast track authority since 1988, when the political context for trade was completely different from the present situation. The vote in 1991 to allow fast track authority for the NAFTA was actually a vote on whether to disallow an otherwise automatic extension of the 1988 authority, a much simpler legislative procedure than a normal trade bill. As the failed effort to obtain new authority in 1997 showed, future efforts to gain trade promotion authority will continue to be a lightning rod for opposition in the United States.

Second, the societal forces that first mobilized to oppose the NAFTA, far from dissipating with the passage of time, have grown considerably stronger. As was demonstrated in the 1997 "fast track" debate, in the 1998 campaign against the Multilateral Agreement on Investment, in the disruption of the 1999 Seattle meeting of the World Trade Organization, and in the protest against the FTAA in Quebec City in 2001, the capacity of the opposition coalition to mobilize political pressure has not waned. The opposition is larger, more capable, more connected, and just as passionate today as it was in 1993.[15] It is also more international, linking environmental, labor, and citizen groups across the civil societies of the Americas. However, it does not have unlimited power, as Congress's 2000 approval of granting permanent normal trading status to China demonstrated. Also, some of the particular features of the NAFTA political landscape are absent in the present context, among them the particular salience of Mexico, including the highly visible border environmental problems and the anxieties about immigration. Nonetheless, the opposition coalition will remain a force with which to be reckoned.

Third, even before the defection of Republican Senator Jim Jeffords returned control of the Senate to the Democrats in 2001, it was difficult to imagine passing trade-promotion-authority legislation over unanimous Democratic opposition. Trade bills have never passed on straight party-line votes. Now, with Democrats chairing committees in the Senate, no trade legislation can move without significant support from Democrats. In many ways, a Republican president complicates matters; Democrats have no ties of party loyalty of the kind that Clinton used in the NAFTA to counter the pressure from labor, environmental, and other groups. It follows, therefore, that there will no fast track legislation without significant Democratic support, and there will be no Democratic support unless fast track is accompanied by significant measures on environment and labor.

There is, however, something of a disconnect between the political realities in the United States and the trajectory of the FTAA talks to date. The

FTAA negotiating structures do not anticipate inclusion of any environmental or labor side agreements (or similar provisions in the text). Moreover, there is strong resistance throughout Latin America to moving in that direction, driven in part by Mexico's experience. Although there is little evidence that the side agreements have negatively affected Mexico, the symbolic significance of being compelled by political necessity to reopen talks with the United States in 1993, and of being forced, alone, to swallow the inclusion of trade sanctions in the agreement, left a very bitter taste. Not surprisingly, other countries in Latin America identify with Mexico.

What, then, should be done to avert a policy stalemate? Clearly the environmental agenda is here to stay. Environmental groups will continue to have enough influence on US domestic politics to insist on continued linkage between trade and the environment. It is not clear, however, that the particular formulation devised for the NAFTA will be necessary or possible in the FTAA. Much of the NAFTA package relating to border issues is irrelevant in the FTAA context. It is also important to recall that the United States did encounter limits to what environmental measures it could impose on its bargaining partners. Those constraints are likely to be far tighter in NAFTA negotiations with a large group of other countries, all less economically dependent on the United States than Mexico and Canada. Trade sanctions as a means of ensuring the enforcement of environmental laws, in particular, seems likely to be outside the realm of possibility.

Clearly some compromise is needed. Logically, one approach would be to de-emphasize state-imposed trade sanctions while increasing the focus on measures to facilitate environmental cooperation and information gathering and to empower forces in the society to bring pressure to bear on environmental problems. Such a compromise would likely be more effective for achieving environmental goals than an agreement that emphasized sanctions, and might well be more politically palatable in Latin America. However, it is imperative that all sides gain better understanding of the political necessity of compromise.

Notes

1. Throughout this chapter, I draw extensively from my account of the side negotiations in Mayer 1998 and from my experience as an aide to US Senator Bill Bradley from August 1992 to December 1993. Other general sources include Audley 1997 and Ayres 1998.

2. Cited in Mayer 1998.

3. For a useful characterization of the political spectrum in the environmental community, see Esty 1994.

4. Cited in Mayer 1998.

5. Ibid.

6. Ibid.

7. Cited in Farnsworth 1993.

8. There is a considerable literature on "two-level bargaining" in political science, most of it spurred by Putnam 1998. Other works in this area include Evans et al. 1993. I provide a more thorough analysis of NAFTA and its side agreements in terms of two-level bargaining theory in Mayer 1998.

9. This perspective runs counter to the realist tradition in international relations. The realist tradition treats countries as the fundamental units of analysis and as rational actors with well-defined and stable interests, thus obviating any need to consider domestic politics.

10. This characterization of trade goes back to Schattschneider 1935. Other classics in this vein include Bauer et al. 1963 and Destler 1992.

11. This increasingly international orientation of business interests is well addressed in Milner 1998.

12. This letter can be found in Magraw 1995.

13. The environmental groups' commitment to sanctions was in some ways largely symbolic. So much of the public discourse, beginning with Mickey Kantor's metaphor of teeth, focused on trade sanctions that it became almost a litmus test for a strong agreement, even with the environmental groups' own memberships. In truth, the environmental groups were much more willing to compromise on the sanctions issue, accepting in the end a process that makes it highly unlikely they will ever be used, than on the degree of independence for the environmental commission.

14. I develop this point more fully in chapter 6 of Mayer 1998.

15. For an elaboration on the nature of the opposition to free trade and its possible implications for the FTAA, see Mayer 2001.

References

Audley, J. 1997. *Green Politics and Global Trade: NAFTA and the Future of Environmental Politics.* Georgetown University Press.

Ayres, J. 1998. *Defying Conventional Wisdom: Political Movements and Popular Contention Against North American Free Trade.* University of Toronto Press.

Bauer, R., I. de Sola Pool, and L. Dexter. 1963. *American Business and Public Policy.* Atherton.

Bentsen, L., and D. Rostenkowski. 1991. Letter to President Bush, March 7.

BNA. 1992a. "Environmentalist, Texas Group Cool to Bush Integrated Border Plan." *BNA International Environmental Daily*, February 26.

BNA. 1992b. "Environmental Issues to Decide Fate of NAFTA, Representative Richardson Predicts." *BNA International Environmental Daily*, June 29.

Bush, G. 1991. Letter to Dan Rostenkowski, May 1.

Business Roundtable et al. 1993. Letter to Mickey Kantor from the Business Roundtable, Council of the Americas, Emergency Committee for American Trade, National Association of Manufacturers, US Chamber of Commerce, US Council of the Mexico-US Business Committee, US Council for International Business, and USA*NAFTA, June 4.

Destler, I. 1992. *American Trade Politics*, second edition. Washington: Institute for International Economics.

Esty, D. 1994. *Greening the GATT: Trade, Environment, and the Future*. Washington: Institute for International Economics.

Evans, P., H. Jacobson, and R. Putnam, eds. 1993. *Double-Edged Diplomacy: International Bargaining and Domestic Politics*. University of California Press.

Farnsworth, C. 1993. "3 Nations Disagree on Trade." *New York Times*, May 22.

Federal News Service. 1991. "News Conference by the Coalition: Mobilization on Development, Trade, Labor and the Environment." May 2.

Gephardt, R. 1991. Letter to President Bush, March 27.

Greenwire. 1991. "Inside Track: Greens Put Their Trade Case to President." *Greenwire*, May 9.

Lee, G. 1991. "'Fast Track's Sprint: Frenzied Lobbying on a Treaty Not Yet Written." *Washington Post*, May 23.

Magraw, D. 1995. *NAFTA and the Environment: Substance and Process*. Washington: American Bar Association.

Mayer, F. 1998. *Interpreting NAFTA: The Science and Art of Political Analysis*. Columbia University Press.

Mayer, F. 2002. "Labor, Environment and the State of US Trade Politics." In *The United States and the Future of Free Trade in the Americas*, ed. M. Lustig.

Milner, H. 1998. *Resisting Protectionism: Global Industries and the Politics of International Trade*. Princeton University Press.

Putnam, R. 1988. "The Logic of Two-Level Games." *International Organization* 42, summer: 427–460.

Schattschneider, E. 1935. *Politics, Pressures, and the Tariff*. Prentice-Hall.

Sierra Club. 1993. Environmental Concerns Regarding the North American Free Trade Agreement.

Workman, W. 1993. Letter to USTR Mickey Kantor, March 8.

II

Environmental Performance of the NAFTA

7

Industrial Pollution in Mexico: Did the NAFTA Matter?

Kevin P. Gallagher

During Mexico's transition to an open economy, Mexican industry experienced compositional and technological changes that were relatively beneficial to the environment. However, these improvements have not been enough to trigger a reduction in overall levels of industrial pollution since the passage of the North American Free Trade Agreement in 1994. Indeed, between 1994 and 2000 industrial air pollution in Mexico, the focus of this study, nearly doubled while Mexico's GDP rose 38 percent (World Bank 2000). And many Mexican industries are alarmingly more pollution intensive than their counterparts in the United States. This chapter seeks to explain how industrial air pollution is outstripping trade-led economic growth in Mexico. It concludes by presenting a framework that would allow the more environmentally positive aspects of Mexico's experience to become more widespread throughout the western hemisphere.

Controversy regarding the environmental implications of trade liberalization flared throughout negotiations for the NAFTA. The launch of negotiations for a Free Trade Area of the Americas has rekindled the debate, but the heated dialogue has been fueled by speculation, emotion and rhetoric rather than data and analysis. Mexico, where trade liberalization began even before the NAFTA's implementation in 1994, provides an ideal laboratory for research on the actual effects of trade liberalization on the environment, particularly on levels of industrial pollution.

Over the course of the past two decades, Mexico has transformed itself from a quintessential example of "import substitution" industrialization to one of the most open economies in the world. In the midst of several macroeconomic crises, four successive phases of trade liberalization were implemented in Mexico: the "Apertura" policy of 1985, accession to the General Agreement on Tariffs and Trade (GATT) in 1986, the "Economic Solidarity

Pact" in 1988, and the NAFTA in 1994 (OECD 1996). (See figure 1.) One of the most significant results of Mexico's trade liberalization is the emergence, since 1985, of manufactured goods as Mexico's chief export, replacing petroleum, the export that fueled both the rise and fall of Mexico's economic prospects.

This chapter develops a clear picture of the differential patterns of industrial air pollution associated with Mexico's manufacturing sector during this period of rapid trade liberalization.

Trade Liberalization and Industrial Pollution in Mexico: Theory and Evidence

The relationship between economic growth and the environmental results has multiple dimensions (Esty 2001; Ros et al. 1996). Three important mechanisms through which trade and investment liberalization can affect the environment have been identified: scale effects, composition effects, and technology effects (Grossman and Krueger 1993).

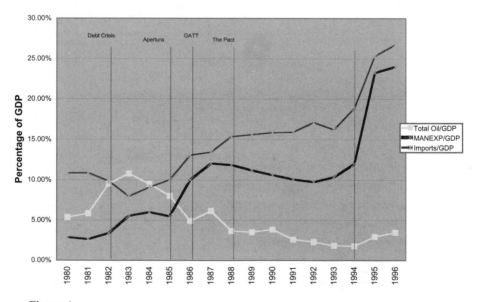

Figure 1
Trade patterns and policies in Mexico, 1980–1996. Source of GDP data: World Bank World Development Indicators, 1998.

Scale effects occur when liberalization causes an expansion of economic activity (output). If the nature of that activity is unchanged but the scale of output is growing, then pollution and resource depletion will increase along with output.

Composition effects occur when changes in trade policy lead countries to specialize in particular sectors, often those where they enjoy a comparative advantage. When comparative advantage derives from differences in the stringency of environmental regulations and/or implementation, then the composition effect of trade will exacerbate existing environmental problems in the countries with relatively lax regulations (i.e., the pollution haven effect). The opposite can be true when comparative advantage is a function of strong environmental controls.

Technique effects—changes in resource extraction and production technologies or processes—can lead to a decline in pollution per unit of output for two reasons. First, the liberalization of trade and investment may facilitate the transfer of cleaner technologies to countries with lower environmental performance (Esty and Gentry 1998). Second, if economic liberalization increases income levels, newly affluent citizens may demand the use of cleaner technologies and production processes.

The Environmental Kuznets Curve hypothesis suggests that environmental degradation may increase during the initial stages of economic development, but that scale, composition, and technology effects can cause those trends to reverse as levels of income increase. The "turning point" when such a reversal is estimated to occur, especially for criteria air pollutants, has been estimated by many economists to be close to US$5,000 GDP per capita in purchasing power parity (PPP) (Stern 1998). Mexico's per capita GDP reached US$5,000 in 1982, and by 1999 it had climbed to more than US$8,000 (World Bank 2000). According to the logic of the Environmental Kuznets Curve hypothesis, trade liberalization should have triggered noticeable environmental improvements in Mexico's economy during this period.

A growing empirical literature on trade and industrial pollution has arisen from these theories. The methodologies employed to examine these relationships vary widely, as do the results. A thorough review of this literature is beyond the scope of this study, but the studies in this realm that are related to Mexico deserve attention.[1]

Of the numerous studies that examine the environmental effects of bilateral trade and investment, one by Grossman and Krueger (1993) found that pollution-abatement costs in the US rarely drove industries to rely on imports from Mexico; instead, traditional economic determinants such as factor prices and tariffs were more important to trade and investment patterns. A more recent study by Eskelund and Harrison (1998) looked at the patterns of US foreign investment in Mexico, Venezuela, Morocco, and Cote d'Ivoire between 1982 and 1994, and also found that the pattern of US foreign investment in any of the recipient countries was not skewed toward industries with high pollution-abatement costs in the United States.

Another line of research has investigated the scale and compositional effects of Mexican industry in general. Since actual data for developing countries have not been collected until recently, most of these studies have relied on useful tools like the Industrial Pollution Projection System (IPPS), which was developed in the early 1990s by the World Bank (Hettige et al. 1994, p. 2).[2]

IPPS provides estimates of pollution intensity by sector and expresses this in terms of pollution per unit of output or pollution per employee. To make estimates, the system merges production and emissions data from a very large sample of firms in the United States in 1987, then calculates changes in pollution attributable to the scale and composition of the industry. This method assumes constant emissions per unit of economic activity (per employee or unit of output) and freezes the technique effect by assumption.[3] IPPS has now been used to estimate intensities in a number of countries, including Brazil, Latvia, Vietnam, and Mexico.

Adriaan Ten Kate was the first to apply the IPPS estimates for pollution per unit of output in the United States to Mexico. Assuming that such intensities did not change over time, Ten Kate found that the proportion of pollution-intensive ("dirty") industries within Mexico's manufacturing sector increased by approximately 50 percent from 1950 to 1970 and by 25 percent from 1970 to 1989. This reflects a marked change in the composition effect as Mexican manufactures moved toward more polluting subsectors over the period. In terms of total pollution (scale effect), Ten Kate found that manufactures were producing 20 times as much pollution in 1989 as in 1950 (Ten Kate 1993).

More recently, the World Bank, in conjunction with Mexico's National Institute of Ecology (INE), created a new database of intensities based on actual Mexican pollution levels. The new database provides estimates for pollution intensity per employee by specific industry sectors in Mexico.[4]

The Mexican estimates are only available for a single year and therefore cannot estimate changes in technique over time.[5]

The World Bank's new estimates were used in a recent study that evaluated the air pollution intensity of Mexican manufactures (Jenkins 1998). The conclusions of this study, conducted by Rhys Jenkins, conflicted with the findings of earlier studies based on IPPS estimates. For example, whereas Ten Kate found a 25 percent increase in the pollution intensity of Mexican manufactures from 1970 to 1989, Jenkins, using the Mexico intensities, found no general increase in pollution intensity over that period. Interestingly, when he resorted back to the IPPS coefficients to re-examine Mexican export-oriented manufactures, Jenkins found that air pollution intensity *increased* from 1979 (when manufactures were still highly protected) to 1990 (ibid.).

Limitations of Existing Research

A lack of sufficient post-NAFTA (that is, post-1994) data has frustrated efforts to examine the effects of Mexico's full transition to liberalized trade. Analysts have had to rely on proxies rather than on actual Mexican data. Though new data-gathering efforts and databases are under development, it is important to understand the four limitations that have hampered efforts to document industrial pollution in Mexico during the era of trade liberalization: the assumption that pollution intensities in developing countries are equal to those in the United States, the practice of holding pollution intensity constant over time, the use of pollution per unit of employment as a measure of economic activity, and the insufficient post-NAFTA data.

First, the World Bank and many of the users of IPPS data make note of the limitations of using US-based pollution intensities as proxies for pollution intensity in developing countries (Hettige et al. 1994). In particular, the United States is known for having some of the most stringent environmental regulations in the world, and most developing countries are just beginning to enact comparable policies. The World Bank's new databases on actual pollution intensities in Mexico (described above) and China are a marked improvement on the earlier IPPS data. They show the wide discrepancy between pollution levels in the United States and those in developing countries (table 1). On average, Mexico's pollution intensity is twice

Table 1
Pollution intensity for sulfur oxides ("SOX") in China, Mexico, and the United States.
Source: World Bank, New Ideas in Pollution Regulation (www.worldbank.org/nipr).

	China	Mexico	US
Food, beverages, and tobacco manufactures	15.89	1.71	0.50
Textiles, wearing apparel, and leather tanneries	11.50	1.56	0.35
Wood and wood products manufactures	43.04	0.68	0.36
Paper, printing, and publishing products	26.63	5.67	2.33
Chemicals, petroleum, coal, rubber, and plastic	84.52	2.80	3.16
Non-metallic mineral products	44.38	1.96	6.19
Basic metals (iron and steel) and non-ferrous metals	n.a.	1.12	11.89
Fabricated metals, machinery, and equipment	4.30	0.08	0.15
Other manufacturing	9.67	0.15	0.03

that of the United States, while China's intensity for sulfur oxides (SOX) is 68 times that of the United States. The new data on pollution intensity also enable researchers to examine the range of pollution intensity within developing countries.

The second limitation of existing research, and the hardest to alleviate, is the fact that often the available data are for only one year, which forces researchers to employ an assumption that pollution per employee remains constant over time.[6] The US pollution-intensity estimates were also collected during a period when pollution-control measures had been in place for quite some time, whereas in developing countries such measures have been largely ignored until recently. With either US-based or Mexico-based estimates it is safe to assume that analysts underestimate pollution intensity in earlier years (Ten Kate 1993).

A third limitation is the use of employment as the measure of economic activity. In the technical papers describing IPPS, the World Bank recognizes that the "volume of output would be the ideal unit of measurement," but data on that volume often are not available (Hettige et al. 1994).[7] The use of employment data poses a problem, particularly for countries undergoing trade liberalization, because employment levels can vary independent of output levels. During transition or crisis, firms can shed workers to maintain productivity without changing production techniques in any way that affects emissions. Thus, estimates based on employment may be tracking large employment losses rather than actual changes in economic activity. (For further details, see the technical appendix to this chapter.)

Analyzing Economic Integration and Industrial Pollution in Mexico

This section will show that significant improvements in industrial air pollution have occurred in some Mexican industries, but not to the extent that they are triggering a reduction in overall levels of industrial air pollution in Mexico. The following analysis works to improve upon existing assessments of industrial pollution in Mexico and move beyond the limitations outlined above by developing and using three sets of data: actual Mexican air emissions intensities, pollution per unit of output for the measuring economic activity, and economic data in Mexican manufactures from 1988 to 2000—the full period of Mexico's transition to open trade.[8] Using the 1997 World Bank air pollution intensities for Mexico (described above), the first subsection of this section will estimate air pollution per unit of output and total pollution (measured in tons of emissions) in Mexican manufactures for 1997, three years after the passage of the NAFTA. Using the same data, the second subsection will estimate the scale and composition effects in Mexican industry from 1988 to 2000. To estimate the technique effect, the third subsection will calculate a "harmonization index" that compares air pollution in Mexico with data from the United States for four air pollutants: particulate matter, sulfur oxides, carbon monoxide, and nitrous oxides.[9]

Pollution Intensity in Mexican Industry

This subsection outlines air pollution intensity and total air pollution for specific sectors in Mexican industry. Table 2 lists the twelve largest sectors in Mexican manufacturing for 1997 (measured in output terms). These firms represent 89 percent of all of Mexican manufacturing. The two largest sectors are transport and food manufactures, representing 23 percent of the top twelve total.[10]

Pollution intensities for these twelve sectors, converted from pollution per employee to pollution per unit of output, are depicted in figure 2. Food manufactures is the most pollution-intensive industry for all of the air pollutants except sulfur oxides. The paper industry is the most pollution intensive for sulfur oxides, followed by food manufactures. For carbon monoxide, electrical machinery and textiles are also among the most pollution intensive.

Table 2
The twelve largest sectors in Mexican manufactures, 1997. (Numerals indicate
International Standard Industrial Classification.)

311	Food manufacturing
312	Food processing
313	Beverage industries
321	Textiles
341	Paper
351	Industrial chemicals
352	"Other chemical, pharmaceuticals, etc."
371	Iron and steel
372	Non-ferrous metals
382	Machinery
383	Electrical machinery
384	Transport

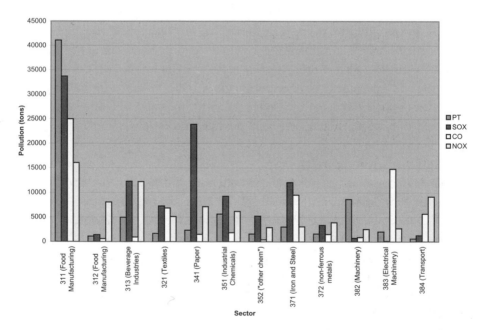

Figure 2
Total pollution in the twelve largest three-digit Mexican manufactures sectors.

Food manufactures, beverages, paper, and non-ferrous metals are the most pollution intensive for nitrous oxides. For particulate matter, machinery is the second most pollution intensive. These results are consistent with previous studies except in the case of textiles. Although many studies consider the textiles sector to be one of the cleaner industries in the world economy, it is near the top of the list for sulfur oxide and carbon monoxide intensity in Mexico.

Total pollution for each of these twelve sectors is shown in figure 3. Food manufactures, the second-largest sector in Mexico, is the largest polluter in all categories. In fact, as a proportion of total pollution in the twelve sectors, food manufactures contributes 55 percent of all particulate matter, 30 percent of all sulfur oxides, 36 percent of all carbon monoxide, and 20 percent of all nitrous oxides. Following food manufactures, paper is the largest sulfur oxide emitter, machinery is the largest carbon monoxide emitter, and the beverage industry is the second-largest emitter of nitrous oxides. These data are used to estimate the scale, composition, and technique effects in the next two subsections.

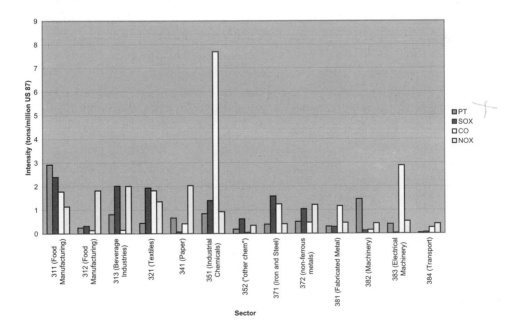

Figure 3
Pollution intensity of the twelve largest Mexican manufactures sectors, 1997.

Scale and Composition Effects in Mexican Industry, 1988–2000

This subsection estimates the composition and scale effect in Mexican manufactures over time.[11] In terms of the composition effect, air pollution intensity in Mexico decreased by approximately 20 percent from 1988 to 1994, but did not change after the NAFTA was passed. In terms of the scale effect, total levels of industrial air pollution decreased from 1988 to 1994 along with the compositional changes. Owing to changes in overall output in the manufacturing sector, however, industrial air pollution has almost doubled since the NAFTA came into force.[12]

From perspective of air pollution, the majority of the trade-related compositional change in Mexico's economy occurred before the passage of the NAFTA. Figure 4 shows pollution intensity calculated biannually from 1988 to 2000 for the nine Mexican industrial categories listed in table 3 with their classification codes. One can observe an approximate 20 percent dip in pollution intensity from 1988 to 1994. Since pollution per unit of output is held constant, these fluctuations in pollution intensity, which occurred during the period immediately following Mexico's Economic Solidarity Pact, reflect changes in the composition of industries during that time—in this case, a shift toward cleaner industries. Though overall output dropped from 1988 to 1994, it dropped less dramatically in cleaner industries. Figure 4 shows that the composition of Mexican industry did not change very much at all after the NAFTA was signed in 1994.

Pollution intensity allows us to observe the composition effect over time. Total pollution levels allow us to observe the scale effect. Total pollution is calculated by multiplying total output by pollution intensity. Figure 5 shows

Table 3
Two-digit International Standard Industrial Classification manufactures industries.

31	Food, beverages, and tobacco manufactures
32	Textiles, wearing apparel and leather tanneries
33	Wood and wood products manufactures
34	Paper, printing, and publishing products
35	Chemicals, petroleum, coal, rubber, and plastic
36	Non-metallic mineral products
37	Basic metals (iron and steel) and non-ferrous metals
38	Fabricated metals, machinery, and equipment
39	Other manufacturing

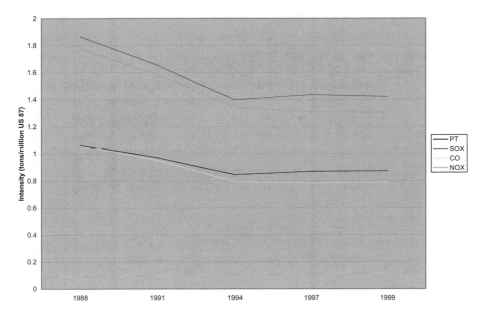

Figure 4
Pollution intensity of Mexican manufactures, 1988–1999.

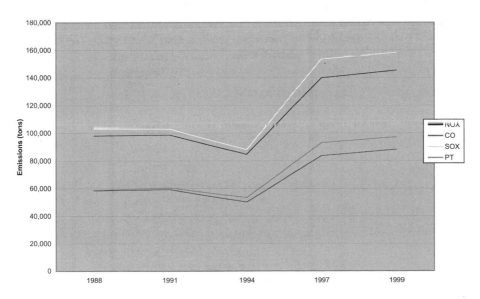

Figure 5
Estimates for air emissions in Mexican manufacturing, 1988–1999.

total air pollution from Mexican manufactures from 1988 to 2000. From 1988 to 1994, overall emissions dropped nearly 10 percent. From 1994 until the present, however, total pollution almost doubles, as a result of strong growth in manufacturing output during the period.

It is important to note that the lack of yearly pollution-intensity data means that pollution intensity has to be held constant at 1997 levels. This means that findings regarding the scale effect can only indicate general trends. As was discussed above, earlier studies reveal that pre-1997 levels of pollution could have been significantly higher than indicated in figure 5. Even if figure 5 underestimates the pollution levels, the evidence remains alarming.

Although there are indications of reductions in pollution intensity in a handful of Mexican industries, such indications are not evident across Mexican industry as a whole. Data on levels of carbon dioxide, available for Mexico on an annual basis, support this claim. Though the carbon intensity, as well as the energy intensity, of Mexican industry has declined in the past 15 years, overall carbon dioxide levels have increased dramatically.[13] The increasing scale effects for each of these criteria pollutants are a cause for considerable concern. Most of Mexican industry is concentrated in a few regions (the US-Mexico border, Mexico City, Monterrey, Guadalajara). Rising levels of air pollution are exacerbating existing trends in regions that already suffer the adverse human health effects of pollution.

Technological Change and Air Pollution in Mexican Industry

This subsection estimates the technique effect on industrial air pollution in Mexico's economy. In the absence of annual data on levels of air pollution for specific Mexican industries, an estimate of the technique effect is calculated by creating an index that compares air pollution intensity in various Mexican industries (using 1997 data) with their counterparts in the United States (using 1987 data). By comparing intensity between the two countries we can examine whether Mexican industry is approaching US levels of environmental technology, policy, and performance (see figure 6). Those industries in figure 6 with a ratio greater than 1 are dirtier in Mexico, and those with a ratio less than 1 are cleaner in Mexico.

It is striking that Mexico is "cleaner" than the United States in three significant industries: a portion of the chemicals industry, iron and steel, and non-ferrous metals. The investment and trade liberalization process, among

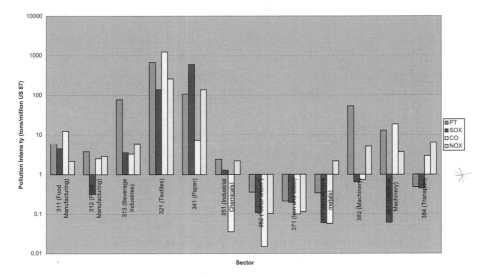

Figure 6
Comparing pollution intensities: Mexico/United States (log scale).

other factors, has led to plant-level upgrades in these industries that have cut down on energy use and therefore levels of air pollution. Unfortunately, however, these significant improvements do not offset the overall picture.

Most Mexican industries, for which pollution is more a function of plant-level processes that require end-of-pipe solutions than of plant-level upgrades, are alarmingly "dirtier" than their US counterparts, in some cases despite increasing levels of foreign and domestic investment. Air pollution intensities in Mexico are greater than those in the United States in nine of the twelve industries in figure 2. Most striking are the ratios for textiles (which are from 137 to 1225 times as dirty in Mexico depending on the pollutant), and paper (which are from 70 to 592 times as dirty). If those two are taken out of the sample, the ten largest Mexican industries are on average only 6.23 times as dirty as US industries. In addition to textiles and paper, the food and beverages industries range from 2 to 10 times as bad as those in the United States. In terms of particulate matter, Mexico's machinery industry is close to 100 times as dirty as machinery in the United States. Electrical machinery is more than 10 times as dirty for particulate matter and carbon monoxide.

A study by the Institute for New Technologies at the United Nations University (Bartzokas and Yarime 1997) provides great insight into the relative differences in pollution intensity between industries in the United States and Mexico. The authors place industrial pollution into two categories: pollution that is a function of waste in energy consumption and pollution that is a function of by-products of the production process. Upgrading energy-combustion technologies and switching fuel inputs can significantly reduce pollution resulting from waste in energy use. By-product wastes can be alleviated by installing end-of-pipe technologies or by making incremental changes in the production process.

The three industries that are cleaner in Mexico per unit of output than their US counterparts (other chemicals, iron and steel, and non-ferrous metals) are all sectors where pollution is in large part a function of energy use and plant vintage. During the process of economic integration, these sectors have been the recipients of large amounts of both domestic and foreign investment. A portion of these investments has gone to new plant equipment ("core technology") that has reduced their energy intensity and thus their air pollution levels.

From 1985 to 1994 (the period for which data are available for both domestic and foreign direct investment by sector in Mexico), investment in the chemicals and primary metals sectors was approximately 35 percent of all investment in Mexican manufacturing. During that period, the iron and steel industries attracted 13.06 percent of total investment, surpassed only by the automobile sector, which received 15.11 percent (Moreno-Brid 1999).[14] From 1994 (when the NAFTA went into effect) until 1998, Mexican manufacturing received more than US$26 billion in foreign direct investment (FDI). (Data are not available for domestic investment during this period.) Of that amount, iron and steel received just over 7 percent of the total, and other chemicals received 4 percent. Automobiles received the most FDI (just over 15 percent), followed by electronic machinery and beverages (10 and 9 percent respectively). Interestingly, these three leading recipients of FDI were dirtier than their US counterparts for many pollutants.

In the "other chemicals" and "primary metals" sectors, it is known that some of the investment went into new plant equipment leading to a reduction in energy use and pollution. The Mexican steel sector provides the most striking example. As part of the economic integration process, the Mexican government and the World Bank brokered a US$400 million in loans to restructure Mexico's steel sector, of which US$170 million was dedicated to modernizing plant equipment. In addition, an agreement was signed between the Mexican government and domestic and foreign investors regarding environmental protections and liabilities in the steel sector (Gentry and Fernandez 1998).

From 1988 to 1998, energy intensity in Mexico's steel sector (measured as energy use per unit of output) decreased by 36 percent, and that sector is now "cleaner" than the US steel sector.[15] This is in part the result of the diffusion of new, "cleaner" electric arc technology to replace the "dirtier" blast furnace technology that remains prevalent in the United States (Reppelin-Hill 1999). Investment spurred the adoption of new technologies and reductions in energy intensity in Mexico's other "cleaner" industrial sectors too. In addition to the 36 percent drop in the energy intensity of steel, energy intensity fell by 43 percent in the aluminum industry and by 25 percent in the chemical industry. Energy intensity declined at a much faster pace in these three industries than in the other sectors of Mexican manufacturing.

Overall, from 1988 to 1998, Mexican manufacturing achieved a general decrease in energy intensity of 9.2 percent. This overall reduction in pollution intensity shows that the combined efforts of the Mexican government, the World Bank, and private industry can assist an environmentally sound transition to openness. For example, recent government policy in Mexico has stimulated manufacturers of chemical fibers to initiate environmental management systems that have led to significant environmental improvement (Dominquez-Villalobos 2000). Such efforts were complemented by a number of private-sector environmental initiatives during the 1990s, including the creation of Mexico's National Council of Ecological Industrialists (CONIECO) in 1992,[16] the creation of the Latin American chapter of the World Business Council for Sustainable Development in 1993, and the creation of the Center for Private Sector Studies for Sustainable Development (CESPEDES) in 1994 (Barkin 1999). The overall impact of these schemes on Mexico's environment remains to be seen. Some Mexican analysts are optimistic (Torres 2000). Others caution that voluntary private-sector initiatives, thought useful, cannot be relied upon to address the massive environmental challenges confronting Mexico (Barkin 1999).

The relative performance of the three industries outlined above is cause for optimism; however, most industries in Mexico are alarmingly "dirtier" than their US counterparts. In the majority of these "dirtier" industries (especially textiles, food, paper, and machinery), pollution is a function of by-product waste and thus of the lack of end-of-pipe technologies rather than of the age of the plants. Such pollution-reducing technologies are not mandatory in Mexico, and many firms do not have the financial means to acquire them.

There are several explanations for the extraordinarily high pollution ratios in the textiles and paper industries. The high pollution intensity of Mexico's textile industry may be a reflection of differences in the composition of US and Mexican textiles manufacturing for areas on which no data are available.[17] In addition, Mexico's textile industry concentrates on more pollution-intensive forms of textile production. Mexico, for example, has had a competitive edge in synthetic fiber production, considered to be one of the more pollution-intensive aspects of textiles production (Botella et al. 1991; Bartzokas and Yarime 1997). A recent study of the textiles industry in Mexico found that, in its sample, the majority of firms have not made the necessary investments to offset the negative environmental impacts of their production processes (Brown 2000). This can be explained in part by inadequate financing. From 1985 to 1994, the textiles industry received only 5 percent of all domestic and foreign investment, and from 1994 to 1998 the industry received less than 1 percent of FDI in Mexican manufacturing (Moreno-Brid 1999). Further research is needed, however, to determine if the differences in the composition of "clean" and "dirty" industries within the textiles sector are attributable to trade policies between Mexico and the United States. (Methodologies for such analysis have been developed—see Gallagher and Ackerman 2000.)

The paper industry reveals a more complex situation. Air emissions in the paper industry are determined by both core technology and by pollution-control technology. No significant wave of innovation in core technologies in paper production has occurred since the middle of the twentieth century (Smith 1997). Controlling air emissions has been left to the development of "end-of-pipe technologies" and their imposition by government or by civil society (Smith 1997; Hartman et al. 1997). Like the textile industry, the paper industry receives little investment: only 0.61 percent of domestic and foreign investment in Mexico from 1985 to 1994, and only 1.7 percent of FDI from 1994 to 1998 (Moreno-Brid 1999).

The experience of the food and beverages industry and the manufacturing industry highlights how industries that rely on end-of-pipe technologies and on core technologies respond to different stimuli to promote environmental improvement. Mexico's food and beverages and machinery industries were two of the largest recipients of domestic and foreign investment during the period of Mexican trade liberalization, but both remain considerably "dirtier" than their US counterparts. The food and beverages industries differ from the textiles and paper industries in that they are labor intensive and process intensive, not energy intensive. Thus, newer plants will not bring the significant environmental gains found in more energy-intensive sectors. The lack of environmental improvement in the food and beverages and machinery industries calls into question the assumption that investment automatically brings cleaner technologies to developing countries as part of the institutional and legal integration process.

Investment best triggers environmental innovations when coupled with the help of governments and international institutions, particularly for industries that rely on end-of-pipe technologies for environmental improvements. The negative environmental performance of the Mexican paper and textile industries highlights the lack of sufficient government commitment to effective regulation of air pollution. The food and beverages industry is the most air-pollution-intensive industry in Mexico and also produces Mexico's highest level of total emissions. Yet it is not even regulated by the Mexican government as a fixed source of air pollution (Baker and McKenzie 2000). Environmental compliance in Mexico also depends on institutional commitment to enforcement of existing regulation (i.e., the frequency of regulatory inspections) as well as the extent of public scrutiny (Dasgupta et al. 1997). During the period when plant-level inspections were at their highest rate in Mexico, inspectors visited only 6 percent of all establishments in manufacturing (figure 7). Although Mexico established a formal system of citizen complaints in the early 1990s, it has not been able to respond adequately to the rising number of submissions. In 1992, 1,281 complaints were filed; in 1997, there were 5,644 (OECD 1998).

Lessons for the FTAA

Industry is clearly a significant contributor to air pollution in Mexico, and it accounts for more than 90 percent of emissions for such pollutants as sulfur oxides in some cities. Growth in manufacturing output and exports over the past decades of increasing openness has exacerbated this already acute problem. By using new data and focusing on changing patterns of industrial

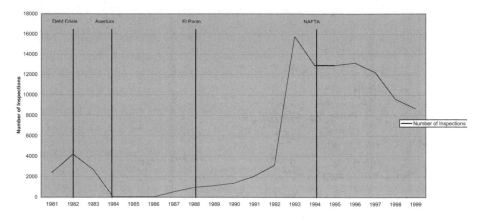

Figure 7.7
Plant-level environmental inspections by Mexican authorities, 1981–1999. Source:
PROFEPA 2000.

air pollution in Mexico during its transition to an open economy, this study
moves beyond the stale political debates regarding whether economic liberalization is "good" or "bad" for the environment. It begins the critical
process of undergirding the trade and environment debate with data and
empirical evidence. This chapter has shown that where certain conditions
prevail, environmental improvements can be achieved through creative management of the liberalization process. Where these conditions are absent,
environmental degradation is likely to dominate. Left to its own devices,
unbridled economic integration can have disastrous environmental results.

This study highlights that where pollution is in large part a function of
plant vintage, new investment and technological changes have led to
marked improvements in levels of air pollution intensity in Mexico.
However, in those industries where air pollution is a function of whether or
not firms employ end-of-pipe technologies, levels of pollution are alarmingly high because neither the NAFTA's environmental package nor
Mexican regulatory systems effectively monitor or enforce their use.
Conditions that could trigger environmental improvements in Mexico
include increases in government-led and citizen-led regulatory enforcement
initiatives, the replacement of outdated combustive technologies with newer
technologies in energy-intensive industries, and the adoption of end-of-pipe
technologies and process-level changes in industries where pollution is a
function of by-product waste. A major disappointment associated with the

NAFTA environmental package and related environmental initiatives is that they do not appear to be adequate to facilitating these processes or addressing the environmental costs of trade-related economic growth in Mexico. Future trade and investment agreements must establish funding mechanisms that give incentives for investors to upgrade core technology for capital-intensive industries and to purchase end-of-pipe technologies where necessary.

Technical Appendix

Problems with Pollution-per-Employee Estimates

The World Bank converts its data into economic indicators as follows:

$$\frac{P}{E} = \frac{P}{Y} * \frac{Y}{E},$$

where P is pollution, Y is output, and E is employment. Then Y/E is productivity. If productivity is growing, then constant pollution per unit of output implies growing pollution per employee (or, constant pollution per employee implies falling pollution per unit of output). Constant pollution per unit of output implies holding technological change constant as well. Seen in this light, using pollution per unit of employment can skew one's picture of pollution because such estimates may be tracking large employment losses rather than actual changes in economic activity. The World Bank data can be converted into a more useful economic indicator as follows:

$$\frac{P}{E} = \frac{P/E}{Y/E}.$$

Comparing Industrial Air Pollution Intensity in the US and Mexico

Pollution intensity is best expressed in terms of pollution per unit of output. Using the 1997 Mexican intensities supplied by the World Bank, we can determine pollution intensity in Mexican manufactures (measured by pollution per unit of output), total pollution in Mexican manufactures measured by tons of emissions, and a pollution-intensity ratio for Mexico relative to the United States. Taken in full, the methodology for obtaining the pollution-intensity ratio can be expressed as follows:

$$R_{MU} = \frac{P_M/Y_M}{P_U/Y_U} = \frac{(P_M/E_M)/(Y_M/E_M)}{P_U/Y_U},$$

where the same notation as in equation 1 is used and where subscript M indicates Mexico, subscript U indicates the United States, and R is the ratio.

Output is measured in 1987 dollars, requiring both deflation and currency conversion. The Mexican National Institute of Statistics, Geographics, and Informatics (INEGI) supplies output data for Mexican manufactures in thousands of current pesos.[18] Using Mexican producer price indexes, output is converted into 1992 pesos and then into 1992 US dollars, which are then converted to 1987 dollars with the GDP deflator (to be consistent with the published US intensities). The year 1992, considered a stable currency year, makes the figures comparable to earlier results. These output figures in each sector are then divided by the number of employees. The World Bank's pollution-per-employee estimates are then divided by the productivity measure. Mexican pollution intensity expressed as pollution per unit of output is then divided by the US intensities expressed in pollution per unit of output to determine the Mexico/US ratio. To determine total pollution, pollution intensity is multiplied by total output.

Composition and Scale Effects over Time

Using two-digit ISIC classification codes for Mexican industry from 1988 to 2000, we can calculate pollution intensity, and therefore total pollution, as follows:

$$\frac{P_m}{Y_m} = \frac{\sum (P_i / Y_i * Y_i)}{\sum Y_i},$$

where the same notation as the previous equations is used, the subscript m indicates all manufactures, and the subscript i indicates industries within manufactures.

To calculate total pollution, pollution per unit of output is multiplied by output in each two-digit industry and then added together. That total is then divided by the sum of the total output for two-digit sectors to determine pollution intensity for the entire manufactures industry.

Notes

1. For thorough reviews of the literature, see Dean 1992, Fredriksson 1999, and Jayadevappa and Chhartre 2000.

2. This literature can be found on the World Bank's "New Ideas in Pollution Regulation" web site (http://www.worldbank.org/nipr).

3. ISIC codes and "digits" are used interchangeably in this chapter and refer to International Standard Industrial Codes (ISIC), a standardized categorization of the world's industrial sectors developed by the United Nations.

4. The World Bank's Mexican Pollution Intensities can be found at http://www.worldbank.org/nipr.

5. The Mexican pollution intensities merge data collected in Mexico over the period 1987–1997 with the bulk of the data compiled in the period 1993–1995. The pollution intensities are for the following pollutants: particulates (PT), sulfur oxide (SOX), carbon monoxide (CO), nitrogen oxide (NOX), and hydrocarbons (HC) (World Bank 2000).

6. Thus, what appear to be sector-wide intensity changes are actually changes in the composition effect.

7. If the data are available, they often require a number of conversions that make them more difficult to manipulate.

8. The data used in this analysis are from several sources. The Mexican intensities are derived from the World Bank estimates. The original units of measurement for employment intensities are tons per employee. The pollution intensities are at two-digit and three-digit ISIC (version 2) levels of industry. Statistics for output and employment for two-digit and three-digit industries are from the Instituto Nacional de Estadistica, Geographica, y Informatico (INEGI), Mexico's national institute of statistics. Exchange rate and price deflator indicators for Mexico are from the Bank of Mexico. GDP deflators for the US are from the Bureau of Economic Analysis. See http://www.bea.org.

9. The methodology for these calculations is shown in the technical appendix to this chapter.

10. Based on earlier studies, the six dirtiest industries in the world economy are Iron and Steel, Petroleum (353 and not included in this data set), Non-Ferrous Metals, Food Manufactures (311), Industrial Chemicals, and Paper (Mani and Wheeler 1999).

11. This part of the analysis is conducted using two-digit classification codes for Mexican industry for the period 1988–2000.

12. The methodology for determining the scale and composition effects is shown in the technical appendix to this chapter.

13. Data on carbon intensity can be obtained from Mexico's Secretaría de Energía (http://www.energia.gob.mx).

14. Data on Mexican FDI can be found at http://www.economia.gob.mx.

15. Data for these calculations is drawn from the statistics gathered by Mexico's Secretaría de Energía at http://www.energia.gob.mx.

16. CONIECO is an organization of manufacturers and retailers of environmentally sensitive products.

17. For example, no data are available for activities that are classified in the three-digit code category.

18. See http://www.inegi.gob.mx.

References

Baker and McKenzie. 2000. *Environmental Law and Policy in Latin America.*

Barkin, D. 1999. The Greening of Business in Mexico. Discussion Paper 110, UNRISD.

Bartzokas, A., and M. Yarime. 1997. Technology Trends in Pollution-Intensive Industries: A Review of Sectoral Trends. UNU-INTECH Discussion Paper 9706, United Nations University, The Netherlands.

Botella, O., G. Enrique, and G. Giral. 1991. "Textiles: Mexican Perspective." In *US-Mexican Industrial Integration*, ed. S. Weintraub et al. Westview.

Brown, F. 2000. "Environmental Performance and Trade Liberalization in the Mexican Textile Industry." In *Industry and the Environment in Latin America*, ed. R. Jenkins. Routledge.

Dasgupta, S., H. Hettige, and D. Wheeler. 1997. *What Improves Environmental Performance? Evidence from Mexican Industry.* Washington: World Bank.

Dean, J. 1992. "Trade and the Environment: a Survey of the Literature." In *International Trade and the Environment*, ed. P. Low (World Bank Discussion Paper 159).

Dominquez-Villalobos, L. 2000. "Environmental Performance in the Mexican Chemical Fibres Industry in the Context of an Open Market." In *Industry and the Environment in Latin America*, ed. R. Jenkins. Routledge.

Eskelund, G., and A. Harrison. 1998. Moving to Greener Pastures: Multinationals and the Pollution Haven Hypothesis. Policy Research Paper 1744, World Bank.

Esty, D., and B. Gentry. 1998. *Foreign Direct Investment and the Environment.* Paris: Organization for Economic Cooperation and Development.

Esty, D. 2001. "Bridging the Trade-Environment Divide." *Journal of Economic Perspectives* 15. no. 3: 113–130.

Frederiksson, P. 1999. "Trade, Global Policy, and the Environment: New Evidence and Issues." In *Trade, Global Policy, and the Environment*, ed. P. Frederiksson. Washington: World Bank.

Gallagher, K. 2000. Structural Reform and Trade Patterns in Mexico: Does Trade Liberalization Matter? GDAE Discussion Paper, Tufts University.

Gallagher, K., and F. Ackerman. 2000. "Trade Liberalization and Industrial Pollution in Developing Countries: A Partial Equilibrium Approach." In *Assessing the Environmental Effects of Trade Liberalization Agreements*, ed. D. Andrews. Paris: Organization for Economic Cooperation and Development.

Gentry, B., and L. Fernandez. 1998. "Mexican Steel." In *Private Capital Flows and the Environment: Lessons from Latin America*, ed. B. Gentry. Elgar.

Grossman, G., and A. Krueger. 1993. "Environmental Impacts of a North American Free Trade Agreement." In *The US-Mexico Free Trade Agreement*, ed. P. Garber. MIT Press.

Hartman, R., M. Huq, and D. Wheeler. 1997. Why Paper Mills Clean Up: Determinants of Pollution Abatement in Four Asian Countries. World Bank Policy Research Working Paper 1710.

Hettige, H., P. Martin, M. Singh, and D. Wheeler. 1994. The Industrial Pollution Projection System. World Bank Policy Research Working Paper 1431.

Jayadevappa, R., and S. Chhatre. 2000. "International Trade and Environmental Quality: A Survey." *Ecological Economics* 32: 77–95.

Jenkins, R. 1998. "Globalizacion y Contaminacion Industrial en Mexico y Malasia." *Comercio Exterior* 48, no. 12.

Levinson, A. 1996. "Environmental Regulations and Manufacturers' Location Choices: Evidence from the Census of Manufactures." *Journal of Public Economics* 62, no. 1–2: 15–29.

Mani, M., and D. Wheeler. 1999. "In Search of Pollution Havens? Dirty Industry in the World Economy." In *Trade, Global Policy, and the Environment*, ed. P. Frederiksson. Washington: World Bank.

Margulis, S. 1996. "Back-of the Envelope Estimates of Environmental Damage Costs in Mexico." In *Pricing the Planet: Economic Analysis for Sustainable Development*, ed. P. May. Columbia University Press.

Moreno-Brid, J. 1999. "Reformas Macroeconomicas e Inversion Manufacturera en Mexico." Serie Reformas Economicas 47, ECLAC, Santiago, Chile.

OECD. 1996. *Trade Liberalisation Policies in Mexico*. Paris: Organization for Economic Cooperation and Development.

OECD. 1998. *Mexico: Environmental Performance Review*. Paris: Organization for Economic Cooperation and Development.

Reppelin-Hill, V. 1999. "Trade and Environment: An Empirical Analysis of the Technology Effect in the Steel Industry." *Journal of Environmental Economics and Management* 38: 283–301.

Ros, J., J. Draisma, N. Lustig, and A. Ten Kate. 1996. "Prospects for Growth and the Environment in Mexico in the 1990s." *World Development* 24, no. 2: 307–324.

Smith, M. 1997. *The US Paper Industry and Sustainable Development*. MIT Press.

Stern, D. 1998. "Progress on the Environmental Kuznets Curve?" *Environment and Development Economics* 3, no. 2: 173–196.

Ten Kate, A. 1993. Industrial Development and the Environment in Mexico. Policy Research Working Paper 1125, World Bank.

Torres, B. 2002. "The North American Agreement on Environmental Cooperation: Rowing Upstream." In this volume.

Van Tongeren, J., S. Schweinfest, E. Lutz, Maria Luna, and G. Martin. 1993. "Integrated Environmental and Economic Accounting: A Case Study for Mexico." In *Toward Improved Accounting for the Environment*, ed. E. Lutz. Washington: World Bank.

World Bank. 2000. *World Development Indicators Database*.

8

Zea Mays: Effects of Trade Liberalization of Mexico's Corn Sector[1]

Alejandro Nadal

Nowhere in the world is corn so intimately related to the social and cultural fabric of a country as in Mexico. Corn is Mexico's primary staple food crop, and it provides livelihoods to millions of Mexicans (Ortega Paczka 1973).[2]

In considering the environmental sensitivity of the NAFTA, one must look beyond the obvious environmental issues and consider those environmental effects associated with the process of trade liberalization itself. In particular, one must try to forecast the transitional effects from changed economic circumstances, analyze their potential environmental effects, and devise adjustment strategies to mitigate potential adverse effects.

In this chapter I compare the anticipated effects of liberalizing the Mexican corn sector under the North American Free Trade Agreement with the actual environmental and socio-economic outcomes. Many existing studies in the trade and environment literature focus on trade liberalization as somehow disconnected from other elements of macroeconomic policy making. I argue that, in the case of Mexico's corn sector under the NAFTA, the Mexican government's failure to properly evaluate the interaction between trade policy and other economic policies (such as policies to reduce inflation, balance fiscal budges, deregulate the banking sector, and privatize and deregulate agricultural markets) led to a serious misunderstanding of how the process of economic restructuring fueled by trade liberalization would unfold.

The Socio-Economic Significance of Corn in Mexico

Since its earliest cultivation in Mexico some 5,000 years ago, zea mays (corn in US English, maize in UK English, maíz in Spanish) has become one of the three most important staple food crops in the world. Nowhere is corn more deeply entwined with the social and economic fabric of a nation, or more genetically diverse, than in Mexico, its country of origin. Mexico boasts thousands of corn varieties (Wilkes and Goodman 1995; Hernández Xolocotzi 1987). These varieties are derived from intense interactions between genotypes and a diverse range of agro-ecological environments, as well as from historical processes of manipulation of germ plasm by Mexican corn growers. Mexican corn producers rely heavily on this rich reservoir of genetic diversity for coping with drought, blight, and other adverse environmental conditions and changes. Agricultural genetic diversity—such as that found in Mexico—promises to play a crucial role in meeting the challenge of food demand in the twenty-first century (Crucible Group 1994).

In most mountainous areas of Mexico, environmental heterogeneity results in a rich array in cultivable productive spaces that are thus exploited in different ways. Corn is a plant endowed with extraordinary adaptive capabilities and can be grown in a very wide range of climatic and soil conditions representing different chemical properties. Producers are widely distributed throughout Mexico and operate under very different social, economic, technical, and environmental conditions. An estimated 60 percent of Mexican producers (1.8 million) use locally adapted corn varieties (SARH 1994). These landraces (traditional varieties developed and used by producers) cover approximately 80 percent of the total area under corn cultivation (ibid.). Sowing several different varieties, with different abilities to withstand a range of environmental conditions (rainfall, humidity, frost, pests, winds), offers some degree of insurance that there will be an adequate harvest. Striking the right balance in the planting of seed varieties requires sophisticated handling of a complex set of relationships involving seed characteristics, soil quality and conservation, surrounding topographic features, weather and climate, etc. In many agro-ecological systems, producers normally sow at least two varieties of corn, one of which matures early but provides low yields and one of which provides greater yields but takes longer to mature. In some communities, at least eight varieties are regularly employed (Cadena Iñiguez 1995, p. 84).[3] Many subsistence producers plant early-maturing varieties to supplement household consumption when the stock from their main harvest is exhausted (Ortega Paczka 1997). The ability to select an appropriate combination of seed varieties and the date of sowing has been considered to be the most powerful technological resource available to these producers (García Barrios et al. 1991, pp. 174–175).[4] These practices rely on knowledge distributed through social networks and cultural practices (Crucible Group 1994; Cuevas Sánchez 1991; Ortega Paczka 1973; Hernandez Xolocotzi 1985).

At the time of the NAFTA negotiations, corn accounted for approximately 60 percent of Mexican land under cultivation and a similar proportion of Mexico's total agricultural output (INEGI 1991). In monetary terms, corn production continues to account for more than two-thirds of the gross value of Mexican agricultural production, well above beans and wheat (12 percent and 11 percent respectively) and followed by rice, sorghum, and soybeans. Other crops, including horticultural crops, account for only 6 percent of the total gross value of agricultural production (SAGAR 2000). In terms of its contribution to employment generation, corn is Mexico's most important commodity. Although the share of Mexico's total GDP generated by agriculture dropped from 15 percent in 1960 to less than 7 percent in 1998, agriculture still employed more than 22 percent of Mexico's total labor force in 2000 (ibid.). Corn employs 40 percent of people working in the Mexican agricultural sector (ibid.).

According to 1991 Mexican government agricultural survey, there were just over 3 million corn-producing units in Mexico (INEGI 1991). Since the average size of a rural family in Mexico is six, the number of people depending directly on corn production for all or a major part of their livelihood is likely to be more than 18 million (INEGI 1990). This figure rises substantially if people depending indirectly on corn production (e.g., through transportation, storage and trade) are taken into account.

The Mexican Government's Agenda for Mexican Corn

For Mexico, corn was by far the most important crop included in the NAFTA. Mexican government officials hailed the liberalization of Mexico's corn sector as a cornerstone of the NAFTA (Téllez 1992). The NAFTA, in turn, was a major policy tool in the Mexican government's efforts to implement a major restructuring of Mexican agriculture.

Mexican government officials believed that their existing agricultural system—dominated by corn—was not only sapping the country's fiscal reserves through subsidies to corn producers but also preventing realization of Mexico's comparative advantage. Mexico's NAFTA negotiators argued that two-thirds of Mexico's corn producers should shift from corn production to the cultivation of other crops, or to non-agricultural sectors of the economy, where Mexico was considered to enjoy a comparative advantage (Telléz 1992). Using the NAFTA, the Mexican government hoped to reallocate productive resources (e.g., land and labor) to more labor-intensive crops and to non-traditional commercial crops (sugar cane, coffee) and land use patterns (forestry and livestock) (ibid.). To this end, under the NAFTA's chapter VII on agriculture, Mexico opened its major market for corn products to imports from the United States.[5] In return, the United States and Canada were to open their markets for horticultural and other labor-intensive crops (fruits, nuts, vegetables, coffee). Mexico was thought to enjoy competitive advantages in these products (due to its labor surplus and lower costs) (de Janvry 1996, p. 6; Hufbauer and Schott 1993, p. 47).

The NAFTA provides 15 years for the final alignment of Mexican domestic corn prices with international (principally US) prices.[6] This long transition period for liberalization reflected recognition of the strategic and economic importance of corn production to the livelihoods of millions of Mexicans. At the beginning of the 15-year transition period, the existing tariff and import permit system was transformed into a tariff-rate quota (TRQ) regime that was to be gradually phased out.[7] The tariff-free quota (initially set at 2.5 million tons per year for Mexican exports to the United States or US exports to Mexico) was to expand at a constant rate of 3 percent per year, while the applicable tariff for imports exceeding the quota would be reduced from 206 percent in 1994 to zero by 2008.[8]

Mexican corn producers received assurances that adjustment-assistance policies would be implemented during the 15-year transition period. Levy and van Winjbergen (1992) emphasized the need for an adequate level of public investment to help transform rain-fed land into irrigated land and to counter any negative wealth effects brought about by reduced prices. A second significant study used by the Mexican government went so far as to quantify the public investments required (Levy and Winjbergen 1995). That study established that, for the first 5 years of the transition period (with a gradual elimination of tariffs and subsidies), 16 billion (constant 1989) pesos would be required.[9] To accompany the NAFTA, the Mexican government proposed the gradual removal of price support mechanisms that had existed for 40 years and their replacement by a series of less trade-distorting adjustment-assistance measures (including direct income support mechanisms, credit, infrastructure investments, and agricultural research and development).

The Mexican government believed that the NAFTA's agenda for agricultural trade liberalization would decrease Mexican corn output due to NAFTA-induced decreases in domestic corn prices, would free labor, land, and capital for reallocation to productive activities that could better utilize Mexico's comparative advantage in abundant low-cost labor, would drastically reduce Mexico's rural population but generate more profitable employment opportunities elsewhere in the agricultural sector or in non-agricultural activities (Téllez 1992), would reduce inflationary pressures through cheaper imports, would decrease the need for subsidies (relieving pressure on fiscal policies), would reduce tortilla prices for Mexican consumers as the use of cheaper imported corn increased, and would improve environmental performance since it was assumed that marginal land vulnerable to erosion would be left fallow (Levy and van Wijnbergen 1995, 1992).[10]

Prospects for the realization of these objectives depended upon a set of fundamental assumptions embedded in studies used to justify the inclusion of corn in the NAFTA (Levy and van Winjbergen 1992, 1995; de Janvry et al. 1995, 1997) and to inform the Mexican government's official stance concerning the trade agreement (Téllez 1992).

The assumptions listed in table 1 led Mexican policy makers to anticipate that, overall, any negative social and environmental impacts of trade liberalization under the NAFTA would be minimal.

Despite the strategic importance of the agricultural sector to Mexico's labor force, natural resource base, and national economy, the Mexican government failed to conduct an in-depth analysis to test and verify these beliefs. By restricting the range of people with whom it consulted and the kinds of questions it asked researchers to address, the Mexican government forfeited the possibility of properly understanding liberalization's potential effects on social conditions and coherence, rural livelihoods, and the environment.[11] Some rough estimates and general models were developed, but these calculations served more as a post hoc rationalization than as a serious effort to examine prospective implications of a major policy decision.

Table 1
Assumptions and logic informing Mexico's corn-sector trade policy.

Assumption	Logic
Frictionless resource reallocation through price signals	Frictionless reallocation of productive resources (land, labor, capital) to more competitive crops through lower corn prices. Producers read market signals correctly and reorient their activities to more profitable activities where they enjoy competitive advantages. The studies used by the Mexican government to assess the impact of the NAFTA on corn product fail to acknowledge cross price elasticities.
Investment in public works to ensure smooth adjustment period	An adequate flow of public resources would facilitate a smooth transition to greater efficiency in Mexican agriculture.
Neutral technical change	The Mexican government's pre-NAFTA studies all employed an unrealistic static model in which technical change does not involve factor substitution. The studies thus assumed certain agricultural activities (such as horticultural production) would be more labor intensive than was realistic.
Unlimited expansion possibilities for Mexico's share of the North American horticulture market	The area of horticultural cultivation would double through the expansion of Mexican exports to the US market. Low labor costs in Mexico would facilitate competitive exports to the United States.
Automatic crop substitution and reallocation	The Mexican government assumed that corn producers would easily reallocate resources and switch from corn to other kinds of agricultural production.
Subsistence producers not affected by lower prices	Subsistence corn producers would not be affected by NAFTA-induced lower corn prices because they do not generate marketable surpluses.
Lower tortilla prices	Cheaper corn imports would lead to lower final consumer prices. Tortillas would sell at a lower price and consumer welfare would increase. The increase in consumer welfare would take place across all levels of rural and urban populations. It would also have the effect of relieving costly programmes to subsidize consumers, a recurring problem in public finance. It was argued that providing price support systems for poor, inefficient corn producers artificially raised the price of the final product.

Table 1 (continued)

Assumption	Logic
Adequate performance of the Mexican economy and employment generation	Studies used by officials to estimate the effects of the NAFTA relied on the assumption of full employment in their modeling. It was argued that social dislocation following trade liberalization would be resolved through market forces. Displaced laborers would be able to find adequate employment opportunities, either in the rural sector or in the urban markets. The fall in the rural wage rate was expected to ensure that excess labor was eliminated. A significant portion of that labor was expected to be reabsorbed through direct interventions such as investment in public works for irrigation. Final equilibrium would be restored through migration.

The NAFTA and Corn: A Case of Unanticipated Outcomes

The NAFTA corn experience has not unfolded as anticipated. The Mexican government's main strategic objectives—particularly in terms of transferring Mexican land and labor out of corn production—have not been achieved. Sadly, the unexpected trends that have emerged are accompanied by severe social and environmental problems.

In the years immediately after the NAFTA came into force, Mexico's planned 15-year transition period for the corn/agricultural sector was compressed to roughly 30 months. Between 1994 and 1998, increased imports of cheap corn into Mexico led to much deeper and faster reductions in the price of corn than originally anticipated. Between 1993 and 1999, US corn shipments to Mexico grew by a factor of 15, to 5.6 million tons, and accounted for 25 percent of Mexico's corn consumption, versus 2 percent before the NAFTA (*Los Angeles Times* 2000). The real inflation-adjusted market price of corn in Mexico, which had begun to fall in the late 1980s, declined by 46.2 percent between 1993 and 1999 (BBC 2000). Mexican corn prices thereby converged with international market some 12 years earlier than the NAFTA had anticipated. Importantly, the predicted drop in Mexico's domestic corn output did not take place. Rather total corn production remained at historically high levels (about 18 million tons per year) (SAGAR, various years). (See table 2.) Similarly, the NAFTA-induced

Table 2
A scorecard for NAFTA's corn regime.

Original plans and objectives	Actual effects
Cheaper imports to induce corn price reductions and lower tortilla prices.	Corn imports caused significant (45%) reductions in the price of corn, but not tortillas.
Tariff-free quota to expand by 3% while over-tariff quota gradually phased out.	Corn imports exceed tariff-free quota and the prescribed tariff is not applied.
Fifteen-year transition period for alignment of domestic prices with international prices.	Transition period truncated; domestic prices aligned with international prices in thirty months.
Resource allocation: domestic corn output to be reduced and land and labour released for non-maize agricultural activities (e.g., horticultural sector).	Corn output stable at highest historical levels of 1994 (18 million tons). Area of corn cultivation expands, yields drop.
GDP in agricultural sector recovers significant growth rates.	Agricultural GDP stagnates during NAFTA's first five years.
Trade balances for agricultural sector to generate a significant trade surplus as comparative advantages develop their trade potential.	Accumulated deficit in agricultural trade balance between 1992 and 1999 is $2.6 billion.
Adequate adjustment policies (e.g., direct income support and other policy instruments) to assist producers during transition period.	Main income support instrument, PROCAMPO, diminished by 40% in real terms and other support mechanisms are reduced. Other public investments fail to materialize as fiscal policy becomes more restrictive.
Tortilla prices to decrease as cheaper imports are used in domestic industries.	Tortilla prices increase by 483%.
Subsidies decline, both to growers and to industries using corn as main input.	Subsidies for industrial corn flour producers continue.

reductions in corn prices stimulated an increase, rather than a decrease, in the area of Mexico's land under corn cultivation (SAGAR, various years). (See table 3.) A review of the post-NAFTA statistics on total corn production and land under cultivation reveals a drop in average yield per hectare (SAGAR, various years; Nadal 2000). (See table 4.)

What is going wrong?

First, the NAFTA liberalization process was implemented far faster than anticipated. Official pre-NAFTA studies had a series of flaws. In particular, Mexican officials failed to consider the range of external political macroeconomic policy pressures that might impinge on Mexico's ability to carry out its intended agricultural reform strategy. In 1994–1995, an economic crisis emerged in Mexico.[12] Rising inflation and instability put Mexico under pressure from the international financial community—particularly the World Bank and the International Monetary Fund—to control and reduce prices and to reduce fiscal expenditures. The 1995 economic crisis made it impossible to allocate fiscal resources in the amount required for a smooth transition of corn growers into other crops or other jobs. The Mexican government, moreover, did not implement the NAFTA's corn tariff and quota provisions as originally planned. Corn imports to Mexico have regularly exceeded the quota set out under the NAFTA. Instead, beginning in 1994, the Mexican government exempted all corn imports from tariff payments, arguing for the need to lower corn prices to reduce the inflationary pressure. In addition, Mexican government transition efforts were largely abandoned. The lack of revenue from anticipated tariff payments combined with a general policy of fiscal restraint resulted in a decline in Mexican support for agriculture and a failure of adjustment assistance to materialize as promised. As a result, the corn producers did not receive the "breathing space" they had been promised to allow them to adjust to a more open trading regime. To assist producers during the anticipated transition period, the Mexican government did establish PROCAMPO (a direct income-support mechanism) and Alianza para al campo (a program designed to increase productivity and competitiveness) (World Bank 2001). These two sets of instruments were to accompany the phasing out of CONASUPO, the state body formerly responsible for support policies, marketing, and distribution of basic grains) and its price-support activities.[13] But these policy instruments rapidly lost most of their effectiveness or were prematurely terminated. In real terms, owing to a combination of reduced budgets and

Table 3
Mexico's corn output by state, 1991–1997 (tons). Source: Centro de estadísticas agropecuarias, Secretaría de agricultura y ganadeía (SAGAR).

	1991	1992	1993	1994	1995	1996	1997
Nacional	14251500	16969724	17961730	18235826	18352856	18023626	18085405
Sinaloa	821000	1080913	2442387	2762275	2027474	1696177	2743653
Mexico	1755997	1901211	1233449	1561746	2146471	2250753	2309412
Jalisco	2310590	2414243	2379659	2125336	2231290	2328157	2074195
Chiapas	983415	1607369	1593298	1096254	1696001	1543675	1485970
Veracruz	797570	852557	761335	929953	1104281	1182712	1186726
Michoacán	979195	919481	1039760	1042268	1293058	1130533	1102230
Puebla	1020398	1161287	944037	881146	1063857	1182504	869402
Guerrero	786516	989801	886835	765736	1112254	1072124	867374
Chihuahua	739955	948238	880086	487031	303627	412303	688196
Sonora	393714	295566	455175	542981	457480	834116	641216
Oaxaca	422014	512818	547654	623953	720714	683624	625462
Hidalgo	383867	484984	362081	453166	406140	427970	465226
Guanajuato	532760	777912	1255706	1020245	824005	757368	558236
Tamaulipas	443304	746921	1108759	1355550	818609	230338	262950
Nayarit	177992	170275	181221	317063	225790	224996	242109
Durango	239127	248487	289217	325088	291280	288146	238427
Zacatecas	216683	243900	243814	277618	296450	302291	229651

Tlaxcala	262051	376081	263250	310065	297076	328046	217356
Campeche	55565	111163	82135	115314	54889	133041	189481
Querétaro	60640	137515	111906	168409	186173	169207	161791
Tabasco	74294	67025	71255	125365	99995	140937	154920
Yucatán	131844	154166	116300	94582	73136	45049	143180
San Luis Potosí	210361	149713	88618	193209	160989	169285	120785
Morelos	67511	102928	94950	97599	115943	100732	98534
Baja Calif. Sur	77843	85203	89562	97492	40484	85065	89267
Colima	65372	57709	76546	90568	90654	94318	70147
Nuevo León	91140	92629	99730	159112	54759	43347	64661
Aguascalientes	47420	73188	65994	74037	85562	77249	63085
Quintana Roo	16227	33546	16848	6616	10410	37778	49917
Coahuila	62955	130384	104002	96172	44855	31851	49446
Distrito Federal	22168	16599	16071	16216	12826	12758	15696
Baja California	2012	25912	60090	23661	6324	7176	6704

Table 4
Changes in post-NAFTA corn production, 1990–1997. Source: author's calculations, based on Anuario Estadístico del Sector Argícola (SAGAR) over several years.

	Cultivated area (ha)	Harvested area (ha)	Production (tons)	Yield (tons/ha)
National total	727497.7	−131880.7	1330367.3	−0.01
Jalisco	38878.2	−34136.8	−208214.3	−0.45
Mexico	−8156.3	3918.5	476741.5	0.80
Sinaloa	161671.2	163068.8	1187407.8	0.82
Chiapas	176856.5	73009.3	143888.8	−0.18
Puebla	17187.8	−82811.2	−168240.0	−0.31
Michoacan	38442.2	−20181.7	68794.5	−0.02
Guerrero	27183.7	−72880.5	−26542.5	−0.016
Veracruz	81550.2	81377.0	294520.2	0.26
Tamaulipas	−153767.0	−177634.0	−592364.8	−1.10
Guanajuato	−42450.0	−135544.5	−288984.2	−0.53
Chihuahua	38117.8	26089.8	55752.3	−0.06
Oaxaca	81278.7	34302.0	78775.8	−0.01
Hidalgo	−5904.7	−28919.3	43491.5	0.18
Sonora	28434.3	29131.5	264298.3	1.23
Tlaxcala	6124.7	−18727.7	−84001.2	−0.50
Zacatecas	574.0	−63390.0	−59997.8	−0.18
Durango	21603.7	−46528.0	−32854.5	−0.26
Nayarit	8947.3	11724.3	39206.5	0.18
San Luis Potosi	59316.2	−2381.3	−57837.7	−0.37
Queretaro	16321.0	−7983.5	33334.5	0.11
Yucatan	11943.5	19244.0	28552.2	0.12
Morelos	6433.2	3732.8	2769.2	−0.20
Nuevo Leon	25277.5	17543.5	−28424.2	−0.48
Tabasco	35929.2	37881.8	66579.0	0.20
Campeche	50279.8	56789.5	104160.3	0.40
Coahuila	658.5	−4507.2	−31353.2	−0.57
Colima	3195.0	2317.5	−5941.2	−0.37
Aguascalientes	−2374.7	−33407.7	−6986.0	−0.07
Baja Calif. Sur	3156.7	2916.2	19304.0	0.30
Baja California	−5720.2	−5118.5	−13806.0	0.21
Quintana Roo	20294.0	40557.7	30247.5	0.29
Distritio Federal	−1534.8	−1333.2	−1909.2	0.12

inflation, the value of PROCAMPO's direct payments to individual producers decreased by 40 percent between 1994 and 2000 (*Houston Chronicle* 2000). Finally, by late 1998, CONASUPO, which was to have been gradually phased out, was completely dismantled. At the same time, other promised transition assistance programs failed to materialize. Access to farm credit, which could have offered corn producers some capacity to make transition investments, fell to extremely low levels and public investment in agricultural projects (including infrastructure such as irrigation) was curtailed severely (SAGAR 2000; *Houston Chronicle* 2000). The crop insurance system ceased operating except in the most profitable cases.

Second, several factors prevented the significant cuts in the price of corn from being passed onto consumers of corn products such as tortillas—part of the staple diet of Mexicans. At the end of 1998, Mexico ended price controls on tortillas and ceased subsidizing tortilla mills (Associated Press 1999). Within a year, consumer prices for tortilla dough in Mexico had risen 22 percent and tortilla prices in Mexico City by 50 percent (*Washington Post* 1999; NPR 2000). Market imperfections in the Mexican tortilla industry (arising, in part from locally isolated markets, market segmentation, and the monopolistic behavior of cartels) also prevented the cut in the price of corn from being passed onto the consumers of tortilla products (Nadal 2000).

Third, the comparative advantages that Mexico's negotiators assumed to have existed have turned out to be founded on poor analysis. Government expectations related to the capacity of non-corn agricultural production—particularly horticulture and fruit—to absorb Mexican labor from the corn sector have been overinflated. There is no doubt that there has been growth in both the Mexican horticultural and fruit sectors. From 1975 to 1997, production of Mexico's eight most important horticultural crops maintained an upward trend, growing from 1.2 million tons to 3.2 million tons (SAGAR, various years). Similarly, in the fruit sector, total output increased from 1.6 million tons in 1975 to 8 million tons in 1997 (SAGAR, various years). However, neither fruit or horticultural production are absorbing the amount of labor or land anticipated by the Mexican government (Málaga and Williams 1999; Gómez Cruz and Rinderman 1993; Grammont 1999; Thrupp 1995; de Janvry 1995). Even though horticulture and fruit production are more labor intensive than corn and other basic grains, more efficient use of inputs has led to greater productivity and higher yields without the anticipated increases in employment (Grammont 1999;

Málaga and Williams 1999). Without adequate transitional assistance and government assistance, the NAFTA has not spurred corn producers to shift their land and production toward fruit and horticulture on the scale expected by the Mexican government (Málaga and Williams 1999; Nadal 2000). Some analysts also highlight the constraints that stringent US public health and product quality requirements, as well as the cost of processing packaging and transportation, could place on the continued expansion of the sector through export growth (Thrupp 1995).

Fourth, the Mexican government undervalues Mexico's corn producers as custodians and curators of Mexico's (and the world's) genetic diversity in corn and thus as crucial contributors to rural livelihoods in Mexico. Preservation of the genetic variability of corn and careful selection of the varieties best suited to local growing conditions are important virtues associated with the "low-efficiency" cultivation of Mexico's poor corn growers. But this "environmental service" is undervalued in today's market economy and by the Mexican government.

Fifth, by assuming that only one variable (market price) would determine changes in the behavior of Mexican corn producers, pre-NAFTA government projections (SARH 1994) critically misjudged the responses of Mexican producers to liberalization of the corn sector under the NAFTA. Producers' decisions are made in the light of many factors, including market prices for alternative crops, wage costs, interest rates, and social as well as cultural preferences (i.e., to maintain their existing lifestyles or remain living in their existing communities). To fully understand the full potential effect of the NAFTA on the production decisions and the resource management capabilities of corn producers, Mexico's government ought to have better analyzed the behavior of different groups of corn producers.

Contrary to the Mexican government's expectations, the negative environmental and social implications of trade liberalization in the corn sector have not been minimal. Absent an effective transition strategy, Mexico's attempt at corn restructuring is generating extreme economic hardships and increased rural poverty for corn producers (Coote 1999; World Bank 2000). As producers they have faced lower prices for their produce, and as consumers they have faced increased prices for agricultural inputs and for corn products such as tortillas (SAGAR 2000; Nadal 2000).

Mexican corn producers have been forced to absorb the shock of rapid trade liberalization without the necessary adjustment programs and investment to enable them to make the shift from corn to other kinds of agricultural production or to seek non-farm employment. Instead, many marginally profitable corn producers and subsistence corn producers have been forced to either maintain corn products in spite of its dwindling economic viability or migrate to urban areas or to the United States in search of off-farm sources of cash income (Hinojosa and Robinson 1992; Tuirán 1998; Salas 1997).

On the environmental front, economic hardship is generating disturbing trends. For those impoverished rural corn producers who do not migrate, economic pressures intensify the struggle to maintain (and in some case to increase) output. Poor growers rely on poor soils, so the yields are low and the risk of crop failure is high. The pressure to produce is driving these farmers to further expand their area of cultivation to even more marginal lands, resulting in soil erosion, deforestation, and encroachment on biosphere reserves and other protected areas across Mexico (Gordillo et al 1995; SEMARNAP 2000; Nadal 2000).

Economic pressures are also diminishing the capacity of rural producers to conserve and develop genetic resources. Local landraces that have undergone a long and systematic process of adaptation to Mexico's diverse agro-ecosystems are a critical element of the world's endowment of genetic resources. The combined pressure of poverty and migration is having a corrosive effect on social institutions, collective community actions, and traditional knowledge systems and practices that have historically been central to resource management and conservation of genetic diversity in many rural communities (Villar Sánchez 1996; Cerda Bolinches 1994; Turrent 1997; Ortega Paczka 1997). If current trends continue, this system of in situ conservation will gradually disappear.

US exports to Mexico of corn seeds containing transgenic material are also generating environmental concern. The environmental implications of the potential genetic flow between traditional varieties and transgenic corn are likely to remain unknown for some time.

Conclusions and Policy Recommendations

Implementation of the NAFTA was central to the Mexican government's strategy for agricultural reform. Yet liberalization of the corn sector under the NAFTA has failed to generate the expected economic, environmental and social benefits. Contrary to the expectations of the NAFTA's negotiators, there has been little incentive and little opportunity for farmers to reallocate productive resources away from corn to other crops, especially since investments and support programs for the transition are being cut. In spite of a sharp drop in corn prices and an increase in imports, Mexico's production has remained stable. In fact, the cultivated surface devoted to corn has expanded, while yields have dropped. In addition, the expected drop in consumer prices for corn products has failed to materialize. Tortilla prices have risen.

This study has shown how the failure to assess trade policy options within the broader the policy, economic, and social landscape can translate into a rapid and poorly planned liberalization process with severe social and environmental impacts. The challenge for Mexico is to develop economic strategies that will improve the livelihood of its rural population in ways that are compatible with long-term environmental sustainability. Trade liberalization can be a part of this policy package only if it is devised with proper consideration of the interaction of a range of different national economic and social policy instruments, trends, and priorities and if it is implemented according to schedule with appropriate support mechanisms for adjustment and sustainability.

Notes

1. This chapter draws from a longer study (Nadal 2000) that involved extensive field work and interviews with rural producers, analyzing technology profiles, social characteristics of producers, resource endowment, and profitability levels.

2. Corn is linked to a range of final uses in Mexico and plays an important cultural role. Typical uses include production of tortillas, tlacoyos (cooked corn pancakes with beans), pozole (corn grain in broth), tamales, atole (a beverage made from maize dough), and pozol (a beverage made in Chiapas from a blend of cocoa and maize grains). Different uses require different corn size, texture, color, flour/starch content, viable storage time, etc. In many cases, traditional cultivars are favored because of their overall performance and contribution to final uses: grain for human consumption, cobs for animals, ear covers for tamales, etc. Also important are certain maize varieties for the preparation of ritual beverages (tezhuino in the Huicot region and bitter atole in the northern Puebla sierras) and the ramified ear corns used in fertility rituals in Chiapas (Ortega Paczka 1973).

3. Local landraces (traditional varieties developed and used by producers) are more frequently used in the southern and central highland states. Penetration of hybrid varieties (developed experimentally through open pollination and other

methods) has been very low in regions where upland production of corn prevails, where poor quality soils are frequent, and where a range of environmental risks is encountered.

4. Ortega Paczka (1997) identified nine factors used by farmers to select seed: soil type, drought resistance, wind resistance, response to inputs, vulnerability to weeds, fertility, yield, end uses, and dietary considerations.

5. In the agricultural sector the NAFTA comprises two bilateral agreements (Mexico-Canada and Mexico-US), whereas US-Canadian trade continues to be regulated by the previously adopted Canada-US Free Trade Agreement. The NAFTA also establishes certain trilateral commitments regarding sanitary and phytosanitary standards, safeguards, and rules of origin.

6. The corn varieties produced in the US and in Mexico are not strictly the same commodity. The US is the largest producer of yellow corn, normally used as animal feed. On the other hand, Mexico is one of the largest producers of white corn varieties that have a finer texture and higher flour content, making them more suitable for direct human consumption.

7. Tariff rate quotas establish a quota for tariff free imports and a tariff rate for imports that exceed the tariff-free quota.

8. See NAFTA, Annex 302.2 in Schedule of Mexico, tariff item 1005.90.99. The starting point was set at 206.9% in 1994, to be reduced during the first 6 years of the agreement by 29.6%. The remaining tariff was to be phased out progressively over the following 9 years until a zero tariff was achieved for all imports.

9. It was argued that this amount of investment would allow for irrigation of approximately 8% of the total land resources and improved productivity of corn producers (Levy and van Winjbergen 1992).

10. By far the most detailed calculations on resource allocation after trade liberalization are found in Levy and van Winjbergen 1992. In another study (1995), the same researchers elaborate on the need to carry out a gradual transition using a general equilibrium model.

11. On the one hand, businesses using corn to manufacture flour, packaged foods, and feed for cattle and poultry were actively involved in the negotiations (Hogenboom 1998). On the other hand, corn producers—particularly representatives of poor, rural communities, and indigenous peoples—were not adequately represented in negotiations, and were excluded from the process leading to final agreement (ibid.).

12. For detailed analyses of the 1994 crisis, the stabilization program, and the lack of progress in terms of recovery, see Nadal 1996, 1997, 1998.

13. CONASUPO regulated the price of corn and beans through direct market intervention, purchasing a significant percentage of the total crop from farmers at a guaranteed price (Burfisher et al. 2000). This allowed producers to rely on a guaranteed price for their output. In 1993 the Mexican government introduced new policies that eliminated CONASUPO's guaranteed floor prices for corn and reduced its prior obligations to buy staple crops (*Business Mexico* 1997). CONASUPO was eliminated in 1998.

References

Associated Press Worldstream. 1999. "Mexican Government Tries to Calm Discontent Over Tortilla Prices." January 7.

BBC. 2000. "Mexico: Senator Says US Agricultural Subsidies Mean NAFTA Should Be Renegotiated." British Broadcasting Corporation, November 4.

Burfisher, M., S. Robinson, and K. Therfielder. 2000. "North American Farm Programs and the WTO." *American Journal of Agricultural Economics* 82, no. 3: 768–774.

Business Mexico. 1997. "The Corn Conundrum; Corn Import Debacle Plays Up Weaknesses in Agricultural Policy." July 1.

Cadena Iñiguez, P. 1995. Del azadón a la labranza de conservación. La adopción de la labranza de conservación en dos comunidades de la Sierra Madre de Chiapas. Maestría en ciencias thesis, Colegio de Posgraduados, Montecillo, Mexico.

Calva, J. ed. 1993. Alternativas para el campo mexicano. Mexico, D.F. Distribuciónes Fontamara, UNAM-PUAL.

Cerda-Bolinches, A. 1994. "The Responses of Abandoned Terraces to Simulated Rain." In *Conserving Soil Resources European Perspectives*, ed. R. Rickson. Oxford University Press.

de Janvry, A. 1996. "NAFTA and Agriculture: An Early Assessment." Paper presented at "NAFTA and Agriculture: Is the Experiment Working?" (Trinational Research Symposium), San Antonio, Texas.

de Janvry, A., E. Sadoulet, and B. Davis. 1995a. "NAFTA's Impact on Mexico: Rural Household-level Effects." *American Journal of Agricultural Economics* 77: 1283.

de Janvry, A., E. Sadoulet, and G. Gordillo. 1995b. "NAFTA and Mexico's Maize Producers." *World Development* 23, no. 8: 1349–1362.

FIDA. 1993. "Reformas del sector agrícola y el campesinado en México." Informa de la Misión especial de programación a la República de los Estados Unidos Mexicanos. Informe Número 0435-ME, Fondo Internacional de Desarrollo Agrícola.

García Barrios, R., and L. García Barrios. 1992. "Environmental and Technological Degradation in Peasant Agriculture: A Consequence of Development in Mexico." *World Development* 18, no. 11: 1569–1585.

García Barrios, R., L. García Barrios, and E. Alvarez Buylla. 1991. *Lagunas. Deterioro ambiental y tecnológico en el campo semiproletarizado.* México City: El Colegio de México.

Goméz Cruz, M., and R. Rindermann. 1993. "El Sistema hortofrutícola en México frente al TLC." In *Alternativas para el campo mexicano*, ed. J. Calva. Mexico, D.F. Distribuciónes Fontamara, UNAM-PUAL.

Gordillo, G., et al. 1994. El Sector ejidal en la agricultura mexicana: impacto de los reformsas. Mimeograph.

Grammont, H. 1999. "La modernización de las empresas horticolas y sus efectos sobre el empleo." In *Agricultura de exportación en tiempos de globalización: El caso de las hortalizas, frutas, y flores*, ed. H. Grammont et al. Mexico City: CIES-TAAM/UACH, IISD/UNAM, CIEAS y Juan Pablos Editor.

Grammont, H., M. Goméz Cruz, H. González, and R. Rindermann, eds. 1999. *Agricultura de exportación en tiempos de globalización: El caso de las hortalizas, frutas, y flores*. Mexico City: CIESTAAM/UACH, IISD/UNAM, CIEAS y Juan Pablos Editor.

Hinojosa, R., and S. Robinson. 1992. "Diversos escenarios de la integración de los Estados Unidos y México: enfoque de equilibrio general computable." *Economía Mexicana* 1, January-June: 71–144.

Houston Chronicle. 2000. "A New Sun Poverty; A People in Want; Poverty Stalks the Nation, But Nowhere Is It Worse Than in the Countryside." *Houston Chronicle*, November 26.

Hufbauer, G., and J. Schott. 1993. *NAFTA: An Assessment*. Washington: Institute for International Economics.

INEGI. 1990. *XI Censo General de Población y Viviendo*. Mexico City: Instituto Nacional de Estadistica, Geografia e Informatica.

INEGI. 1991. *National Agricultural and Livestock Census*. Mexico City: Instituto Nacional de Estadistica, Geografia e Informatica.

Levy, S., and S. van Wijnbergen. 1992. "Maize and the Free Trade Agreement between Mexico and the United States." *World Bank Economic Review* 6, no. 3: 481–502.

Levy, S., and S. van Wijnbergen. 1995. "Transition Problems in Economic Reform: Agriculture in the North American Free Trade Agreement." *American Economic Review* 85, no. 4: 738–754.

Los Angeles Times. 2000. "Growing Troubles in Mexico." January 17.

Nadal, A. 2000. *The Environmental and Social Impacts of Economic Liberalization on Corn Production in Mexico*. Gland and Oxford: Oxfam GB and World Wide Fund for Nature.

National Public Radio. 2000. "Rise in Tortilla Prices." *Morning Edition*, January 20.

OECD. 1997. *Review of Agricultural Policies in Mexico: National Policies and Agricultural Trade*. Paris: Organization for Economic Cooperation and Development.

Ortega Paczka, R. 1997. Maíz en el Tratado de Libre Comercio: Implicaciones para el medio ambiente, recursos genéticos. Mimeograph, Consultoría sobre recursos genéticos preparada para la Comisión de cooperación ambiental, México.

SAGAR. 2000. Propuesta de Desarrollo Rural y Agroalimentario 2001–2006. Mexico, D.F.: Secretaría de Agricultura, Ganandería y Desarrollo Rural.

Salas, C. 1997. Corn and NAFTA: Potential and Effective Migration in Mexican Agriculture. Mimeograph.

SARH. 1991. Encuesta nacional sobre rentabilidad y productividad, maíz. Mexico City: Secretaria de Agricultura y Recursos Hidraulicas.

SARH. 1994. Encuesta nacional sobre rentabilidad y productividad, maíz. Mexico City: Secretaria de Agricultura y Recursos Hidraulicas.

Stedman-Edwards, P. 1997. Socioeconomic Root Causes of Biodiversity Loss: The Case of Calakmul, Mexico. Mimeograph, WWF-Oficina program, México.

Suppan, S., and Lehman, K. 1997. Food Security and Agricultural Trade Under NAFTA. Minneapolis: Institute for Agriculture and Trade Policy.

Téllez, Luis. 1992. La modernización del sector agricola en Mexico. Mexico: Fondo de Cultura Economica.

Turrent, A. 1997. Maíz en el TLC: Implicaciones para el medio ambiente. Consultoría sobre calidad de suelos. Mimeograph, Comisión de Cooperación Ambientall.

Villar Sánchez, B. 1996. Erosionabilidad de suelos y su impacto en lad productividad del maíz en el trópico mexicano. Doctor en ciencas thesis, Colegio de Posgraduados, Montecillo, Mexico.

Washington Post. 1999. "Tortilla Price Hike Hits Mexico's Poorest." January 12.

World Bank. 2000. "Rural Poverty." Available at http://wbln0018.worldbank.org/external/lac/lac.nsf/.

World Bank. 2001. *Mexico: A Comprehensive Development Agenda for the New Era.* Washington, D.C. World Bank.

Statistical Sources

SAGAR, various years. Anario Estadístico del Sector Agrícola. Mexico City: Secretaría de Agricultura, Ganadería, Desarrollo Rural, Pesca y Alimentación.

SAGAR, various years. Centro de Esatidística Agropecuaria. Available at http://www.siea.sagarpa.gob.mx.

9

An Investment Regime for the Americas: Challenges and Opportunities for Environmental Sustainability

Howard Mann and Mónica Araya

The experience with the North American Free Trade Agreement's investment provisions provides important evidence about how investment policies and environmental sustainability relate. The lessons learned from the NAFTA are especially relevant because the approach, content and scope of the NAFTA investment model have considerable gravitas in the context of negotiations for a Free Trade Area of the Americas (FTAA).[1]

Over the past decade, foreign direct investment (FDI) has emerged as a popular tool for financing development in Latin American and Caribbean countries.[2] As FDI grows in importance, so does the necessity of understanding FDI flows, the international agreements that help facilitate them, and their implications for growth and environmental sustainability.[3] Choosing a "business as usual" approach to investment threatens to erode the already fragile public support for moving the hemispheric trade and investment agenda forward. A carefully considered relationship between investment provisions and the demands of environmental sustainability might, in contrast, enhance public support for further liberalization.

This chapter surveys the shortcomings of the NAFTA approach to investment. It argues that the FTAA investment regime needs to move beyond the NAFTA's one-dimensional focus on protecting foreign investors, and it offers suggestions of how FTAA negotiators can avoid the pitfalls of the NAFTA approach.

Investment and Environmental Sustainability

The role of FDI in achieving or obstructing improvements in environmental sustainability is inadequately understood (Esty and Gentry 1997;

OECD 1999a). While FDI can provide much-needed financing for infrastructure development, technology transfers, and capacity building such as technological, managerial, and environmental training, there is no guarantee that FDI will substantially improve overall environmental performance.

To date, the literature examining the relationship between investment and environmental sustainability suggests the following[4]:

• There is no overarching conclusion yet regarding the balance of environmental sustainability benefits and risks posed by FDI flows (Schmidheiny and Zorraquin 1996; OECD 1999a). There are negative, positive, and neutral effects, and the outcome depends on context (Zarsky 1999; Gentry 1998).

• In a highly competitive global economy, the ambition of many developing countries to attract foreign investors has affected local/national environmental standards and their enforcement. Although there has not been a broad-based "race to the bottom" (Wheeler 2000), there is some evidence of "political drag" (Esty 1994), and a "race to the top" has become more difficult to achieve, particularly in the absence of a global environmental regulatory framework (Vogel 1995; Esty 1995).

• Although the existence of "pollution havens" is difficult to prove empirically (Low and Yeats 1992; Fredricksson 1999, Neumayer 2001); pollution patterns based on differences in the income or education of local communities (rather than on differences in national environmental standards) are discernible (Zarsky 1999).

What is clear is that economic growth alone, whether driven by trade, by FDI, or by other domestic means, provides no guarantee of gains in environmental quality (Nordström and Vaughan 1999; Esty 2001). Effective environmental regulation is necessary to achieve good micro and macro environmental performance—and regulatory approaches must evolve in parallel with economic integration efforts (Dua and Esty 1997). In fact, empirical studies from leading economic institutions such as the World Bank and the World Trade Organization have shown that in the absence of good environmental management, economic liberalization could have negative effects on the environment (Fredriksson 1999; Nordström and Vaughan 1999). A recent comprehensive study (World Bank 2000) clearly highlights that even though the pace of economic growth remains important to environmental performance, the quality of economic growth is more critical.[5]

Investment Provisions and the Government's Role in Protecting the Environment

The objectives, provisions, and language of investment agreements can have a crucial influence on whether foreign investment becomes a driver for or against environmental sustainability. This section reviews the investment provisions in the NAFTA's chapter 11 and assesses their effects on government efforts to promote higher levels of protection of the environment and public health.

Experience with the NAFTA's chapter 11 that certain investment provisions can unduly constrain the ability of governments to protect the environment. In particular, corporations have used the provisions of chapter 11 to lobby against the promulgation and implementation of domestic laws (including environmental or human health protection laws) and have sought redress (through government compensation) for negative effects on their profitability due to the adoption of such laws (Mann 2000; Mann and von Moltke 1999; Rugman et al. 1999; Peterson 2001). (Table 1 presents a list of environment-related disputes that arose under chapter 11 between January 1994 and December 2000.)

At least two important reasons can be suggested for these developments. First, the purpose and use of investment agreements has evolved. Investment agreements were originally designed as a protection, or shield, for foreign investors against the threat of discriminatory treatment and nefarious host government activities such as expropriation of assets or limitations on the repatriation of profits. Investment "disciplines" were intended as recourses of last resort for companies that had suffered unfairly. In fact, only in the 1990s did mandatory investor-state arbitration processes emerge as an alternative to local legal remedies (Sacerdoti 1997; Comeaux and Kinsella 1997). Chapter 11 has formalized and sharpened this "internationalist" approach. Second, in keeping with a general trend toward expanding investor protection, chapter 11 provides investors with a broader combination of rights and remedies than all previous bilateral investment treaties (Horlick and Marti 1997). While environmental advocates were the first to point out the unintended consequences of the NAFTA's investment provisions, trade liberalization proponents have also begun to articulate similar concerns (Hufbauer et al. 2000; Graham 2000).

Table 1
Environment-related disputes under NAFTA's chapter 11, January 1994–December 2000.
Sources: Mann and von Moltke 1999; Hufbauer et al. 2000; Morrison and Warn 2000.

Initial filing date	Petitioner (country of origin)	Defendant	Subject of dispute; claim	Status or outcome
September 1996	Ethyl Corp. (US)	Canada	Import ban on MMT gasoline additive for environmental reasons. Claim: US$250 million	Case settled out of court in July 1998 for US$13 million; import ban lifted.
October 1996	Metalclad Corp. (US)	Mexico	State decision (with support of national government) preventing location of hazardous waste facility; zone in question later declared an ecological reserve. Claim: US$65 million	Case settled in August 2000 in favor of Metalclad; Mexico ordered to pay US$16.7 million in damages.
December 1996	DESONA de C.V. (US)	Mexico	Alleged breach of contract to operate a landfill and seizure of property. Claim: US$19 million	Resolved in Mexico's favor in November 1999.
June 1998	USA Waste ("Acaverde") (US)	Mexico	Contract dispute over payments for solid waste collection and disposal. Claim: US$60 million	In June 2000, arbitration panel dismissed case due to lack of jurisdictional authority; USA Waste re-filed claim in September 2000.
June 1998	S.D. Myers (US)	Canada	Temporary ban on PCB waste exports. Claim: US$20 million	Case settled in November 2000 in favor of S.D. Myers; determination of monetary damages forthcoming.
December 1998	Sun Belt Water, Inc. (US)	Canada	Alleged biased treatment by provincial government because of decision to prevent bulk water exports. Claim: US$220 million	Consultations ongoing.

Table 1 (continued)

Initial filing date	Petitioner (country of origin)	Defendant	Subject of dispute; claim	Status or outcome
December 1998	Pope & Talbot Inc. (US)	Canada	Alleged discriminatory implementation of the Canada-US Softwood Lumber Agreement. Claim: US$130 million	In June 2000, arbitration panel dismissed two of four of Pope & Talbot's claims; remaining two still pending.
June 1999	Methanex Corp. (Canada)	US	State ban on use of MTBE gasoline additive. Claim: US$970 million	Commencement of tribunal process in September 2000.

The following discussion focuses on five issues that have emerged from jurisprudence on chapter 11 which require both careful rethinking by the NAFTA countries and re-consideration by negotiators of future investment agreements.[6]

Purpose of the Investment Provisions
Properly identifying the purpose and objectives of investment agreements has emerged as a critical factor in their interpretation. Preambulatory language stating the general objectives of an agreement is a crucial component of any international agreement. But such hortatory provisions are not substitute for specific statements of goals within the substantive chapters of trade and investment agreements. The recent *Metalclad* case illustrates this point well (Metalclad 2000). In this dispute, the US-based Metalclad Corporation petitioned the government of Mexico, contesting Mexican state and municipal decisions preventing the opening of a hazardous waste landfill. The NAFTA tribunal ordered Mexico to pay Metalclad $16.7 million in damages. In its decision, the tribunal referred expressly to three general purposes expressed in the NAFTA's preamble and objectives that it believed were relevant to interpreting the provisions of chapter 11 (Metalclad, paragraphs 70–75):

- transparency (see NAFTA Objectives)
- substantial increases in investment opportunities (see NAFTA Preamble)
- ensuring a predictable commercial framework for investors (see NAFTA Preamble)

The tribunal added an interpretational spin to these objectives, arguing that it was the goal of the NAFTA and hence of chapter 11 to ensure the successful implementation of investment initiatives (paragraph 75).

By identifying only investor-focused objectives in a complex environmental case, the tribunal's decision illustrates the high risk of imbalanced interpretations of the main provisions of chapter 11. In the absence of specific preambular language for the investment chapter, the tribunal developed its own interpretations of the chapter's objectives that ignored the NAFTA's preamble calling for a focus on environmental protection and sustainable development alongside economic objectives.

In the more recent *S.D. Myers* case, the tribunal's decision established a further precedent in this regard by incorporating the broader objectives of the NAFTA directly into its interpretation of chapter 11's investment provisions. This dispute involved the US-based company S.D. Myers, Inc. (petitioner) and the Government of Canada. The subject of the dispute was an export ban on PCB (polychlorinated biphenyls) waste established by the Canadian government, which S.D. Myers claimed caused it to suffer losses of US$20 million. The tribunals's interpretation of national treatment under chapter 11 is infused with references to the NAFTA's overarching principles regarding the preference for the use by governments of least-trade restrictive rules, despite the complete absence of any text in chapter 11 that endorses the application of this principle to the investment arena (S.D. Myers 2000, paragraphs 238–257).

These two cases highlight the need to put specific environmental or sustainable development principles and objectives into any investment provisions that may emerge in future trade agreements, and to separate the interpretation of investment provisions from other trade-related obligations and objectives. The need for economic and environmental balance must not be left to a vague overarching perambulatory statement of objectives and principles for the trade agreement as a whole.

National Treatment and Most Favored Nation Treatment
The rationale for national treatment and most favored nation treatment disciplines in investment agreements is to ensure foreign investors receive no less favorable treatment than other domestic investors or investors from other countries. In this context, several important environmental issues warrant consideration.

First, investment agreements should ensure appropriate sensitivity to national and local environmental laws and decision-making processes that set thresholds for water and ambient air quality. When standards change, new investors may face more stringent environmental requirements than existing investors. Such differential treatment should not be seen as per se discrimination. Indeed, ambient air quality often decreases with additional polluters. Later polluters can inflict higher marginal costs on society and thus quite legitimately can be held to higher standards.

Second, governments need to retain their ability to hold investors liable for environmental damages. Governments may, for example, want to require foreign investors whose activities could create environmental risk to put up bonds or provide insurance guarantees to address potential liabilities from pollution damage. Again, while this practice could be seen as differential treatment of domestic and foreign investors, it should not be condemned to the extent that the investors are not similarly situated.[7]

Third, the interpretation of key terms can have important environmental implications. Chapter 11 stipulates that when foreign investors share "like circumstances" with domestic investors, governments must treat them in an equally favorable manner. That is, governments must accord national treatment to investors in "like circumstances." The definition, interpretation, and application of terms such as "like circumstances" are thus critical to determining when principles of national treatment must be applied, and should not be left ambiguous or ill defined.[8]

Minimum International Standards
Chapter 11, like most bilateral investment agreements, contains provisions requiring host countries to treat foreign investors in a manner that complies with minimum international standards. These provisions were designed to provide a treatment floor, regardless of whether domestic firms were being treated "equally badly" by the government in question. However, neither the term "minimum international standards" nor the accompanying terminology has received satisfactory definition.[9] These NAFTA provisions have important environmental implications.

The *Metalclad* decision, with some support from the more recent *S.D. Myers* decision (paragraphs 258–269), and a further decision in the *Pope and Talbot v. Canada* case, raises important questions as to the scope of the minimum international standards provision (article 1105 of the NAFTA).[10]

In the Metalclad case, "minimum international standards" were held to include, among others, procedural requirements for full transparency of all national and subnational laws as well as extensive rights of access to the decision-making process (Metalclad 2000, paragraphs 75–100).

Given the potential for expansive interpretation of provisions such as those on minimum international standards, negotiators ought to consider the scope they wish such a provision to have. In so doing, they should consider whether their own domestic law and regulation making standards fall within such a scope, and define the obligations accordingly.

Performance Requirements Prohibitions

Chapter 11's investment rules incorporate provisions to regulate the use of performance requirements (article 1106). The purpose of these performance requirements provisions is to prevent a host government from imposing conditions on investors that might compromise the efficiency of the investor's operation. The type of provisions that might run afoul of this discipline include requirements related to minimum or maximum levels of imported or exported materials, incorporation of a given level of domestic inputs or services, rules governing the use of particular technologies or personnel, or mandatory technology transfer provisions (Brooks 1997). Requiring investors to use a specific domestically patented air emissions technology is one example of an environment-related performance requirement that might be unacceptable.

The way in which, to date, chapter 11's provisions on performance requirements have been interpreted by tribunal's decisions extends the original intent of the article in a very significant way (Mann and von Moltke 1999). Recent interpretations enable private companies to challenge the imposition of any general trade measure that affects its production inputs or product sales. Although, three environmental-related chapter 11 cases have involved challenges to performance requirements, to date, no such case has been decided in favor of the investor.[11] But the potential for abuse by foreign investors remains.

Expropriation

The NAFTA's provisions on expropriation have critical implications for environmental lawmaking. Clear analysis of chapter 11's expropriation provisions is again frustrated by ambiguous and unclear definitions of key terms such as "expropriation," "creeping expropriation," and "measures tantamount or equivalent to expropriation."

An additional unresolved question for policy makers concerns the question of the relationship of the NAFTA's expropriation provisions to the normal exercise of a government's "police powers" to protect its environment and the health and well being of its public. In traditional investment law, the activities that fall under a state's legitimate police powers can not be limited by an expropriation discipline or subject to any compensation (Mann and von Moltke 1999, section 3.7). In practice, however, at least two of the chapter 11 cases (*Metalclad* and *Pope and Talbot*) have failed to recognize or consider the importance of police powers.

In the *Metalclad* case, one of the environmental measures that provoked legal proceedings was a measure that withdrew land from potential use as a waste management facility or other industrial purposes to create an ecological reserve. In its decision, the tribunal stated expressly and concisely that it "need not decide or consider the motivation or intent of the adoption" of the environmental measure in question (Metalclad 2000, paragraph 111). Had the tribunal held that such changes of land use, while in the public interest, normally require payment of compensation to landowners, much less controversy would have arisen. By arguing that the purpose is simply not relevant, the tribunal established a precedent that could be interpreted as broadly limiting a state's authority to engage in environmental protection.

The legal test that the *Metalclad* tribunal did apply is one of the *scale of impact* of a measure on a business. The question here is whether there was a "significant" impact on "the use or reasonably-to-be-expected economic benefit of property, even if not to the obvious benefit of the host State" (Metalclad 2000, paragraph 103). The *Pope and Talbot* decision repeats this focus on the scale of impact (with extensive reference to US international law doctrine as a basis for this approach) (Pope and Talbot 2000, paragraphs 96–105). The specific threshold for "significant" impact is not clear in either case, though *Pope and Talbot* notes that a "negligible" impact of a measure on an investor's business will not qualify. What is more critical, however, is that both cases focus on the scale of impact rather than the purpose behind a measure as the main test for considering whether the governmental action in question falls within the scope of chapter 11's expropriation provision.[12]

Investment Negotiations in the FTAA

Existing case law under chapter 11 consistently places the rights of investors ahead of the ability of governments to protect the environment. In effect, this jurisprudence turns the "polluter pays" principle on its head—creating a "pay the polluter" rule of law. In the FTAA context, there is a critical need to avoid establishing hemisphere-wide investment provisions that could constrain the ability of governments to protect the environment and the public.

The inclusion of investment in the FTAA agenda is highly significant, from both a regional and global perspective. Many aspects of the FTAA project were envisioned in the halcyon days immediately after the Uruguay Round Agreement of 1994, and before the debacle of the Multilateral Agreement of Investment (MAI) in October 1998.[13] The lukewarm attitude of developing countries toward the inclusion of investment issues in the World Trade Organization (which prevailed at the WTO Ministerial Meetings in Singapore 1996 and Seattle in 1999) has deterred FTAA proponents. Prospects for action on the trade-investment interface are greater in the western hemisphere than in most of the other international settings.[14]

The basic mandate for the investment negotiations—set out in the 1998 Declaration of the Second Summit of Americas in San José—calls on the negotiators to "establish a fair and transparent legal framework to promote investment through the creation of a stable and predictable environment that protects the investor, his investment and related flows, without creating obstacles to investment from outside the hemisphere." While many bilateral investment treaties in the Americas already exist, proponents of the inclusion of investment provisions in the FTAA hope that the added presence of a hemispheric-wide accord, with strong rules establishing arbitration procedures for disputes between a foreign investor and the host state, will offer a powerful incentive for attracting new foreign investment to the region. The focus is on promoting investment through provisions that will protect the investor by enforcing a predictable legal framework for investment and a stable macroeconomic and institutional context.

This one-dimensional economic focus neglects the essential qualitative dimension of a successful economic growth and development strategy. FTAA negotiators need to balance the objectives of investment stability, predictability, and risk management with other important public purposes and considerations, such as the environmental sustainability of FDI and its distribution within and between countries. (See box.) From a legal perspective, by focusing narrowly on protecting investors and their investments, the current FTAA strategy fails to match investor rights with appropriate obligations.

The FTAA negotiating group on investment officially released its first draft negotiating text for the FTAA in July 2001.[15] Its approach mirrored the scope and content of investment provisions contained in the NAFTA (in chapter 11) and in other bilateral investment treaties. The range of issues under consideration fall within three clusters.

Cluster 1: Basic definitions, national treatment, most favored nation treatment, fair and equitable treatment

Cluster 2: Scope of application, key personnel, transfers, performance requirements

Cluster 3: Expropriation and compensation, compensation for losses, general exceptions and reservations, and dispute settlement.

The Growth and Distribution of FDI in Latin America

In the 1990s, global capital flows experienced very pronounced shifts. While official development assistance flows continued to decline in real terms, private capital flows set new records for foreign direct investment, portfolio investment, and commercial bank lending. Globally, FDI reached a record level of US$1,118 billion in 2000 (Thomsen 2000; UNCTAD 2000). This dramatic increase represents a doubling in just three years; only a decade ago, total annual flows were about US$200 billion (UNCTAD 2000).

While developing countries have successfully attracted FDI inflows in the 1990s, reaching $207 billion in 1999 (compared to a annual average of $46 billion in the period 1988–1993), they still lag behind developed countries. The United States alone receives more FDI flows (34.7 percent) than the entire developing world (24 percent). Within developing countries, three quarters of the inflows go to eleven economies alone.

In Latin America and the Caribbean, FDI flows to the region have increased substantially, reaching a new record level of $90 billion in 1999. This represents approximately 10 percent of the global FDI flows. The uneven distribution of the flows, however, is striking: more than 80 percent of all FDI inflows into LAC is concentrated in only four economies, Brazil, Mexico, Argentina, and Chile (UNCTAD 2000; OECD 2000).

The main players in the investment talks to date have been the United States, Canada, Mercosur, Chile, and Mexico. Only the United States has publicly indicated its negotiating goals and concerns on this issue, but has provided little in the way of detail. One can glean, however, that the United States is aware of problems associated with the unintended impacts of the language adopted in the NAFTA's investment provisions, as well as public concerns over the secretive nature of the associated investor-state dispute settlement mechanism, but no specific proposal has been submitted to the FTAA on investor-state provisions.[16]

Canada has also developed an FTAA negotiations web site, but it does not contain any statements on the investment negotiations.[17] On the Latin American side, Chile appears to concur that some concerns about the NAFTA's investment provisions deserve consideration in the FTAA context. Mexico, on the other hand, appears to stand behind the NAFTA investment provisions as the appropriate model for the FTAA (although the Fox administration might change this position).

United States negotiators are already beginning to experience aggressive lobbying from business groups anxious to not only keep but expand the NAFTA's investment provisions in the FTAA context, including those that have caused governmental and public concern to date (*Inside US Trade* 2001a). Reports from the press confirm that the United States has met strong opposition to preliminary proposals that governments be prohibited from relaxing environmental and labor standards to attract investment, and provisions that would require an environmental impact study for projects of a certain size (*Inside US Trade* 2001b).

While the limited available information indicates little innovation or "thinking outside the box" at the present time, specific provisions have yet to be negotiated, leaving open the possibility of broader alternative approaches and more specific recommendations for each of the existing issue clusters.

Incorporating Sustainability Considerations into the FTAA Investment Provisions

This subsection outlines several priorities that ought to inform the FTAA investment negotiations including language that averts unintended negative effects on government regulatory abilities and mechanisms for transparency in the dispute resolution processes.[18] Specifically, we suggest consideration of the following:

• Language promoting sustainability-related goals and principles within the investment chapter's preamble or objectives to guide later interpretation, especially in a dispute resolution context.

• Provisions setting out governmental responsibilities and obligations regarding the substance and enforcement of environmental law. The minimum provision would place a bar on the lowering of environmental standards to attract or maintain investments and oblige signatories to enforce effectively their environmental laws.

• Mechanisms that establish investor responsibilities and obligations alongside their rights. Among items for consideration should be provisions for the undertaking of environmental impact assessments in accordance with the host country's, the home country's, or the World Bank's standards before the investment was made and for maintaining a certified environmental management system (ISO 14001 or equivalent) during the life of the investment.

• Requirements to act in a manner consistent with the obligations assumed by both the host country and the home country in any international environmental agreements.

• Commitment to investment principles and objectives that balance environmental and economic issues and make clear that economic growth goals do not take precedence over environmental aims.

• Definitions of key terms, including what constitutes "like circumstances" and the appropriate scope of comparison for national treatment purposes.

• Expropriation provisions that recognize explicitly the police powers of each party and affirm the ongoing right of governments to protect the environment and human health (including the strengthening of standards) without fear of having to pay compensation for so doing.

• Transparency, public disclosure of all documents, and a commitment to modern judicial procedures in all investor-state dispute resolution process.[19]

Conclusion

The high profile of investment-related issues in the public perception of trade and investment liberalization cannot be escaped. In many ways, the issues raised by the substance of chapter 11 of the NAFTA and the process of the MAI negotiations exemplify public fears about the negative effects of globalization. Poorly designed investment provisions and unclear language can limit legitimate environmental regulation. The FTAA negotiations on investment provide an opportunity to move beyond the narrow focus of the NAFTA's investment chapter as well as its substantive and procedural failures.

Notes

1. For an overview of investment issues in the FTAA context, see Roberts 2001.

2. For information and case studies about FDI flows to Latin America and the Caribbean, see UNECLAC 2000.

3. Foreign investment includes portfolio investments in stocks and other capital assets, banking activity with securities and other types of financial transactions. The main subject of this chapter, however, is foreign investments in the form of the ownership and management of businesses in a foreign country. This is referred to as foreign direct investment. Traditionally, FDI has been defined as an investment involving management control of a resident entity in one economy by an enterprise in another economy (UNCTAD 2000). International agreements protecting foreign investment apply, however, to a broader range of foreign capital investments. For a discussion of capital flows in general and their environmental repercussions, see O'Connor 2000.

4. Zarsky (1999) also finds that the evidence is still poor compared to the research needs. Additionally, she offers a useful conceptual framework for FDI-environmental linkages. For reviews of empirical work on the intersection of FDI and environment issues, see OECD 1999a and chapter 10 of UNCTAD 1999.

5. The World Bank study identified four core elements that can affect the *quality* of development: the distribution of opportunities, environmental sustainability, the variability of global risks, and the framework of governance. An OECD report (1999b) also highlights the role of policy coherence (i.e., strong social safety-nets, strong investment in human capital, and good governance) in a development strategy, highlighting that trade and investment opening is a necessary condition for developing countries to grow and reduce poverty but by no means a sufficient one.

6. Much of this section is a concise version of a broader review of the legal issues found in Mann and von Moltke 1999, supplemented by the cases that were decided on these issues since June 1999. This jurisprudence is reviewed in detail, including individual case summaries, in IISD/WWF 2001. There is no attempt here to provide an exhaustive legal review of each issue, but rather to summarize the main issues and suggest an appropriate path for addressing these issues in the negotiating process. Citations refer to the main decision in each case, unless otherwise noted, which are listed in the references. A comprehensive database of all NAFTA chapter 11 decisions can be found at http:www.naftalaw.org.

7. To our knowledge, no litigation under investment agreements to date has raised this or the preceding issue.

8. The *S.D. Myers* case (2000) highlights the serious potential for interpretive confusion. The tribunal had to choose how it would compare the circumstances of foreign and domestic investors to decide whether the foreign investor shared "like circumstances" with domestic investors. In this case, the tribunal chose a vast scope of comparison for establishing when foreign investors ought to enjoy national treatment.

9. Similarly ambiguous associated terminology in the NAFTA includes "treatment in accordance with international law," "fair and equitable treatment," and "full protection and security."

10. The Pope and Talbot dispute involved the US-based company Pope and Talbot Inc. (the petitioner) and the government of Canada. The dispute is related to the US-Canada Softwood Lumber Agreement, which the petitioner claims has been implemented in a discriminatory fashion.

11. The three cases are *S.D. Myers* (paragraphs 289–300), *Pope and Talbot* (paragraph 74), and the *Ethyl v. Canada* Decision on Jurisdiction (paragraphs 62–64). There is no jurisprudence to the contrary. *Methanex Corp. v. United States of America* also raises questions about chapter 11's provisions regarding performance requirements.

12. Express reference to expropriation not normally covering government regulation is found in *S.D. Myers*, but the decision creates a genuine ambiguity. The tribunal concludes that the main reason the particular measure in question does not fall within the scope of article 1110 of NAFTA is that it is a temporary measure. Hence, the influence of this case, though potentially constructive, is self-limited (S.D. Myers 2000, paragraphs 279–287).

13. The MAI negotiations were later dropped.

14. In fact, Sauvé (2000) argues that, in the face of the ongoing doubts as to the appropriateness for its inclusion in other global forums, "the FTAA countries can do much to reenergize the sagging spirit of trade and investment liberalization that has taken root in many quarters of the world." For an overview of current investment agreements in the Americas, see OAS 1999.

15. The official text of the draft investment agreement can be found at http://www.alca-ftaa.org/ftaadraft/eng/draft_e.asp.

16. USTR released a public summary of the US position related to the FTAA's negotiations in early 2001. See USTR 2001. The summary is also available at http: www.ustr.gov.

17. Only sporadic statements of concern made by the Minister for International Trade over the current text of chapter 11 of NAFTA can serve as a guide to Canadian intentions, but these are too imprecise to afford a basis for any detailed comment. See the department's web site (http:// www.dfait-maeci.gc.ca/trade/ menu-e.asp).

18. Drafting alternatives include direct provisions in the text, interpretive notes as an add-on to the text, appropriate preambular paragraphs and objectives provisions located specifically in the investment chapter of an FTAA. Some of these recommendations are influenced by approaches suggested by the Environment Division of the OECD Secretariat in the context of the MAI negotiations (OECD 1997).

19. In particular, governments should allow public access to proceedings, ensure publication of all panel decisions, and establish appropriate conditions and formal provisions for the *amicus curiae* submissions. For example, a chapter 11 tribunal hearing the *Methanex* case against the US recognized the importance of allowing *amicus* participation (Methanex 2001).

20. All these cases are available at http://www.naftalaw.org.

References

Brooks, J. 1997. "Performance Requirements." In *Proceedings of the Special Session on the Multilateral Agreement on Investment Held in Paris on 17 September 1997*. Paris: OECD.

Comeaux, P., and S. Kinsella. 1997. *Protecting Foreign Investment under International Law. Legal Aspects of Political Risk*. Oceana.

Dua, A., and D. Esty. 1997. *Sustaining the Asia Pacific Miracle: Environmental Protection and Economic Integration*. Washington: Institute for International Economics.

Esty, D. 1994. *Greening the GATT: Trade, Environment, and the Future*. Washington: Institute for International Economics.

Esty, D. 1995. "Private Sector Foreign Investment and the Environment." *Review of European Community and International Environmental Law* 4, no. 2.

Esty, D. 2001. "Bridging the Trade-Environment Divide." *Journal of Economic Perspectives* 15, no. 3: 113–130.

Esty, D., and B. Gentry. 1997. "Foreign Investment, Globalization and Environment." In *Globalization and the Environment*, ed. T. Jones. Paris: Organization for Economic Cooperation and Development.

Fredriksson, P. ed. 1999. *Trade, Global Policy, and the Environment*. Washington: World Bank.

Gentry, B., ed. 1998. *Private Capital Flows and the Environment: Lessons from Latin America*. Elgar.

Graham, E. 2000. *Fighting the Wrong Enemy: Antiglobal Activist and Multinational Enterprises*. Washington: Institute for International Economics.

Horlick, G., and A. Marti. 1997. "NAFTA Chapter 11B: A Private Right of Action to Enforce Market Access Through Investments." *Journal of International Arbitration* 14, no. 1: 43–54.

Hufbauer, G., D. Esty, D. Orejas, L. Rubio, and J. Schott. 2000. *NAFTA and the Environment: Seven Years Later*. Washington: Institute for International Economics.

IISD/WWF (Institute for International Sustainable Development and World Wildlife Fund-US). 2001. *Investors Rights and Private Rights, Public Problems: A Guide to NAFTA's Chapter on Investor Rights*. Winnipeg: IISD.

Inside US Trade. 2001a. "US Industry Pushes NAFTA Investment Model in Future Agreements." April 27.

Inside US Trade. 2001b. "FTAA Vice Ministers Fail to Agree on Negotiating Timeframes." *Inside US Trade*, February 2.

Low, P., and S. Yeats. 1992. "Do dirty industries migrate?" In *International Trade and the Environment*, ed. P. Low. Washington: World Bank.

Mann, H., and K. von Moltke. 1999. *NAFTA's Chapter 11 and the Environment. Addressing the Impacts of the Investor-State Process on the Environment.* Winnipeg: International Institute for Sustainable Development.

Mann, H. 2000. "Assessing the Impact of NAFTA on Environmental Law and Management Processes." In *First North American Symposium On Understanding the Linkages between Trade and Environment*. Montreal: Commission for Environmental Cooperation.

Morrison, S., and K. Warn. 2000. "Canada Loses First NAFTA Suit." *Financial Times*. November 14.

Neumayer, E. 2001. *Greening Trade and Investment: Environmental Protection without Protectionism.* Earthscan.

Nordström, H., and S. Vaughan. 1999. "Trade and Environment." *Special Studies* 4. Geneva: World Trade Organization.

O'Connor, D. 2000. Global Capital Flows and the Environment in the 21st Century. Technical Paper 161, Organization for Economic Cooperation and Development.

OECD. 1997. What Would an MAI With High Environmental Content Look Like? Internal working document.

OECD. 1999a. *Foreign Direct Investment and the Environment*. Paris: Organization for Economic Cooperation and Development.

OECD. 1999b. *Trade, Investment, and Development: Policy Coherence Matters.* Paris: Organization for Economic Cooperation and Development.

OAS. 1999. *Investment Agreements in the Western Hemisphere: A Compendium.* Washington: Organization of American States.

Peterson, L. 2001. "Changing Investment Litigation, Bit by Bit." *BRIDGES Monthly* J, no. 4: 11–12.

Roberts, M. 2001. "Investment." In *Toward Free Trade in the Americas*, ed. J. Salazar-Xirinachs and M. Roberts. Washington: Brookings Institution and Organization of American States.

Rugman, A., J. Kirton, and J. Soloway. 1999. *Environmental Regulation and Corporate Strategy: A NAFTA Perspective*. Oxford University Press.

Sacerdotti, G. 1997. Bilateral Treaties and Multilateral Instruments on Investment Protection. Recueil des Cours 269, Academy of International Law, The Hague.

Sauvé, P. 2000. "Making Progress on Trade and Investment: Multilateral versus Regional Perspectives." In *Services Trade in the Western Hemisphere. Liberalization, Integration, and Reform.*, ed. S. Stephenson. Washington: Brookings Institution and Organization of American States.

Schmidheiny, S., and F. Zorraquin. 1996. *Financing Change: The Financial Community, Ecoefficiency and Sustainable Development*. MIT Press.

Thomsen, S. 2000. Investment Patterns in a Longer-Term Perspective. Working Paper on International Investment 2000/2, World Bank.

UNCTAD. 1999. *World Investment Report*. Geneva: United Nations Conference on Trade and Development.

UNCTAD. 2000. *World Investment Report*. Geneva: United Nations Conference on Trade and Development.

UNECLAC. 2000. *Foreign Investment Flows in Latin America: 1999 Report*. Santiago: United Nations Economic Commission for Latin America and the Caribbean.

USTR. 2001. Negotiating Group on Investment: Public Summary of US Position. Washington: US Trade Representative Office.

Vogel, D. 1995. *Trading Up: Consumer and Environmental Regulation in a Global Economy*. Harvard University Press.

von Molkte, K. 2000. *An International Investment Regime: Issues of Sustainability*. Winnipeg: International Institute for Sustainable Development.

Wheeler, D. 2000. *Racing to the Bottom? Foreign Investment and Air Quality in Developing Countries*. Washington: World Bank.

World Bank. 2000. *The Quality of Growth*. World Bank and Oxford University Press.

Zarsky, L. 1999. "Havens, Halos, and Spaghetti: Untangling the Evidence about Foreign Direct Investment and the Environment." In *Foreign Direct Investment and the Environment*. Paris: Organization for Economic Cooperation and Development.

NAFTA Chapter 11 Cases[20]

Ethyl v. Canada 1998: Award on Jurisdiction in the NAFTA/UNCITRAL Case between Ethyl Corporation and the Government of Canada, June 24.

Metalclad 2000: *Metalclad Corporation v. United Mexican States*, Award, International Center for Settlement of Investment Disputes (Additional Facility), Case no. ARB (AF)/97/1, August 30.

Methanex 2001: *Methanex Corporation v. United States of America*, Decision of the Tribunal on Petitions From Third Persons to Intervene as "Amici Curiae," January 15.

Pope and Talbot 2000: In The Matter of an Arbitration under Chapter Eleven of the North American Free Trade Agreement Between Pope and Talbot Inc. and the Government of Canada, Interim Award by Arbitral Tribunal, June 26.

S.D. Myers 2000: In A NAFTA Arbitration Under UNCITRAL Arbitration Rules, *S.D. Myers, Inc. v. Government of Canada*, Partial Award, November 13.

III
The NAFTA's Environmental Provisions

10

Trade and Environment in the FTAA: Learning from the NAFTA

Richard Fisher

As we begin the new century, negotiators are looking ahead to one of the most demanding trade negotiations in history: the goal is a comprehensive agreement linking 34 countries and 800 million people from the Arctic Ocean to Tierra del Fuego. The proposed Free Trade Area of the Americas (FTAA) will address every major industry, commodity, and trade issue: agriculture, goods, services, intellectual property, electronic commerce, and more. The countries of the western hemisphere have discussed the idea of a regional free trade zone on innumerable occasions. Any student of the western hemisphere's history can recite the free trade proposals of Simon Bolivar, James G. Blaine, and Benito Juarez. None of these earlier initiatives developed beyond the point of preliminary discussion.

At the 1994 Miami Summit of the Americas, leaders from across the western hemisphere took the concept from dream to vision, committing themselves to creating a single hemispheric trade zone that will deepen regional economic relationships that already absorb more than 50 percent of the goods exported from Brazil and roughly 46 percent of those exported from the United States (USTR 2001). The FTAA process is now well underway. At the Second Trade Ministerial, held in Cartagena in 1996, the hemispheric leaders directed trade officials to begin formal negotiations for a FTAA. In Quebec City in April 2001, negotiators presented the first draft of an actual text of the agreement.[1]

The FTAA is an extraordinarily ambitious, complicated, and important initiative. It brings continental giants, including the United States and Brazil, together with some of the smallest countries in the world. Its realization will be no simple task, and one should underestimate neither the technical nor the political challenges negotiators will face. The protests in the streets of Quebec City—and wherever else trade liberalization is on the agenda—

have made clear the range of views and values that must be addressed. One of the most pressing questions is how to build appropriate environmental sensitivities into the agreement. The North American Free Trade Agreement and its environmental side agreement—the North American Agreement on Environmental Cooperation (NAAEC)—provide both precedents from which to learn and environmental success stories on which to build. As the FTAA negotiations begin, it is crucial that negotiators from all countries in the hemisphere reflect on the NAFTA trade and environment model and experience with an eye to garnering lessons for the FTAA context. That said, we can and should also improve on both the NAFTA process and its substantive approach to linking trade and the environment.

In this chapter, I offer an assessment of the trade-environment balance in the NAFTA from the perspective of someone who has faced these issues as a US negotiator. I review what was done well and what could have been done better. Finally, I advance several ideas about lessons that have been learned for the FTAA process and submit several recommendations to the Bush administration as it grapples with the question of how best to foster an environmental dimension to regional trade.

The NAFTA Experience

In many respects, the NAFTA was a bold experiment. It was the first major trade negotiation where environmental issues played a central role, both in terms of challenges and opportunities, throughout the negotiations (Esty 1994a). Concern about the possible environmental impacts of the agreement, particularly along the US-Mexico border, prompted policy makers to "think outside the box."[2] Thus the US, Canadian, and Mexican governments sought to avert potential environmental problems and, more significantly, to build environmental sensitivity into the NAFTA.

In the NAFTA negotiations, US trade officials worked with US Environmental Protection Agency officials to an unprecedented degree (Esty 1994a, pp. 68–69). For the first time, the Office of the US Trade Representative conducted an environmental review of a trade agreement while it was under negotiation, and used its conclusions to create a better agreement (USTR 1992). NGOs and other interested stakeholders also played an influential role in the development of US negotiating positions and the NAFTA negotiations.[3] Though members of environmental and other NGOs may hold differing views on the results of the negotiations, their participation made a difference.[4] It is fair to say that the NAFTA negotiations were a crucial stimulant to ongoing analysis of the relationships between trade liberalization, the environment, and the protection of human health (Audley 2001; Esty 1998).

Because of these efforts, the NAFTA emerged a better agreement. And the proof is in the record of the last 8 years: the NAFTA's environmental dimensions have helped the United States and its North American trading partners to improve the environment and the quality of life across the continent as well as to advance other basic values such as a common commitment to the rule of law and more accountable governance. The NAFTA achieved these environmental improvements in two ways. First, the NAFTA stimulated the economic growth that is necessary for enhanced public- and private-sector investments in the environment. Second, the incorporation of innovative environmental provisions into the NAFTA's main text and the institutions created by its environmental side agreement enhanced the ability of the NAFTA countries to protect their environments against potential negative effects of expanded trade and concomitant economic expansion, to improve cooperation and working relationships on the environmental front, and to guard against specific kinds of environmental harm (Hufbauer et al. 2000).

The NAFTA's Text

A significant—and often overlooked—aspect of the NAFTA environmental package is the environmental language in the main text of the NAFTA itself. The preamble of the agreement sets the tone for what follows, committing governments to undertake the implementation of the NAFTA "in a manner consistent with environmental protection and conservation."[5] The preamble also incorporates commitments to sustainable development and the "development and enforcement of environmental laws and regulations." In addition, in several of its subsequent sections, the NAFTA incorporates strong principles relating to environmental protection (Johnson and Beaulieu 1996).

To understand the significance of these environmental provisions, it is important to recall that one of the central purposes of trade agreements is to ensure that the national laws of trading partners follow certain "trade disciplines" that require them to treat imported products in a manner no less favorable than the manner in which they treat those produced domestically (Esty 1994b). The NAFTA's important innovation is that it allows governments to make exceptions to these disciplines to protect the environment and human health and to conserve natural resources (Audley 2001, p. 8). For example, article 712 explicitly recognizes the right of parties to adopt, maintain, or apply sanitary or phytosanitary measures for the protection of human, animal, or plant life or health, including measures more stringent than an international standard.[6] Article 904 recognizes similar rights to establish other standards-related measures (including those that exceed international technical product and food safety standards.[7] In both instances, the NAFTA allows countries to establish environmental standards so long as they do not constitute a "disguised barrier to trade."[8] Finally, in the case of a dispute involving environmental legislation, the NAFTA's provisions for dispute settlement place the burden of proof on the challenging party to show that the environmental provisions in question restrict trade unnecessarily.[9] (For further exposition of the environmental provisions contained in the NAFTA's main text, see appendix A.)

While there have been no direct challenges to human health or environmental laws under NAFTA to date, there is no evidence to date that compliance with these provisions has caused economic harm to any of the three countries. On the contrary, the NAFTA's language has offered regulatory agencies considerable flexibility to pursue domestic measures to protect the environment (particularly in comparison with WTO rules). Some claim that the only exception to this general rule is the unintended environmental consequences associated with the NAFTA's Investment chapter (chapter 11). Article 1114 clearly addresses the question of pollution havens by stipulating that the parties should not waive or derogate domestic health, safety, or environmental measures to encourage investments in their territories, and provides a right to request consultations should a party consider another party to have offered such encouragement (Esty 1994a).[10] Concerns about the implications of the NAFTA's expropriatory provisions for environmental policy making, however, may warrant further exploration (Mann and von Moltke 1999).

The NAFTA's Environmental Side Agreements
When the Clinton administration turned its attention to negotiating the side agreements, it sought to achieve a delicate balance. On the one hand, the administration wanted to put in place mechanisms that would help the

United States to restore and protect the environment. At the same time, negotiators were mindful that the United States would have to live with anything it asked Canada and Mexico to accept.

The two supplemental agreements that emerged—the North American Agreement on Environmental Cooperation and an additional bilateral cooperation agreement between the United States and Mexico—struck that balance.[11] They provided needed assurance that all the NAFTA partners would enforce their environmental laws and regulations, by committing the countries to strengthening their own administrative and judicial procedures.[12] The agreements also created mechanisms through which both countries, as well as private individuals and organizations, can challenge a pattern of non-enforcement of environmental regulations in another country.[13] None of the institutions that the NAFTA established usurps the right of each country to set its own laws. Nor can they replace federal, state, tribal or local authorities in the enforcement of their own laws.

Perhaps the most remarkable achievement of the NAFTA's environmental side agreement is the establishment of a trilateral North American Commission on Environmental Cooperation (NACEC) with a mandate to "facilitate cooperation and public participation to foster conservation, protection and enhancement of the North American environment for the benefit of present and future generations, in the context of increasing economic, trade and social links between Canada, Mexico and the United States."[14] Article I of the NAFTA's environmental side agreement presents public participation as a fundamental goal. This article supports general commitments to make environmental information publicly available and the specific commitments to ensure private access to remedies for environmental matters and fair, open, equitable procedures at the domestic level (Johnson 2000). The environmental side agreement also creates a Joint Public Advisory Committee (JPAC) to the NACEC designed to provide input from NGOs and the private sector to the NACEC's governing council. Two other NAFTA-related institutions—the Border Environment Cooperation Commission (BECC) and the North American Development Bank (NADBank)—were established by the US-Mexico bilateral agreement to assist in the development of environmental projects along the US-Mexico border.[15]

What have these institutions achieved?

First, through a combination of research, newsletters, and new tools for environmental monitoring and data collection, the NACEC has improved understanding of trade and environment issues, environmental reporting, and the availability of information about the state of the environment and environmental laws in North America. This increased environmental knowledge is a prerequisite for improvement of environmental cooperation in the region (Johnson 2000). The development of an inventory of North American Pollutant Releases and Transfers provides a case in point (NACEC 1998).

Second, the NACEC has initiated a range of initiatives that support greater environmental cooperation among the NAFTA countries.[16] Cooperative actions fall in several areas. The NACEC facilitated an agreement among the NAFTA partners on the conservation of North American birds (NACEC 1999). It has also helped the NAFTA partners devise regional action plans, scenarios, and proposed strategies to address several environmental issues, including phasing out or sound management of toxic substances (including DDT, chlordane, PCBs, and mercury), joint implementation of the United Nations Environment Program's Global Program of Action for the Protection of the Marine Environment from Land-Based Activities, and reduction of exposure to the persistent organic pollutant lindane (Johnson 2000). NACEC efforts have also helped develop constructive responses to some environment disputes and promote efforts to harmonize environmental standards in specific areas (Torres 2002). Specific projects to improve Mexico's environmental capacity have been a critical component of the environmental cooperation agenda (Rosenberg 2000). The NACEC has developed model cooperation projects—such as one on the San Pedro water management area (NACEC 1999).[17] In addition, the North American Fund for Environmental Cooperation (NAFEC) has extended small grants to an array of projects in Mexico focused on improving environmental protection, natural resource management and local livelihoods. And, in the years since the NAFTA was signed, the US Environmental Protection Agency has trained hundreds of Mexican environmental official, and Mexico has substantially increased its budget resources and inspections related to environmental law compliance.

Third, the NACEC has made important contributions to improving the enforcement of environmental laws by devising tools and mechanisms to promote pollution control, investigation, and reporting as well as public participation (NACEC 1998). The NACEC's coordination of collaborative work on monitoring and environmental enforcement is helping the NAFTA partners reduce the costs of environmental protection. The United States and Canada, for example, have established protocols for the coordinated review of certain new pesticides (such as those that are designed to replace older, more risky pesticides). By sharing data review responsibilities, joint reviews lower regulatory costs, expedite registration of safer pest-control tools, increase the efficiency of the registration process, and provide farmers across North America with more nearly equal access to pest-management tools.[18] The NACEC has also made important efforts to promote enforcement of environmental laws under the auspices of articles 13–15 of the NAAEC. Article 13 permits the NACEC to initiate reports on any environmental issue within the scope of the work plan approved by its council (Johnson 2000).[19] Article 14 enables the NACEC to develop factual records following a private party's submission regarding an alleged non-enforcement of environmental law in one country. As of April 2002, the article 14 provisions have already attracted 33 citizen submissions.[20]

Finally, the NACEC has worked admirably to achieve its mandate regarding public participation. By consistently working to seek public input and incorporate the insights and expertise of civil society into its activities and projects, the NACEC's initiatives have been greatly enhanced. Of course, there remain issues of which the NACEC needs to make further progress. Some officials in the trade community have worried about the NACEC getting too deep into trade and environment issues. Environmentalists have complained that the NACEC has not gone far enough in this direction (Carlsen and Salazar 2002; Sierra Club 2001).

The NADBank and the BECC have also achieved considerable success— and revealed some weaknesses. Together, they are working with more than 100 communities throughout the US-Mexico border region to address their environmental infrastructure needs. Since 1994, these two institutions have invested US$940.5 million to aid the development of at least 44 environmental infrastructure projects related to water, sewage, and municipal waste in communities on both sides of the US-Mexico border, benefiting almost 7 million border residents (White House 2001). With the assistance of the BECC and NADBank, the city of Juarez, for example, recently completed its first wastewater-treatment plant. The results will be better health and cleaner water for a million people in Juarez, another million in El Paso, and for towns and villages all along the upper Rio Grande. The NADBank has experienced considerable difficulty operating as a lending agency—with a large share of its resources not yet deployed. Part of the issue arises from the fact that many border entities lack the resources to pay off loans, forcing the Bank to rely more heavily on grant support to develop and implement important infrastructure projects (GAO 2000).

In environmental improvement, as with the reduction of barriers to trade in goods and services, the NAFTA is incomplete—it remains a work in progress. The NAFTA—hailed by the *Dallas Morning News* (1999) as the "greenest commercial pact ever"—has represented a significant step forward in terms of the recognition of the environmental aspects of trade. The US, Canadian, and Mexican environments are better off with the NAFTA than without it. Governments still need to address many environmental challenges, but the NAFTA and its side agreements put us in a better position to deal with them.

Trade and Environment Lessons for the FTAA

The NAFTA's environmental provisions and side agreements demonstrate that countries can explicitly deal with environmental issues, both within and around the context of trade negotiations, without impeding their mission of trade liberalization. FTAA negotiators have much to learn from the NAFTA and post-NAFTA experience of working to make trade policies and environment policies mutually reinforcing.

First, the NAFTA highlights the importance of taking the environmental implications of the negotiations into account from start to finish of negotiation processes. This means that we need discussions and negotiations not only to ensure that environmental agreements "do no harm" but also to take advantage of positive opportunities to enhance environmental performance. In addition, the NAFTA experience instructs us that trading partners must equally share a commitment to handling the environmental aspects of trade. Much is made about the economic might of the United States, the subtext being that the United States should be able to get whatever it wants on the environmental front. However, the issue is not that easy. Even if the United States does get what it wants in an agreement, positive results depend on the degree to which its trading partners see environmental protection as being squarely in their national interest. Awakening Latin

Americans to their own self-interest in the trade-environment linkage remains a significant challenge within the FTAA.

Second, governments across the region—including the US government under the new Bush administration—must bring to the FTAA negotiations the lessons that the NAFTA taught about the benefits of working closely with environmental agencies, non-government organizations, and other interested parties before and during negotiations as well as in the implementation phase of any trade liberalization effort. The development of the NAFTA's side agreement and environmental provisions in the main NAFTA text can be attributed—in important respects—to unprecedented levels of engagement of environmental regulatory agencies in domestic trade policy making processes and the negotiations themselves.[21] The NAFTA taught us that an informed public and open debate can help to raise key issues, ensure that non-trade values are folded into the negotiations, and build public support for the trade agreement that emerges. By engaging their parliaments and the public in the formulation of national trade policy objectives, trade negotiators can develop trade initiatives with a clear sense of the standards and benchmarks which legislators and the public expect them to meet. The NACEC's independent factual reports on the enforcement of environmental provisions show how public engagement can stimulate government action on environmental protection.[22] In recognition of the lessons regarding transparency in negotiations and the need for public input, the Clinton administration exerted strong leadership in creating the FTAA's Committee of Government Representatives on Civil Society and working to give the committee a meaningful role in the negotiating process—even in the face of opposition from a number of Latin American countries. At the national level, the United States continues to solicit public comment to help shape negotiation objectives. Bush administration negotiators need to continue to strive for transparent negotiations, to seek the input of the USTR Trade and Environment Policy Advisory Committee, and to maintain a dialogue with all elements of civil society through various means throughout the negotiations. Governments across the Americas would do well to also explore ways to systematically engage environmental agencies and the public in the FTAA negotiations and national trade policy making processes. Simply put, environmental issues are unlikely to be avoided in a regional trade agreement (Dua and

Esty 1997). It is better to tackle the issues that will inevitable arise in an open and straightforward manner.

Third, the NAFTA and post-NAFTA experience demonstrates that environmental cooperation can be a complementary outcome of trade negotiations that benefits all parties. The NAFTA's parallel agreements show that "countries can think creatively to develop mechanisms for sharing technology, experience in environmental protection, and the financial resources necessary to build environmental infrastructure where it did not exist previously" (Audley 2001, p. 11). Environmental cooperation imposes costs on trading partners. The creation of the BECC and NADBank established the important precedent of financing to support environmental cooperation and improvement. That said, the mixed performance of the NADBank and the BECC underscores the importance of maintaining high levels of interest in, and support for, the environment and trade agendas of the NAFTA's institutions and any future environmental mechanisms governments may establish in the Americas (Kelly et al. 2001). The US-Jordan Free Trade Agreement, recently passed by the US Congress, has already tried to build on the NAFTA experience.[23] While this agreement does not guarantee new resources to Jordan, it creates a Joint Forum on Environmental Technical Cooperation to advance environmental protection in Jordan with the potential to channel resources from the US Agency for International Development toward improving Jordan's environmental capacity.[24] Trade negotiations can focus attention on the need to build environmental capacity and infrastructure among trading partners.

Environmental reviews are clearly a key component in this effort. They enable governments to consider environmental issues fully during the course of the negotiations, to understand where the need to build additional environmental infrastructure, and to provide a useful tool for involving the public in the trade policy making process. The US experience with the NAFTA provided inspiration for President Clinton's 1999 strategy to ensure that US efforts to expand trade and promote development reflected a strong commitment to achieving environmental protection in the United States and abroad. As part of this initiative, in 1999, President Clinton signed an executive order requiring careful assessment and written review of the potential environmental impacts of major trade agreements to ensure that US positions in trade negotiations incorporate environmental considerations.[25] President Clinton also issued a White House Policy Declaration on

Environment and Trade, outlining a set of principles to guide US negotiators and to ensure that trade supports sustainable development globally.[26] The groundwork for an environmental review of the FTAA has already been laid, and the Bush administration has affirmed its commitment to implementing environmental reviews of trade agreements.[27] An interagency group has developed recommendations on the appropriate methodology for analyzing the potential environmental effects of free trade.[28] A quantitative analysis of the impact of tariff elimination is, however, only one of the necessary components of an effective environmental review. Research on non-quantitative effects as well as regulatory and legal effects is also important. Moreover, these reviews should provide ample time for public input and review of the data, analysis, methodology, and recommendations.

A final, crucial lesson learned from the NAFTA experience is that countries can make the greatest progress in improving their environments when their economies are on the right economic path. Expanding trade need not—and should not—spell concurrent environmental doom. The evidence is clear: a strong economy and a clean environment go hand in hand. The United States provides a case in point. By mid 2000, the opening of world markets had helped spark a 62 percent expansion of American exports since 1992 (USTR 2001). Together with, and inseparable from, domestic policies including fiscal discipline, deregulation, and investment in education and job training, as well as private-sector adjustment to the new economic paradigm of the Information Age, the opening of world markets has contributed to a remarkable economic record. The United States has experienced the longest economic expansion in its history, an expansion in its manufacturing industry, increases in real wages for non-supervisory workers, and broadly shared benefits.

The NAFTA has contributed to this prosperity by creating fairer and more open markets. During the NAFTA's first 7 years, goods exports from the United States to its NAFTA partners increased 104 percent to more than US$247 billion (USTR 2001). Today, Canada is the United States' largest trading partner (in terms of two-way trade flows), and the success of the NAFTA has been a significant factor in stimulating Mexico to become our second-largest trading partner, surpassing Japan. The easiest way to summarize the weight of these two countries on the "sell side" of our trade equation is this: a quarter of everything the US sells abroad goes to Canada and almost 15 percent goes south to Mexico. The NAFTA partners thus account for 40 percent of US exports. In 2001, the United States sold more to Canada than to the European Union. By 2004, the United States is likely to sell more to Mexico than to the European Union.

Although any causal relationship is difficult to establish, it is clear that at the same time as the NAFTA has expanded trade, US environmental performance has strengthened. In early 2000, the White House released a report from the Council on Environmental Quality highlighting dozens of administration initiatives over the past seven years to improve public health, restore endangered wildlife, promote "green" business, protect oceans and coasts, strengthen environmental enforcement, and combat global warming (CEQ 2000). The same pattern can be found in Canada and Mexico. Data for the NAFTA era show improved air and water quality, accelerated toxic cleanups, dramatic reductions in toxic releases, and increased environmental protection for millions of acres across America (CEQ 2000). Since 1993, the number of Americans breathing clean air has grown by 44 million, the number receiving clean drinking water has grown by nearly 34 million. Superfund cleanups have more than tripled, environmental technology exports have more than doubled, and spending on environmental priorities has risen dramatically (CEQ 2000). While the United States still faces significant environmental challenges, there is no doubt that a strong economy will contribute to future environmental gains. Higher productivity and living standards generate resources and political will for environmental improvement. A stronger regional trading system will strengthen the prosperity of countries in the western hemisphere and thus strengthen their ability to achieve shared environmental goals and address common environmental responsibilities.

Conclusion

The NAFTA is a dynamic agreement. Like the FTAA, it is a work in progress. In fact, the NAFTA will not be completely implemented until 2008. All three countries are learning from experience thus far, using it to improve the agreement as it goes into force. Using the cooperative framework built through the NAFTA, countries have undertaken the challenge of advancing not only their mutual interests in freer trade but also their shared goal of an improved North American environment. The United States can be very pleased with the record of the NAFTA six years after its passage. Both the Bush and Clinton administrations predicted that the NAFTA agreement would mean growth, better and more jobs, rising standards of living, and a higher quality of life. For the United States, the agreement is keeping these promises. The United States has more jobs, higher wages, and a much stronger economy than it did at the time the NAFTA was concluded. The NAFTA governments are working more closely and accomplishing more than ever before on environmental protection, workplace safety, and all the other issues that affect the daily lives of our citizens. And—most important of all—the prospects are better than ever before of passing on to our children, the invaluable legacy of peace, cooperation, and progress on the North American continent that we have inherited from past generations.

Each new trade negotiation is different. The handling of the environmental aspects in the NAFTA was strongly shaped by the borders the United States shares with its NAFTA partners, as well as certain other factors unique to those countries. FTAA negotiators will need to develop their own environmental strategy, not just because of the great number of countries and the multiplicity of common borders, but also because the times have changed, and we now know more about how best to build environmental sensitivity into a trade agreement. The point is that context matters. Policy makers need to think about environmental issues in terms of the specific context of each negotiation. The US-Jordan Agreement illustrated this point and the potential for approaches to making the environment-trade link evolve over time into palatable formats. The United States and Jordan negotiated a dispute resolution process that compels parties to resolve trade tensions in a cooperative, respectful manner to replace the threat of sanctions to punish non-enforcement of environmental laws (Audley 2001, p. 21). This variation on the NAFTA theme represents an evolutionary step forward, which can be further refined in the FTAA context.

Building practicable environmental sensitivities into FTAA negotiations will help produce a regional trade agreement that works for Latin Americans and that will be accepted by the US Congress and the American people. As the Bush administration proceeds with these important trade negotiations, it must work with Latin American governments to devise strategies for technical cooperation and civil society participation. In this respect the NAFTA provides a great model to build on with the refinements mentioned above.

Notes

1. The full draft text is posted on the FTAA web site (http://www.ftaa-alca.org).

2. For a comprehensive set of positions papers from governments and environmental organizations regarding environmental concerns and recommendations presented over the course of the NAFTA negotiations, see Magraw 1995.

3. From June 1990 until November 1993, US Congress held more than one hundred hearings on environmental issues raised during the NAFTA negotiations, inviting more than 80 members of the US environmental community to offer their opinions before Congress (Audley 1997).

4. For a detailed presentation of the positions taken by US environmental organizations and their participation in the NAFTA negotiations, see Audley 1997.

5. See North American Free Trade Agreement 1993 and http://www.nafta-sec-alena.org/english/index.htm.

6. Article 712 is included in NAFTA's chapter 7 on sanitary and phytosanitary measures.

7. Article 904 is contained in NAFTA's chapter 9 on technical barriers to trade.

8. Despite limitations to the total freedom of parties to adopt trade-restrictive measures, the language of articles 712 and 904 make NAFTA considerably more open to high levels of environmental protection than prior regimes (such as the Canada-US Free Trade agreement and the General Agreement on Tariffs and Trade) (Johnson 2000). Audley (2001, p. 8) writes: "Several of the trade disciplines set forth in NAFTA use language that more clearly provides the flexibility that is important for preserving critical environmental protections. For example, while the WTO Agreement requires that technical regulations not be more trade restrictive than necessary, NAFTA does not include this test. Instead, NAFTA explicitly allows each party to adopt, maintain, or apply, product standards *and any measure* to ensure their enforcement or implementation. (NAFTA articles 712.1: 904.2—emphasis added)" The NAFTA also clarifies several points that are ambiguous in the WTO Agreement. In the area of food safety (and other sanitary and phytosanitary measures), for example, the NAFTA articulates that international harmonization of such measures must be achieved without reducing the level of protection. In addition, Audley explains that the NAFTA improves on the WTO by requiring that food safety measures have a scientific basis rather than a "sufficient" scientific basis (leaving the interpretation of what constitutes "sufficient" scientific basis to dispute settlement panels of legal specialists who are not expert in scientific and regulatory issues. The NAFTA also provides importing countries the right to determine, on a scientific basis, whether the exporting country's food safety measures achieve a standard of protection that is equivalent to that of the importing country (ibid., p. 9).

9. In the WTO context, the General Agreement on Tariffs and Trade holds that the burden of proof falls on the party attempting to uphold the environmental provisions under challenge.

10. In the case of disputes, it is noteworthy that this provision is not accompanied by any remedies (e.g. sanctions or fines) beyond consultations (Johnson 2000).

11. The full text of the 1994 Agreement between the Government of the United States of America and the Government of the United Mexican States concerning the establishment of a Border Environmental Cooperation Commission and a North American Development Bank can be located at: http://www.cocef.org/englishbecc.html.

12. Article 5 of the NAAEC establishes specific commitments of the NAFTA parties oriented toward the persistent and effective enforcement of environmental laws and regulations. The substance of this article is defended in the dispute resolution procedure described in chapter V of the NAAEC.

13. Chapter V (on dispute resolution) provides for such challenges between countries. Canada is not, however, a party to the specific agreement between the US and Mexico that provides for sanctions, after an exhaustive process, in the case of noncompliance. Article 14 of the NAFTA puts in place a mechanism whereby an NGO or an individual can voice their complaints by presenting a submission "asserting that a party is failing to effectively enforce its environmental law." The culmination of this procedure is the drafting of a factual record on the alleged non-enforcement by the NACEC that can then be made public. The process is designed to shine an international spotlight on cases of alleged non-enforcement and facilitates the application of public pressure on and within that country, but no legally binding obligation can be imposed as a result of this particular process (Johnson 2000).

14. The NACEC recognizes "the importance of the environmental goals and objectives of the NAFTA," and commits the parties to "promote sustainable development based on cooperation and mutually supportive environmental and economic policies," to "increase cooperation between the parties to better conserve, protect, and enhance the environment, including wild flora and fauna," and, inter alia, to support the environmental goals and objectives of the NAFTA. See North American Agreement on Environmental Cooperation "Preamble" and "Objectives," http://www.ccc.org/pubs info resources/law treat agree/naaec. For a description of the governance and organization of the NACEC, see appendix A.

15. For complete descriptions of these institutions, see http://www.becc.org and http://www.nadbank.org. Each web site also contains extensive bibliographies from which to select additional readings

16. Environmental cooperation also proceeds outside the NAFTA context. The US and Canada, for example, forged the International Joint Commission to resolve disputes over waters from the Gulf of Maine to the Gulf of Alaska. More recently, the US has worked through our International Boundary Waters Commission with Mexico to fight pollution and provide for the fair allocation and the use of the shared water resources.

17. Under the authority of article 13 of the NAAEC, the Commission produced a report on the San Pedro Secretariat in 1999. The report describes the cooperative efforts to ensure the continued health of the San Pedro watershed, including a growing number of public and private partnerships (NACEC 1999a).

18. For further information about the NAFTA's Program on Pollutants and Health see, http://www.cec.org/programs_projects/pollutants_health/.

19. Additional reports that have been issued to date include a 1995 report on the massive bird kill at the Silva Reservoir in the State of Guanajuato in Mexico, a 1997

report tracing pathways for air pollutants in North America, and a 1999 report assessed water problems in the San Pedro Riparian Conservation area along the Mexico-Arizona border. See the NACEC's publications page at http://www.cec.org/pubs info resources/publications.

20. The NACEC's web site provides a complete list and history of all article 14 submissions. See http://www.cec.org.

21. Environmental agencies involved in the development of US negotiation positions and the NAFTA negotiations included the Department of Interior, the Environmental Protection Agency and the National Oceanic and Atmospheric Administration. In some instances, environmental agency officials chaired the US negotiating teams (Magraw 1995). Before the NAFTA negotiations, most US regulatory agencies were not involved in trade policy negotiations or relied on their international trade divisions that were more focused on promoting US exports than addressing trade impacts on domestic regulatory authority (Audley 1997).

22. In the Cozumel Pier case against Mexico, for example, the submission helped stimulate engagement between the Mexican government and civil society in the development of regulations for the island. The island was ultimately declared a protected area (Hufbauer et al. 2000).

23. See http://www.ustr.gov/regions/eu-med/middleast/US-JordanFTA.shtml for the text of the US-Jordan Free Trade Agreement.

24. See US-Jordan Statement on Environmental Technical Cooperation at http://www.ustr.gov/regions/eu-med/middleast/US-JordanFTA.shtml

25. Executive Order 13141, dated November 16 1999. In July 2000, the Office of the US Trade Representative and the Council of Environmental Quality released draft guidelines on implementation of Executive Order 13141 (Environmental Reviews of Trade Agreements) which mandates procedures for integrating consideration of environmental issues into the trade negotiating process (Department of State 2000). For more information, see: http://www.whitehouse.gov/library/index.html.

26. This Declaration of Principles was also released on November 16 1999. Both the Declaration and the Executive Order were mentioned in a statement by the president on that day and the White House also released an information sheet on the Declaration. See http://www.whitehouse.gov/library/index.html.

27. In December 2000, the Clinton administration released a Notice of Initiation of Environmental Review and Request for Comments on Scope of Review, and Notice of Availability of the Report of the Quantitative Analysis Working Group and Request for Comments in the Federal Register (Federal Register 2000). This signaled the intent to formally initiate an environmental review of the proposed Free Trade Area of the Americas. The new Bush administration affirmed its commitment continuing with a written environmental review of the FTAA (AFFI 2001). See http://www.affi.com/capconnection4-27.asp.

28. The US FTAA Environmental Group, chaired at the Trade Policy Staff Committee level, created an interagency Quantitative Analysis Working Group composed of experts from relevant agencies. The Working Group was charged with providing advice on an analytical methodology for quantifying the environmental effects of hemispheric trade liberalization. The Working Group has released recommenda-

tions regarding the completion of a quantitative analysis in a report to the FTAA Environmental Group. This document, entitled Report of the Quantitative Analysis Working Group to the FTAA Interagency Environment Group, is available on the USTR web site, see http://www.ustr.gov. Work to develop and test methodologies for environmental assessment has been underway for several years at the NACEC. See e.g. Spalding and Stern 1996 and NACEC 1999b. In October 2000, the NACEC held a conference to test the methodologies it has developed. Papers and proceedings of that conference are available at http://www.cec.org/symposium.

References

AFFI. 2001. "USTR to Continue Environmental Review of Trade Agreements." *Public Policy Debate*, April. Available at http://www.affi.com/capconnection4-27.asp.

Audley, J. 1997. *Green Politics and Global Trade: NAFTA and the Future of Environmental Politics*. Georgetown University Press.

Audley, J. 2001. "Back to the Bad Old Days of NAFTA." Paper presented at NWF/Yale Center for Environmental Law and Policy conference on "Trade and Environment in the Americas: Lessons Learned from the NAFTA." April 2000, revised April 2001. Available at http://www.ycelp.org.

Carlsen, L., and H. Salazar. 2002. "Limits to Cooperation: A Mexican Perspective on the NAFTA's Environmental Side Agreement and Institutions." In this volume.

CEQ. 2000. *Earth Day 2000 Report*. Washington: Council on Environmental Quality.

Dallas Morning News. 1999. Editorial, January 4.

Dua, A., and D. Esty. 1997. *Sustaining the Asia Pacific Miracle: Environmental Protection and Economic Integration*. Washington: Institute for International Economics.

Esty, D. 1994a. "Making Trade and Environmental Policies Work Together: Lessons from NAFTA." *Aussenwirtschaft* 49: 59–79.

Esty, D. 1994b. *Greening the GATT*. Washington: Institute for International Economics.

Esty, D. 1998. "Environmentalists and Trade Policymaking." In *Representation of Constituent Interests in the Design and Implementation of US Trade Policies*, ed. A. Deardorff and R. Stern. University of Michigan Press.

Federal Register. 2000. "Guidelines for Implementation of Executive Order 13141: Environmental Review of Trade Agreements." *Federal Register* 65, no. 244: 79442–79449.

GAO (General Accounting Office). 2001. *US Experience with Environment, Labor, and Investment Dispute Settlement Cases*, August 21.

Hufbauer, G., D. Esty, D. Orejas, L. Rubio and J. Schott. 2000. *NAFTA and the Environment: Seven Years Later*. Washington: Institute for International Economics.

Johnson, P. 1999. "Five Windows for the Future of NAFTA's Environment Commission." *Policy Options* 20, no. 5: 27–32.

Johnson, P. 2000. "Trade Liberalization and the Environment: From NAFTA to the FTAA." *ISUMA* 1, no. 1: 1–8.

Johnson, P., and A. Beaulieu. 1996. *The Environment and NAFTA: Understanding and Implementing the New Continental Law.* Island.

Kelly, M., C. Reed, and L. Taylor. 2001. *The Border Environmental Cooperation Commission (BECC) and the North American Development Bank (NADBank): Achieving Their Environmental Mandate.* Texas Center for Policy Studies.

Magraw, D. ed. 1995. *NAFTA and the Environment: Substance and Process.* American Bar Association.

Mann, H., and K. von Moltke. 1999. *NAFTA's Chapter 11 and the Environment: Addressing the Impacts of the Investor-State Process on the Environment.* Winnipeg: International Institute for Sustainable Development.

NACEC. 1998. North American Pollutant Releases and Transfers. Montreal: North American Commission on Environmental Cooperation.

NACEC. 1999a. *Ribbon of Life: An Agenda for Preserving Transboundary Migratory Bird Habitat on the Upper San Pedro River.* Montreal: North American Commission on Environmental Cooperation.

NACEC. 1999b. *Assessing Environmental Effects of the North American Free Trade Agreement (NAFTA). An Analytic Framework (Phase II) and Issue Studies.* Montreal: North American Commission for Environmental Cooperation.

Public Citizen and RMALC. 1996. *NAFTA's Broken Promises: The Border Betrayed, US-Mexico Border Environment and Health Decline in NAFTA's First Two Years.* Washington: Public Citizen.

Rosenberg, R. 2000. "Can NAFTA's Approach to Environmental Cooperation Instruct the FTAA Negotiating Process? Emerging Institutionality and the Struggle for Political Will." Presented at NWF/Yale Center for Environmental Law and Policy conference on "Environment in the FTAA: What Can We Learn from the NAFTA Model?"

Sierra Club. 2001. Sierra Club Position on the FTAA: Statement by Robert Cox.

Spalding, M., and M. Stern. 1996. Potential NAFTA Effects: Claims and Arguments 1991–1994." Report 2, Environment and Trade Series, Commission for Environmental Cooperation, Montreal.

Torres, B. 2002. "The North American Agreement on Environmental Cooperation: Rowing Upstream." In this volume.

USTR. 1992. "Review of U.S-Mexico Environmental Issues." Washington: US Trade Representative.

USTR. 2001. "Testimony of Ambassador Peter F. Allgeier, DUSTR, Before the International Relations Committee, Western Hemisphere Subcommittee of the US House of Representatives - July 12 2001." Washington: US Trade Representative.

White House. 2001. White House Fact Sheet on BECC and NADBank, September 5.

11

The North American Agreement on Environmental Cooperation: Rowing Upstream

Blanca Torres

The North American Free Trade Agreement spurred profound changes in the way bilateral and trilateral issues were handled by Mexico and its new trading partners. Traditional, tacit cooperation between Mexico and the United States was abandoned in favor of explicit, formal cooperation, not only in trade but also in other areas, including environmental protection.[1] Mexican cooperation with Canada, almost non-existent before the NAFTA, was broadened.[2]

Although many analysts consider the provisions of the North American Agreement on Environmental Cooperation (the NAFTA's environmental side agreement) to be groundbreaking developments, both the creation and the implementation of the NAAEC's institutions for environmental cooperation have presented a number of challenges to governments. Many environmentalists have become skeptical of the NAAEC institutions' potential to foster effective environmental cooperation. The performance of the NAAEC institutions has been hampered by the complexity of the environmental issues at stake, the political context, and the character of the side agreement itself. The NAAEC's emphasis on litigious rather than collaborative functions did not bode well for its implementation (Spalding 1997). It also places excessive pressure on Mexico, the most vulnerable of the three new partners, to increase its rate of environmental protection, even though few provisions were introduced by the NAFTA or its environmental side agreement to help Mexico facilitate swifter progress toward better pollution control.

In this chapter, I argue that the NAFTA's environmental commitments have contributed to a greater concern for the environment among relevant state and non-state actors in Mexico. The results of this new attitude have improved environmental training and research as well as the design of environmental institutions and mechanisms for public participation. These improvements will foster better and long-lasting environmental protection in Mexico. I further argue that the NAFTA's institutional accomplishments, albeit modest, are significant, particularly since cooperation between the states faced powerful obstacles.

Challenges to the Implementation of NAFTA's Environmental Side Agreement

The North American Commission for Environmental Cooperation (NACEC) is the institution responsible for implementing the environmental side agreement. Through a collaborative, incremental approach, it has helped to spawn a range of positive environmental initiatives in Mexico. NAAEC's environmental institutions, the NACEC, the Border Environmental Cooperation Commission (BECC) and the North American Development Bank (NADBank), have sought to make valuable environmental contributions.[3] From the outset, the NACEC emphasized collaboration (possibly to de-emphasize the environmental side agreement's controversial features). It focused on tasks such as capacity building and data gathering to create a solid base for cooperation. It also produced several thorough reports (such as the Upper San Pedro River Basin Report) and developed several promising new programs (such as the Silva Reservoir Management Program, the Sound Management of Chemicals Initiative, and the North American Fund for Environmental Cooperation (NAFEC)).[4]

At the same time, the NACEC strove to find its niche among the institutions working for environmental cooperation in the region (such as the International Joint Commission, the International Boundary and Water Commission, the recently established Border XXI Program, BECC, and the NADBank), some of which boasted more experience or resources than the NACEC (GAO 1996).[5] The NACEC's initial attempts to advance very rapidly on collaborative issues postponed the development of a clear strategic vision to guide its work and ultimately undermined its efforts (Independent Review Committee 1998). However, during recent years it has tried to develop greater focus for its work.

Six distinct sets of challenges, discussed below, have thwarted the smooth implementation of the NAFTA environmental side agreement.

Economic Asymmetries

Sharp differences, due in large part to different levels of economic development, exist between the NAFTA partners in terms of environmental priorities, strategies, and the capacity to address problems. Table 1 provides a rough idea of the magnitude of the economic and social asymmetries between the three new partners, the United States, Canada and Mexico.

Reservations about Formal Cooperation

Economic asymmetries between the NAFTA partners also help explain Mexico's historical efforts to avoid formal cooperation with the United States on bilateral issues. Mexico's preference for tacit cooperation is not surprising because weak states tend to perceive formal agreements as more binding on them than on strong countries.[6] Sharing a 3,000-kilometer border with a country as powerful as the United States has increased Mexico's desire for autonomy in domestic and foreign affairs. Thus, avoiding formal accords, particularly bilateral accords, was an essential element of Mexico's strategy for several decades (Krasner 1990).[7] Certainly some of the formal agreements reached with the United States were indeed linked to environmental issues, but in general Mexico preferred a tacit approach to cooperation to ensure its sovereign rights remained intact.[8]

Therefore, it was not surprising that in 1980 and in 1989 Mexico rejected US proposals to negotiate a bilateral free trade agreement (Purcell 1981–82; Garciadiego 1994). Two factors seem to have caused President Carlos Salinas to change his mind and propose negotiating a trade agreement with the United States soon thereafter. First, in the face of increasing protectionism in the United States, Mexico felt the need to ensure access to the US market and recognized the significant economic opportunity created by

Table 1
Source: OECD 2000. (GDP data are in $US. GDP and GDP per capita reflect 1995 prices and exchange rates.)

	US	Canada	Mexico
GDP, 1999	$8,637 billion	$654 billion	$348 billion
GDP per capita, 1998	$32,300	$19,600	$4,300
Population	266.7 million	30.2 million	94.1 million
Inhabitants/km²	28	3	47

its location alongside the United States. Second, Salinas understood the need to attract more foreign capital to modernize Mexico's economy. Negotiating the NAFTA was a way of "locking in" the new model of economic growth, thereby creating a stable investment climate (Garciadiego 1994).

Both US and Mexican trade negotiators initially tried to prevent the inclusion of contentious "non-trade" issues, such as the environment, in the NAFTA negotiations. Their main arguments are well known and continue to be maintained by many Mexican trade officials (although they are not necessarily shared by all members of the Mexican government, especially in the Fox administration):

• There are no clear causal links between free trade and environmental degradation.
• Increased levels of growth fostered by free trade will provide less developed countries with the resources necessary to improve environmental protection.
• Trade will modernize the economy and force the adoption of cleaner technologies.[9]
• Environmental safeguards could be used to mask trade protectionism. (This concern was exacerbated by the US tuna embargo.[10])

Eventually, Mexican and Canadian authorities yielded to the US government on the environment front to win US congressional support for the NAFTA's approval (Mayer 1998). Nevertheless, both Mexico and Canada sought to limit their commitments as much as possible. Reluctant acceptance of some aspects of the NAFTA environmental side agreement continues to color US-Canada-Mexico environmental relations.

A second challenge that the environmental side agreement faced, and Mexico continues to face, is the political need to move from limited environmental cooperation (largely on US-Mexico border issues) to a broader scheme entailing stronger commitments and action. In this process, the Mexican government had two experiences of formal environmental cooperation on which to build. Neither the International Boundary and Water Commission (IBWC) nor the Agreement on Cooperation for the Protection and Improvement of the Environment in the Border Region (commonly known as the La Paz Agreement) provided a satisfactory model.[11] The IBWC was severely criticized by US non-governmental organizations during the NAFTA negotiations because of its capital-intensive approach to projects and the absence of public participation in decisions (Mumme 1991). The La

Paz Agreement was regarded as insufficiently binding and as concerned with only a limited range of problems (Spalding 1999; Sánchez 1995).

During his 1992 electoral campaign, President Clinton promised to deliver an environmental agreement "with teeth" to accompany the NAFTA. Consequently, the US government requested the creation of a trilateral commission with a clearly defined "watchdog" function. The nature of the final environmental side agreement reflects the range of disagreements among, and compromises reached by, several contending forces. While US environmental NGOs focused on defending domestic environmental standards, the Mexican government wanted to avoid excessive interference by its northern neighbors as well as commitments to harmonize environmental standards, which could be used as an excuse to block Mexican products from the US and Canadian markets. The US government was concerned with reconciling opposing pressures from environmental NGOs and big business.[12] Canadian negotiators maintained a largely detached position after reluctantly agreeing to inclusion of the issue in the negotiations.

None of the three NAFTA parties wanted an organization with supranational powers. The US government was, however, more willing than Canada and Mexico to give the trilateral commission power to initiate investigations. For their part, Mexico and Canada worked to limit the secretariat's autonomy. Eventually, Mexico accepted US proposals allowing the imposition of fines and, as a last resort, trade sanctions for noncompliance of either party with its respective domestic environmental laws. Canada, however, did not accept the possibility of trade sanctions, and an alternate recourse was devised.[13]

Given the political necessity of concluding the side agreement negotiations quickly, the NAFTA negotiators papered over their differences. They left loopholes and ambiguities, and they wrote escape clauses or insurmountable complexity into the NAAEC when consensus was not possible. This postponed the task of resolving outstanding differences between the negotiators—differences that would arise once again during the implementation process, provoking disagreement and friction between the parties.

Legacies of the NAFTA Negotiations
The tense negotiating process generated several destructive legacies.

First, the NACEC's valuable work has been overshadowed by three controversial issues that the negotiation process left unresolved: the secretariat's powers, the citizen submission mechanism, and public access to information. The three parties to the agreement still disagree on the powers of the secretariat, and the ambiguities of the environmental side agreement have sparked heated discussions. In the words of the Independent Review Committee (1998), it is not easy to define "the system of management oversight of the secretariat work that most, if not all, international environmental institutions have."

The second legacy is an approach and agenda for trilateral environmental cooperation that reflects US preferences and priorities. Mexican officials believe that the NACEC collaborative agenda primarily addresses US concerns, particularly the views of vocal US environmental groups. The incorporation of a wider range of green issues (such as a biodiversity protection agenda) into recent NACEC consultations has, to some extent, helped to reduce this feeling. However, some officials still believe that Mexican environmental priorities, such as the need to address desertification and land erosion, remain unacceptably neglected.

Third, Mexico felt that it was unfairly targeted in the NAAEC negotiations. Indeed, the fact that the negotiations took place at all left Mexico with a "sense of imposition" (Independent Review Committee 1998).

These legacies have worked to diminish the trust required to facilitate the work of the new institutions charged with managing environmental cooperation between the three countries.

The NAFTA's Litigious Approach

The environmental side agreement's litigious approach presents a fourth challenge. A careful reading of the environmental side agreement's main objectives and the NACEC mandate reveals that, in principle, each of the NAFTA's parties preferred to rely on environmental cooperation. The body of the environmental side agreement, however, adopted a litigious approach to many important issues. There is an imbalance between the number and scope of the provisions promoting collaboration and those involving dispute resolution (which were largely responses to US domestic political pressure from the environmental community). To quote Mark Spalding (1999): "Advocates for creating the Commission for Environmental Cooperation did not put much stock in the agency's "cooperation" aspects. The stronger focus was on the NACEC's litigious functions: dispute resolution, citizen petitions, and access to justice provisions." It is possible that NAAEC's wording of these matters (largely provided by moderate US NGOs) was intended to be a signal that the environment could not be ignored in the trade context, setting a precedent for future trade agreements. Several Canadian scholars see the language as a reflection of the dominant trade perspective of the parties, claiming that "negotiators displayed, throughout the NAFTA and the NAAEC's negotiations, a bias favoring the preservation or the reestablishment of trade equilibria as the main priority" (Johnson and Beaulieu 1996). Whatever its source, it is incontestable that the confrontational approach that pervades the side agreement has been a source of strain.

The two provisions of the environmental side agreement that challenge cooperative endeavors are the most litigious in spirit. These are the threat of trade sanctions and the provisions for citizen complaints against governments. The NAAEC's government dispute mechanism provides for financial and trade sanctions if an arbitration panel finds a pattern of persistent failure of the accused party to effectively enforce its environmental legislation and if, further, the accused party fails to remedy the situation. As a first step, the panel would impose "monetary assessments" on the accused party. If they were not paid, the party that brought the complaint could apply trade sanctions.[14] While the application of trade sanctions requires a lengthy process that has discouraged and reduced the likelihood of their use, Mexican officials still consider these sanctions provisions threatening.

Though no party has used it, some analysts argue that the very existence of the sanctions mechanism discourages Mexican efforts to complete or improve environmental legislation. Some government officials, not all from Mexico, made similar observations when interviewed by an independent committee charged with reviewing the NAAEC (Independent Review Committee 1998). As Johnson and Beaulieu (1999) state: "Regulators might be wary of adopting new environmental laws and regulations if they are not certain to have the resources necessary or a clear commitment from official decision making to adequately enforce the new measures."

The NAAEC's provisions for citizen submissions regarding noncompliance with domestic environmental regulations (whether or not these are related to trade) inspired similar concerns from Mexico.[15] Because resources were scarce and other problems limited the rapid improvement of law enforcement, some Mexican officials believed this mechanism posed great short-term risks to Mexico. It was certainly a riskier provision for Mexico than for Canada or the United States, because Mexico was more likely to fail in its domestic environmental enforcement efforts.[16] Given limited resources, Mexican officials also feared that this mechanism might inhibit the creation of a more coherent agenda of environmental priorities, at least in the short term. They envisioned the possibility that a large number of petitions of varying urgency or importance would distract already overburdened Mexican environmental officials and dilute or exhaust available resources. Moreover, although the penalty for non-compliance with a citizen submission is merely public exposure, some observers, including the NAAEC's Independent Review Committee (1998), have observed that the parties could use these submissions as the basis for launching inter-government disputes. Notably, one party may decide to use a factual record generated by the NAFTA's environmental commission to initiate a dispute resolution process under part V of the environmental side agreement (ibid.).

The NAAEC's citizen submission procedures have produced considerable debate, despite the relatively small number of submissions that have been filed. There could be several reasons for this limited use. Some NGOs complain that the citizen submission procedures are too costly, too bureaucratic, or simply ineffective. The low number of submissions could also be a reflection of the range of domestic avenues and legal remedies open to US and Canadian NGOs and citizens. On the Mexican side, the limited number of cases brought against Mexico is, at least partially, explained by the NGO community's self-restraint. NGOs appreciate the risk of dispersing the Mexican government's and the NACEC's scarce resources too widely and of damaging the environmental commission's reputation by over-reaching.

Though the citizen submissions procedure has been used rarely, the process for handling existing cases has strained relations between the NAFTA's parties and between the parties and the secretariat. According to the Independent Review Committee (1998), it has not been easy for the NACEC, a small organization, to cope with both the "adversarial" process and its cooperative functions. The committee suggested several means of reducing friction, which were immediately adopted, but they failed to prevent ongoing conflict.

The ambiguities of the NAAEC's language dealing with the citizen submission process (in articles 14 and 15), in addition to the different juridical traditions of the NAFTA partners, prompted Mexican officials to request a more precise definition of the secretariat's specific role in the citizen submission process. More recently, the Canadian government showed a similar interest. Environmental NGOs were not, however, sympathetic to their concerns. In June 2000, the NACEC's council took steps to address these ambiguities. It tried to address NGOs' concerns by incorporating the environmental commission's Joint Public Advisory Committee (JPAC) into the process of clarifying the language of articles 14 and 15. A constructive discussion of the precise relationship between the secretariat and the citizen submission process is urgently required.

The Demand for Equal Effort
The fifth challenge for the NAAEC has been its demand for equal effort by all parties. Mexican officials perceive that the NAAEC unfairly demands stronger and more immediate efforts by Mexico to improve its level of current environmental protection than it requires of others. While Mexico successfully opposed proposals to harmonize environmental protection levels or to advance new legislation, the obligation to comply with its existing domestic environmental laws raised distinct challenges [17] US officials and environmental NGOs believed that the 1988 Mexican General Law of Ecological Balance and Environmental Protection (LGEEPA) provided an acceptable level of environmental commitment, or that it was at least a good start. However, the law was relatively new, and many regulations and norms necessary for proper enforcement were still lacking.

Although it was also widely acknowledged in the three countries that Mexico had limited financial, technical, and human resources necessary for achieving a big leap in enforcement, no one introduced a schedule of work to provide for better compliance. Mexico was obliged to meet the commitments to finance and to capitalize the BECC and the NADBank. Mexico was also expected to make significant investments in environmental infrastructure along the US-Mexico border to prevent further environmental degradation and to clean up the existing problems.[18] Despite this hefty list of legal obligations, none of the NAFTA's environmental institutions provides special treatment for Mexico in their programs, their funding, or their lending.[19] There is a lingering view that Mexico's financial contribution to the NACEC budget is too large a burden.[20] It is, however, probably more accurate to see this as a case of missed opportunities than as a case of high

costs. If the programs and specific projects had addressed Mexico's priorities and perception of fairness, Mexican officials' trust in the NACEC might
be greater.

Different Constitutional Arrangements

Last, but certainly not least, the profound differences in constitutional
arrangements and judicial systems in each of the three NAFTA countries
have recently become an important sixth challenge to the environmental
side agreement. Here, it is useful to recall that according to NAAEC article
41 and annex 41, the rights and obligations of the agreement apply only at
the federal level in Canada, except in those instances where Canada's
provincial governments have also accepted the agreement (as of 2001 only
three provinces—Manitoba, Quebec, and Alberta—have joined the
NAAEC).[21] This partial adherence is delaying the introduction of specific
environmental measures. Moreover, although according to the same article
41 rights and obligations apply at federal and non-federal levels in the
United States and Mexico, there are also differences; in particular, US states
have more autonomy on environmental policy making than do Mexican
states. To overcome these problems, the parties devised some temporary
responses. For example, on trans-boundary environmental assessment
issues they agreed to make decisions based on reciprocity: a Mexican state
would accept a commitment if the neighboring US state or states accepted
it. Since such difficulties will continue to emerge, all three parties to the
agreement need to develop innovative, well-designed procedures to deal
with them. At the same time, they must reach a deeper understanding of
each other's positions; otherwise, their efforts are sure to be stymied by the
more controversial issues.

The NAFTA's Positive Environmental Story: Improving Environmental Performance in Mexico

Despite challenges, the NAFTA and the NAAEC commitments have helped
advance Mexico's environmental record. The NAFTA process has helped
the Mexican government focus on the pollution-control and natural-
resource-management challenges it faces. Incremental but steady implementation of NAAEC has encouraged Mexico to invest in data, training,
and other critically important prerequisites for enhancing environmental

performance. However, these improvements also have been prompted by the several multilateral agreements Mexico signed during the 1980s and the early 1990s, and by the growing domestic public awareness and environmental activism during those years. [22]

Other aspects of the NAFTA process have helped to bolster Mexico's environmental awareness and performance.

First, the NAFTA debate strengthened the Mexican environmental NGO community. Mexico's NGOs, which became fragmented in the late 1980s, regained a considerable degree of cohesion after the Rio conference and increased their activities accordingly.[23] After a short delay, some became involved in the NAFTA process, establishing or strengthening networks both with those US NGOs supporting a conditioned NAFTA and with the ones that opposed it. Domestic discussions of the NAFTA favored the increase in the number of NGOs in Mexico. By the mid 1990s, there were more than 400 active environmental groups across the country. As the public gained access to environmental monitoring data and realized the health risks associated with air pollution in major cities, there also was a concomitant increase in public concern and demands for action, although these focused largely on this specific environmental problem.

Second, the NAFTA debate and the NAAEC have worked to foster new and revitalized environmental institutions in Mexico. For example:

• In 1992, two new institutions, the office of the Federal Attorney General for Environmental Protection (PROFEPA) and the National Institute of Ecology were established. The first became Mexico's agency responsible for environmental enforcement; the second provides technical and scientific advice and is responsible for setting standards. There is little doubt that President Salinas created these agencies to demonstrate Mexico's commitment to environmental protection.[24]

• In 1992, Mexico established the National Commission for the Knowledge and Use of Biodiversity (CONABIO).[25]

• In December 1994, a month before the NAFTA came into effect, Mexico established the Ministry for Environment, Natural Resources and Fisheries (SEMARNAP) (renamed the Ministry for Environment and Natural Resources—SEMARNAT—in late 2000).[26] SEMARNAT's mission is to promote sustainable development and to limit environmental deterioration and natural resource depletion. It brings together under one national ministry the policies, programs, and administrative resources for soil conservation and restoration and forests.[27] SEMARNAT's budget increased between 1995 and 1997 from US$803.5 million to US$1068.8 million. Bad economic conditions forced an estimated 20 percent reduction in 1999 (OECD 2000). Total public-sector and private-sector expenditure on pollution abatement and control also started to grow; by 1997, it was estimated to be roughly 0.8 percent of GDP, a figure that is still far from sufficient (OECD 2000). Although there are well-grounded complaints of unsatisfactory inter-secretarial cooperation, SEMARNAT's power relative to other government agencies appears to be increasing slightly as a result of Mexico's environmental commitments under the NAFTA.

• After thorough diagnosis of prevailing conditions, several specific five-year programs have been introduced since 1995, including programs for Natural Protected Areas, for Fisheries and Aquaculture, for Forestry and Soil, for Conservation of Wildlife and Diversification of Rural Production, and for Minimization and Integral Management of Industrial Wastes, and also including the Special Program of Sustainable Development of Beaches, Coasts, Federal Zones, and Lands. More than 50 norms are now in force. The number and the size of protected areas also have increased (SEMAR-NAP 2000).

• In 1996, a substantially amended General Law of Ecological Balance and Environmental Protection (LGEEPA) came into force (SEMARNAP 1997). In an attempt to modernize regulation, the LGEEPA allows for self-regulation, and it incorporates a number of economic instruments. It has also expanded the number of activities for which an environmental impact assessment is required. The LGEEPA provides for further, albeit gradual, return of environmental responsibilities to states and municipalities, an incremental approach that will allow them to increase their capacity to address these new tasks (SEMARNAP 1997). The new LGEEPA also increases public participation in decision making, and it establishes the right of citizens to access environmental information and dispute the actions of environmental authorities. In addition, the Mexican Congress introduced an amendment of the Penal Code that incorporated a chapter on environmental crimes, thereby facilitating the classification of these crimes and increasing the sanctions.

• In 1997, a Forest Law was added to the LGGEPA.

• In May 2000, the Mexican Congress approved a General Wildlife Law.[28]

Many environmental NGOs and scholars argue that some aspects of the new legislation and programs are steps backward rather than forward.[29] However, most would agree that what we have now is preferable to the previous legal vacuum. Progress must be closely monitored because even those who praise Mexico's new policies and programs stress that this legislative and policy framework is still at a very early stage of implementation.[30]

Third, Mexico's new environmental laws and institutions have created new opportunities for public participation in decision making and monitoring. The NAFTA undoubtedly provided an impetus to these developments, though it is difficult to measure its precise influence. Since 1995, a series of consultative councils for sustainable development have been established, with representatives from business, academia, NGOs, government, and the community at large to advise municipal, state, or federal authorities. The are four regional councils, each comprising three or four states; there is also a federal council, and there are numerous local and issue-specific councils. According to official data, 628 public participation organizations with nearly 6,000 participants existed in the year 2000. Monitoring Committees (Comités de Vigilancia Participativa) involved a further 2,685 people (SEMARNAP 2000). While complaints have been voiced about the lack of responsiveness of the Mexican authorities to the demands and proposals raised at these councils and committees, there is no doubt that these forums represent a considerable step forward for public participation in Mexico. Better information systems in public and private institutions have also led to a qualitative change in public participation (Grupo de Reflexión G-25 2000).

Fourth, compliance with environmental regulations has improved. PROFEPA began making enforcement visits to large industries in 1992. The percentage of industries with zero irregularities in their compliance with environmental regulations increased from 18.25 percent in 1994 to 25.00 percent in 2000 (SEMARNAP 2000). The percentage of small irregularities decreased from 78.67 to 73.50 and the percentage of closures from 3.08 to 1.50 (ibid.).

Fifth, the NAFTA's environmental provisions and side agreement are helping to change private-sector environmental attitudes. Since the NAFTA came into force, Mexican big business has been more receptive to discussing, proposing, and adopting new ways of addressing pollution abatement, a new attitude largely due to its increasing interest in exports. Businesses have strongly supported voluntary agreements (such as ISO 14000 product and process certification standards and eco-auditing). With industry cooperation, PROFEPA has implemented a Clean Industry Certificate Program. Voluntary audits by private enterprises in that program increased from 179 in 1995 to 221 in 1998. By the beginning of the year

2000, the number of voluntary audits had reached 968. With financial support from the NACEC, these audits included 30 small and "micro" businesses in 1999 (SEMARNAP 2000). Business organizations have also set up a research center, the Private Sector Studies Center for Sustainable Development (CESPEDES), to encourage better environmental performance in industry. Some of its research projects have been jointly sponsored by the Lower Chamber of Congress in Mexico (CESPEDES 1999a and 1999b).

Big industry's deeper environmental concern is being shared by large semi-state enterprises. In 1992, PEMEX, the Mexican oil company, initiated an environmental program that was triggered by a disaster in Guadalajara and air pollution emergencies in Mexico City.[31] It aims to reduce air pollution in densely populated areas, to produce fuels of higher environmental quality, to reduce emissions of volatile organic compounds, to control discharges of wastewater, and to evaluate marine pollution caused by PEMEX's industrial activities (OECD 2000). The state oil company has also adopted PROFEPA's auditing program. PROFEPA's audits of PEMEX and other semi-state enterprises increased from 12 in 1995 to 97 in 1998 (SEMARNAP 1998).

Only a small percentage of Mexican firms, however, can be classified as "big business." So far little has been done to encourage and enable small and medium-size industries to improve their environmental performance. The NACEC, Mexico's Confederation of Industrial Chambers (CONCAMIN), and the Mexican Foundation for Innovation and Technology Transfer (FUNTEC) have created a fund to finance pollution-prevention projects for these industries (NACEC 1999). A pilot project supporting twelve small enterprises, mostly tanneries in the state of Guanajuato, is designed to demonstrate the benefits of prevention over end-of-the pipe cures.

Finally, there has been a substantial and encouraging increase in Mexico's "environmental market," from US$95 million in 1988 to US$3.761 million in 2000 (La Jornada 2000a). According to the Council of Ecological Entrepreneurs, in the year 2000, US$2.306 million was to be invested in wastewater treatment projects, US$786 million in air pollution, US$610 million in solid and hazardous waste treatment, and US$57 million in soil improvement (La Jornada 2000a).

Conclusion

Mexico's environmental advances are undoubtedly small when compared to the magnitude of the environmental problems facing the country. Still, the NAFTA's environmental commitments can be said to have encouraged the creation of a more accurate diagnosis of environmental problems in Mexico, a prioritized agenda, and a range of government activities on the environment. Even if some of this momentum reflects a strategic calculation rather than a real political commitment to the environment, these are important steps in the right direction.

In view of the dynamics of US politics, the inclusion of environmental issues in the ongoing negotiations for a Free Trade Area of the Americas is inevitable. Based on the NAFTA experience, the parallel agreement approach clearly provides a commendable starting point for negotiators. While some of the environmental challenges that the NAFTA's environmental side agreement sought to address (such as border problems and those involving shared natural resources) are specific to the relationship between the United States and Mexico or Canada and may not apply to the hemisphere as a whole, the NAAEC experience does provide some instructive lessons for those committed to incorporating environmental considerations into the FTAA. These lessons include the importance of thinking carefully about how economic asymmetries and different environmental priorities are likely to play out. Negotiators will also have to consider how to ensure a fair and workable balance between collaborative and dispute resolution mechanisms, specify the scope of the institutions created and the powers of their governing bodies, and address the significant economic, political, and juridical differences between countries that might otherwise impede implementation of the agreement.

Notes

1. Intergovernmental cooperation takes place when the policies actually followed by one government are perceived by its partners as facilitating realization of their own objectives, as the result of a process of policy coordination. See Milner 1992 and Keohane 1984.

2. Several US-Mexican agreements were signed in previous decades, but they usually were very specific and limited. There were even fewer Canadian-Mexican agreements.

3. The last two institutions were provided for by a second bilateral agreement signed between the US and Mexico. As for the BECC's performance, public, scholarly, and official perception has been increasingly positive over the past few years. Despite a slow start, it was able to build on previous bilateral cooperative experiences and decades of high-level interaction between governments and people on both sides of the border (Mumme 1999). Its collaborative, highly focused agenda (water and wastewater treatment issues), though criticized by some, has facilitated its work. Nevertheless, some modifications to the BECC's initial approach had to be introduced to achieve better and smoother implementation (GAO 1996).

4. These publications are available on the NACEC's web site: http://www.cec.org.

5. The International Joint Commission (IJC) was established by the 1909 Boundary Waters Treaty between the US and Canada with the goal of assisting governments to find solutions to problems that affects the rivers and lakes the like along, or flow across, the border between the two countries. See http://www.ijc.org. On the powers and the performance of the IJC, see Munton and Castle 1992 and Holsti and Levy 1976. The IBWC is a century-old commission that initially was in charge of settling all major surface water allocation and distribution problem related to the Colorado and Bravo rivers. In the 1970s and the 1980s, the IBWC's powers were enlarged to address the countries' shared sanitation problems (Szekeley 1992). See also Mumme 1991 and http://www.ibwc.state.gov. For further information on the Border XXI program, see http://www.epa.gov/usmexicoborder/ef.htm.

6. From a realist perspective, when a very deep-seated asymmetry exists, the stronger state would also tend to eschew explicit cooperation as it is seldom willing to be bound by a formal accord when unilateral action is a viable alternative (Krasner 1990).

7. See also Domínguez 1997.

8. For a list of the 20 quite specific agreements and programs involving commitments of environmental cooperation, including the establishment of the Joint Committee for the Conservation of Wildlife (1984), see Szekely 1992.

9. See Jaime Serra, cited in Garciadiego 1994. Many multilateral development bank officials shared similar views. For a full exposition of the range of views on this issue, see Low 1992.

10. The US embargo, introduced in 1991, was applied only to the tuna fished in the Tropical Pacific Ocean, where the Mexican fleet was the largest one. Two reasons why this second US embargo is seen as disguising protectionist interests are (1) the fact that the US has since disregarded Mexico's considerable achievements in reducing incidental death of dolphins and (2) the experience just few years earlier with an embargo on Mexican tuna that was not linked to environmental concerns. See Torres 1993a,b.

11. The La Paz Agreement provides for the negotiation of "annexes" to address specific border environmental problems and for the set up of working groups. The initial groups were Air, Pollution Prevention, Water, Hazardous Waste, Emergencies, and Enforcement Compliance. After NAFTA, three new groups were formed: Environmental Information, Health, and Natural Resources. See Hufbauer and Schott 1992.

12. For a comparative review of the demands of seven large US NGOs and the negotiation positions of the three parties, see Kelly 1993.

13. Mayer's chapter in this compilation further discusses the negotiating dynamics on the issue of sanctions and the eventual agreement between Canada and the US, which allowed for resort to domestic courts (Mayer 2002).

14. Timetables for every step of the process were established. See chapter 5 of the NAAEC.

15. During the side agreement negotiations, Mexican officials unsuccessfully proposed limiting petitions to problems that related to trade. Canadian and Mexican officials "agreed that North American solutions are needed for regional problems ... but they initially refused to accept that trilateral solutions are required for local pollution" (Hufbauer and Schott 1992).

16. Historically, Congress approved very progressive Mexican laws on social issues, even though immediate compliance with them would be very difficult. Subsequent efforts were made to gradually improve their enforcement. This practice helps to explain the large enforcement deficit of Mexican laws, including environmental ones.

17. Articles 3 and 5 of the parallel agreement.

18. Both countries were to share equally the financing and the capitalization of these institutions.

19. At this stage, Mexican negotiators were not interested in proposing differential treatment. Instead, they hoped that equality in financing would translate into equality in decision making, and, eventually, into more even-handed institutions.

20. It is quite common to hear Mexican officials underscoring that their country's contribution to the NACEC is higher than its funding for its own domestic protected environmental areas.

21. At the time of signing the NAAEC, Canada's federal government agreed to be bound only for matters within the federal jurisdiction in Canada. Participation by the provinces is crucial in carrying out the NAAEC since most environmental legislation falls under provincial jurisdiction. A Canadian Intergovernmental Agreement Regarding the NAAEC was subsequently negotiated with the provinces and territories to create a mechanism that would provide for their participation. The NAAEC stipulates that until Canadian provinces representing 55% of GDP become signatories to the CIA, Canada cannot fully benefit from the dispute resolution or arbitral panel process. Further, Canada may only make a request under the dispute resolution and arbitral panel mechanisms provided 55% of the sector which is the subject of dispute is represented by provinces which are included as signatories to the CIA. For further information, see http://www.naaec.gc.ca/english/can/implementation.htm.

22. National and regional Mexican newspapers now publish complaints coming from different groups and communities almost daily.

23. By the mid 1980s there were several dozen environmental NGOs in Mexico, of very different types and size. The largest and best organized were directly involved in conservation efforts; many others were small grassroots organizations. Some established early trans-national links. In 1988, 50 NGOs signed a joint environmental statement that is considered a well-justified analysis of problems and suggestions of

possible solutions. See also Kurzinger 1991, SEDESOL/Instituto Nacional de Ecología 1994, and El Foro Mexicano de la Sociedad Civil para Río 1992.

24. For a summary of the measures taken by the Salinas government to demonstrate Mexico's environmental commitment, see Hufbauer and Schott 1992. Certainly, as I have argued in some other articles, some of them reflected neither Mexican priorities nor were they environmentally or socially sound. See Torres 1993a.

25. CONABIO coordinates activities and research programs in this area, including a national information system listing the biodiversity in Mexico. Recently it published a well-designed national strategy for preserving biodiversity (SEMARNAP/CONABIO 2000).

26. Mexico's Ministry for Urban Development and Ecology was its first ministry related to environment. Established in 1982, it was transformed a decade later into the Ministry for Social Development.

27. SEMARNAT is responsible for coordinating the programs and policies of seven semi-independent agencies: the National Water Commission; the National Institute of Ecology; the Commission for the Knowledge and Use of Biodiversity, the Mexican Institute of Water Technology, the National Commission for Protected Areas, the National Forest Commission and Office of the Federal Attorney General for Environmental Protection. For further information, see http://www.semarnat.gob.mx.

28. Additional environmental laws include the Federal Hunting Law (1952), the Law of the Sea (1986), the Federal Fishing Law (1992), the Federal Metrology and Standardization Law (1992), and the Law on National Waters (1992).

29. For a discussion of the recently approved law for wildlife protection and opposition from NGOs such as Grupo de los Cien and Greenpeace-Mexico, see La Jornada 2000b.

30. See e.g. OECD 2000.

31. An explosion that killed dozens of people and destroyed hundred of houses in April 1992 was attributed to large gasoline spills into the sewer system.

References

Centro de Estudios del Sector Privado para el Desarrollo Sustentable. 1999a. *Competitividad y protección ambiental: Iniciativa Estratégica del Sector Industrial Mexicano*. México: CESPEDES.

Centro de Estudios del Sector Privado para el Desarrollo Sustentable. 1999b. *Retos y oportunidades para una Reforma Fiscal Ecológica en México*. México: CESPEDES//Cámara de Diputados.

Domínguez, J. 1997. "Mexico's New Foreign Policy: States, Societies and Institutions." In *Bridging the Border. Transforming Mexico-US Relations*, ed. R. de la Garza and J. Velasco. Rowman and Littlefield.

El Foro Mexicano de la Sociedad Civil para Río 92. 1992. *Sistematización de una experiencia organizativa. Síntesis documental 3*. México: Debase, A.C./Grupo de Estudios Ambientales.

Garciadiego, J. 1994. *El TLC día a día. Crónica de una negociación*. México: Miguel Angel Porrúa.

GAO (General Accounting Office). 1996. *Environmental Infrastructure Needs in the US-Mexican Border Region Remain Unmet*, Report to the Ranking Minority Member, Committee on Commerce, House of Representatives, GAO/RCED-96-179, July.

Grupo de Reflexión G-25. 2000. *Hacia una política ambiental eficaz para el desarrollo sustentable*. México: Jiménez Editores e Impresores.

Holsti, K., and T. Levy. 1976. "Bilateral Institutions and Transgovernmental Relations between Canada and the United States." In *Canada and the United States, Transnational and Transgovernmental Relations*, ed. A. Fox et al. Columbia University Press.

Hufbauer, G., and J. Schott. 1992. *North American Free Trade: Issues and Recommendations*. Washington: Institute for International Economics.

Independent Review Committee. 1998. Four-Year Review of the North American Agreement on Environmental Cooperation: Report of the Independent Review Committee. Montreal: North American Commission for Environmental Cooperation.

Johnson, P., and A. Beaulieu. 1996. *The Environment and NAFTA. Understanding and Implementing the New Continental Law*. Island.

Kelly, M. 1993. "Review of the NAFTA Environmental Side Agreement: Does it Meet Expectations?." *Frontera Norte* 5, no. 10.

Keohane, R. 1984. *After Hegemony*. Princeton University Press.

Krasner, S. 1990. "México y Estados Unidos." In *Interdependencia ¿Un enfoque útil para el análisis de las relaciones México-Estados Unidos?* ed. B. Torres. Colegio de México.

Kurzinger-Weimman, E., ed. 1991. *Política ambiental en México. El papel de las organizaciones no gubernamentales*. Mexico: Instituto Alemán de Desarrollo-Friedrich Ebert Stiftung.

La Jornada. 2000a, "Se dispara el mercado medioambiental mexicano." March 7.

La Jornada. 2000b. "Especies silvestres, en riesgo si prospera una ley." February 25.

Low, P., ed. 1992. International Trade and Environment. Discussion Paper 159, World Bank.

Mayer, F. 1998. *Interpreting NAFTA: The Science and Art of Political Analysis*. Columbia University Press.

Mayer, F. 2002. "Negotiating the NAFTA: Political Lessons for the FTAA." In this volume.

Milner, H. 1992. "International Theories of Cooperation among Nations." *World Politics* 30: 466–496.

Mumme, S. 1991. "Innovation and Reform in Transboundary Resource Management: A Critical Look at the International Boundary and Water Commission, United States and Mexico." *Natural Resources Journal* 33, winter: 93ff.

Mumme, S. 1999. "NAFTA and the Environment." *Foreign Policy in Focus* 4, no. 26: 1–4.

Munton, D., and G. Castle. 1992. "The Continental Dimension: Canada and the United States." In *Canadian Environmental Policy: Ecosystems, Politics and Process*, ed. R. Boardman. Oxford University Press.

NACEC. 1999. Informe annual. Montreal: NACEC.

Purcell, K. 1981–82. "Mexico-US relations: Big Initiatives Can Cause Big Problems." *Foreign Affairs* 60, no. 2.

Sánchez, R. 1995. "El TLC y el medio ambiente." In *México frente al TLC: costos y beneficios*, ed. V. Bulmer-Thomas et al. Colegio de México.

Secretaría de Desarrollo Social/Instituto Nacional de Ecología. 1994. *Directorio verde: Organismos no gubernamentales*. Mexico: SEDESOL/INE.

SEMARNAP. 2000. *La gestión ambiental en México*. México: SEMARNAP.

SEMARNAP. 1997. *Ley General del Equilibrio Ecológico y la protección al ambiente*. México: SEMARNAP.

SEMARNAP/CONABIO. 2000. *Estrategia sobre biodiversidad de México*. México: SEMARNAP/CONABIO.

Spalding, M. 1997. "Lessons of NAFTA for APEC." *Journal of Environment and Development* 6, no. 3: 252–275.

Spalding, M. 1999. "Governance Issues under the Environmental Side Agreement to the NAFTA." In *Sustainable Development in San Diego-Tijuana*, ed. M. Spalding. La Jolla: Center for US-Mexican Studies and University of California, San Diego.

Szekely, A. 1992. "Establishing a Region for Ecological Cooperation in North America." *Natural Resources Journal* 32, summer: 563ff.

Torres, B. 1993a. " El medio ambiente en la relación bilateral México-Estados Unidos." In *Liberalización económica y libre comercio en América del Norte*, ed. G. Vega Canovas. Colegio de México.

Torres, B. 1993b. "La cuestión del medio ambiente en la agenda bilateral (1991–1992)." In *México, Estados Unidos, Canadá*, ed. G. Vega Canovas and F. Alba. Colegio de México.

12

Limits to Cooperation: A Mexican Perspective on the NAFTA's Environmental Side Agreement and Institutions

Laura Carlsen and Hilda Salazar

The fundamental difference between our proposal and the official proposal is not one of isolationism vs. opening, but over the existence of a national project or the lack of one. The market and the world economy have a role to play in a national project of long-term sustainable development but the definition of that project cannot be left to the market. . . .
—Mexican Action Network on Free Trade

The highest hopes for environmental improvement under the North American Free Trade Agreement were pinned on the institutions created alongside it to enhance environmental cooperation—the North American Commission for Environmental Cooperation (NACEC), the North American Development Bank (NADBank), and the Border Environmental Cooperation Commission (BECC).[1] Although the NAFTA's environmental side agreement—the North American Agreement on Environmental Cooperation (NAAEC)—does not contain powerful mechanisms for direct pressure for environmental improvement, many environmental organizations were hopeful that the environmental side agreement and its related institutions could serve to generate and promote exchange of environmental information, identify common environmental interests, encourage public participation, create pressure for compliance with national environmental laws, and finance badly needed environmental infrastructure in the US-Mexico border region.

In political terms, those who worked to include environmental considerations in the NAFTA package see the results simultaneously as a win and a loss. On the "win" side, the very fact that environmental and labor issues made it to the NAFTA negotiation table is considered a major breakthrough. The NAFTA's approach to the environment set three crucial precedents. First, it formalized the inclusion of environmental aspects within a trade agreement despite the resistance of trade ideologues who insisted that the market would automatically address environmental problems (Audley 1997; Esty 1994).[2] Second, the NAFTA process spawned three international institutions for environmental cooperation: the NACEC, the NADBank, and the BECC. Participants in one prominent meeting of non-governmental organizations concluded that "in spite of the insufficiency of the parallel agreements, they constitute a first draft of the kind of supranational institutionality that must be generated in processes of regional integration" (RMALC 1994a, p. 6). Third, the negotiation of the environmental side agreement mobilized citizen groups across North America[3] and stimulated an ongoing coalition of non-governmental organizations focused on long-term efforts to influence trade policy (Audley and Uslaner 1994; Barba Pírez 1993; Arídjis 1993).[4] On the "loss" side, the actual content of the NAFTA's environmental side agreement is weak, reflecting the fact that its existence owes more to a mixed bag of environmental pressures, political expediency, and business concessions than to a commitment in principle to the environment on the part of trade officials (Mayer 1998, 2002).[5] The environmental side agreement and issues of social justice and environmental quality in general suffer from absolute subordination to the "real issues" of commerce and financial flows. Fundamentally, the environmental side agreement fails to modify or condition the trade agreement itself. In fact, many of the terms and effects of the NAFTA contradict, impede, or fail to sufficiently encourage compliance with the environmental provisions incorporated into the trade agreement itself (listed in appendix A of this volume) or those set out in the environmental side agreement (Seligman 1998).[6]

Did those concerned about the environment really gain anything from the NAFTA's environmental side agreement and related institutions? This chapter responds to that question. The side agreement must be judged on at least two criteria: (1) the adequacy of the institutional structures it created and how these bodies live up to their mandates and (2) what a comprehensive environmental agreement should and could do, not only in the context of a specific trade agreement, but in the context of the broader economic integration between countries.

NACEC: A Trail of Disappointments

The performance of the NAFTA's Commission for Environmental Cooperation (NACEC) has been disappointing on several fronts.[7] The first set of problems stems from the subordination of the environmental side agreement—and thus the commission—to the NAFTA's trade priorities. The effects of this subordination have been threefold:

• Inadequate integration between environmental and trade processes and bodies. The NACEC's prospects for success depend in large part on its ability to work with the NAFTA economic institutions and engage them in effective dialogue at the trade-environment nexus. So far these links have been precisely the NAFTA's weak points. The failure of the trade and environmental ministers of the NAFTA countries to meet, never mind their failure to interact effectively, is an impediment to trade and environment dialogue. The absence of central NAFTA coordinating offices within national governments or a tri-national trade secretariat able to mount a dialogue with the NACEC is notable.[8] A central problem is that the main NAFTA governing body—the Free Trade Commission—has been slow to accept the environmental mandates contained within the NAFTA and the NAAEC.[9] There is no single location for systematic periodic and ongoing reviews of the environmental aspects of the NAFTA agreements and how well the parties are proceeding to meet their obligations. These shortcomings also preclude possibilities for adjusting trade priorities and deadlines to meet new environmental priorities.[10] An internal 1997 evaluation of the NACEC states that, despite the environmental side agreement's call for close coordination between the NAFTA 's Free Trade Commission and the NACEC, "there has been a paucity of contact, communication and in some cases trust, let alone integrated decision making and cooperation, between NAFTA's economic bodies with specified environmental responsibilities and the NACEC itself" (CEC 1997, p. 46).

• Low impact of environmental and sustainable development concerns on trade priorities: Efforts to integrate environmental and sustainable development issues into the NAFTA's trade agenda have met with many obstacles. No institution exists to monitor and evaluate environmental progress or the overall goal of sustainable development. The NACEC has only recently expanded its activities to respond to sustainable development considerations such as sustainable use of primary natural resources (mostly in agriculture), conservation of biodiversity in the trade of wild species, and sustainable tourism in natural areas (CEC 1999). A NACEC-sponsored experts' workshop on sustainably produced Mexican coffee held in Oaxaca in April of 2000 and several small grants to initiatives to promote shade-grown coffee and sustainable tourism are a promising if small-scale start.[11] Much skepticism exists about the NACEC's possibilities to influence or inform trilateral trade institutions on issues of sustainable development.[12]

• Piecemeal political commitment to, and implementation of, environmental mandates. The NACEC has been allocated a long list of responsibilities, including research, exchange of information, environmental cooperation, supervision of the terms of the environmental side agreement, protection of endangered species, and promotion of public awareness.[13] As an institution, however, the NACEC lacks the mechanisms, resources and political commitment to assure proper implementation of these responsibilities.

A second area of criticism of the NACEC concerns its mandate to increase public participation. There seems to be a lack of clarity as to how public consultations should or could feed into the work of the NACEC. The NACEC's Joint Public Advisory Committee itself recently noted that the NACEC "is still 'confused' by this responsibility (public participation)" (JPAC 2000).[14] Governments' limited followup on public consultations reinforces NGOs' fears that consultations play no actual role in the NACEC's policy making (Naumann 2001). In each of the NAFTA's three member countries, the perceived futility of participation has prompted several citizens' groups that were active during the original negotiations on the side agreement to withdraw from ongoing consultation processes (Naumann 2001). Some groups claim that the process for article 14 citizen submissions to the NACEC (to allege non-compliance of the parties with environmental regulations) is so complicated as to discourage such participation.[15] Repeated untransparent and secretive attempts by governments to restrict the citizen submission process provided for by the side agreement for cases of alleged environmental non-compliance has further dampened enthusiasm (National Advisory Committee 2000).[16] In addition, the Mexican NGO community has voiced concern that the public consultation process has not been adequately broad and representative (CEMDA 2000). They have complained of discretional criteria for determining the eligibility of NGOs to participate in the consultations. The JPAC (2000) notes that important sectors of the North American civil society are left out of the consultation processes entirely, especially remote communities and indigenous peoples in Mexico.

Third, inadequate political will and conflicts among the NAFTA member countries have hampered the operation of the NACEC. Political infighting caused intolerable delays before the appointment of a second NACEC Executive Director and undermined the NACEC's efforts to create a forceful environmental agenda. In addition, the NACEC's three-party governing council appears to have made the success of the NACEC a relatively low priority.[17] This lack of commitment limits the NACEC's authority, its decision-making power, and its political leverage as well as diminishing its profile in each of the three countries. Examples of how these shortcomings constrain the operations and effectiveness of the NACEC include the following[18]:

• Restricted scope for NACEC intervention in matters of non-compliance with national environmental regulations. Only issues of non-compliance with existing environmental laws fall within the purview of the NACEC. Failure to regulate is not a consideration. In addition, environmental tensions related to commercial agriculture and exploitation of natural resources are largely excluded.

• Incapacity to carry out independent investigations. NACEC officials do not have the capacity to carry out effective independent environmental evaluations and investigations (such as the factual reports provides for in articles 13–15 of the NAAEC) because they must rely only on publicly released documents or volunteered information.

• Difficulty initiating a process. The NACEC requires a two-thirds vote of its council to initiate a Factual Report to investigate allegations of non-compliance with domestic environmental regulations.

• Limited authority and thus limited efficacy. The process for protecting environmental goals and programs in the NAFTA context is not well developed. The documentation required to initiate environmental disputes far exceeds that required for trade conflicts and the process for resolving them is prohibitively long.

• Conflicts of interest. There are few if any norms to prevent cases of possible conflicts of interest in which NACEC officials may have interests in a company under investigation (RMALC/Public Citizen 1996, pp. 31–32).

Revitalized and meaningful processes of public participation are needed to countervail the tendency of governments to shunt environmental work to the bottom of national and international agendas. The North American Fund for Environmental Cooperation (NAFEC) may in fact be the NAFTA's only environmental institution that directly links trade effects to sustainable development goals.[19] Housed within the NACEC, the NAFEC aims to finance community projects that promote its aims and objectives. The projects must be innovative, sustainable, and replicable, must have access to other sources of funding, must enhance the capacity of local institutions, and must encourage a relationship of cooperation between sectors and countries in the NAFTA region. NAFEC projects in Mexico have emphasized alternative trade and production processes that simultaneously promote environmental protection and the internalization of environmental costs. However, the program receives an average of just US$1 million per year, to be divided among small-scale community projects in all three countries. Projects are so few and so small that they cannot be considered to have a major effect on overall trade and environment trends.

Inadequacies of the BECC and the NADBANK

The NAFTA's other environmental institutions—the BECC and the NADBank—have had to recognize in practice what the trade agreement refused to accept conceptually: that Mexico needs preferential treatment to achieve basic national and region-wide economic, social, and environmental goals.[20] The macroeconomic asymmetries between the two countries received inadequate attention during the NAFTA negotiations and have been largely ignored in assessing environmental needs and priorities. In the area of environmental cooperation, the need for preferential treatment arises not so much from the common assertion made by US environmentalists that Mexico has more environmental problems and less ecological consciousness as from the simple fact of its inability to pay for needed infrastructure.[21] Each of the NAFTA's institutions has found itself ill-prepared to deal with the challenges of effective financing and project development.

First, the focus of the BECC and the NADBank has been on "catching up" on environmental infrastructure needs faced by Mexico. This emphasis has diverted the strategic planning and feasibility process away from analysis of long-term macro-economic considerations and existing developmental trends along the border—including rapid border industrialization, the growth of the maquiladora industry, serious deficits in housing and urban and social services, and enormous population growth due to migration from other regions of Mexico.

Second, the NADBank's focus on loans (debt financing) has proved ineffective in both the United States and Mexico. Recent evaluations of the BECC and the NADBank highlight severe financing problems, including the limited availability of grant funding and structural problems with access to and eligibility for debt financing (NWF 1999; Kelly 1998). By 2001, only 3 percent of the NADBank's paid-in capital has been used for loans; this has generated serious concern about the underutilization of this money (Kelly et al. 2001). A central reason for this is that, at 1 percent *above* the US market rate, NADBank financing is well out of reach to poor border communities (Kelly et al. 2001; Kouros 2000b).[22] Continued external pressures for fiscal austerity in Mexico have consistently constrained the capacity of federal, state, and municipal agencies to take on loans and commit funds to environmental projects. Together with the unequal allocation of funds to states and municipalities and increasing poverty in many communities, Mexico's authorities and communities have a severely limited capacity to take on loans and debt they are unable to pay back (Kelly et al. 2001).

The NADBank's management has acknowledged that "the very few communities able to afford a relevant proportion of debt financing [are] all on the US side [of the US-Mexico border] . . ." (Rodríguez 2000). A 1999 review of 19 environmental projects administered through the BECC and the NADBank shows that between 50 percent and 100 percent of all funding to Mexican public infrastructure projects had to take the form of grants rather than loans because of Mexico's inability to assume more debt obligations (Kelly 1998). The NADBank's 1999 10-year border outlook identified US$2.1 billion in specific water and wastewater projects needed in the border region but estimated that debt financing would be feasible for only 25 percent of the projects (Rodríguez 2000).

Communities on both sides of the US-Mexico border also have a low capacity to pay for construction and services through user fees. The Mexican government's deepening dependence on foreign-financed maquiladora development has precluded higher tax bases and user fees among the maquila businesses (even thought they generate a large portion of industrial wastes along the border) (Kelly et al. 2001, p. 6). Another obstacle to Mexican projects is the limited power of local Mexican governments to develop their own financing strategies—such as local taxation and borrowing money from foreign sources.

Another problem relates to realistically assessing costs. Many wastewater treatment projects, for example, have run into cost overruns due to insufficient planning or implementation. Finally, a survey of border projects reveals that the majority of the projects financed are aimed at treating pollution, not preventing it.[23] Little effective monitoring has been developed, and even fewer projects are designed to identify and restrict the major *sources* of pollution.

Failed Climate of Cooperation

Supporters of the environmental side agreement claimed that, beyond the institutional framework of the agreement, the NAAEC would foster a climate of greater environmental cooperation among the NAFTA countries. In our view, NAAEC's record on this front has been extremely disappointing. Instead, where cooperative responses have been needed most, the NAFTA's environmental institutions have played a negligible or non-existent role (Hufbauer et al. 2001, p. 37). In the pages that follow, we highlight the inadequacies of the NAAEC's response to four significant NAFTA-related environment issues that arose over the past 6 years: the US tuna embargo against Mexico; tensions over a proposed toxic dump site in Sierra Blanca, Texas; the North American Agreement on Trans-Border Environmental Impact Evaluation; and the Metalclad investment dispute.

The tuna-dolphin case is perhaps the most famous trade and environment dispute ever. For almost a decade it has been crying for a cooperative solution.[24] A ten-year US trade ban on the importation of Mexican tuna (imposed because of concerns about the high rate of incidental death of dolphins in Mexican tuna-fishing operations) has caused irreparable damage to an important Mexican export industry (SEMARNAP 1998; González 1999).[25] In response, Mexico has worked to improve fishing practices and adhere to several relevant international environmental agreements (Associated Press 2001). After Mexico succeeded in reducing dolphin deaths, the embargo on tuna was lifted by the US Department of Commerce in April 2000, but the controversy rages on. Several US environmental NGOs mounted legal challenges to the Department of Commerce's decision to lift the ban, and legal proceedings continue on the question of whether the US Department of Commerce acted wrongfully in its decision to weaken the standards required to use a "dolphin safe" label (Defenders of Wildlife 2001). However, the removal of the ban has done little to satisfy Mexican tuna exporters, because much of the catch cannot be labeled dolphin safe for US markets (Associated Press 2001). Mexican producers are convinced that the US tuna ban is not truly a response to environmental concerns but rather a response to commercial concerns of the US tuna industry, and that it reflects the opportunism of some environmental groups (Associated Press 2001; Torres 1993). Surely the ongoing tuna-dolphin controversy is a clear case where an institution such as the NACEC—with the promotion of environmental cooperation as its main mandate—should have been empowered to carry out a tri-national scientific investigation that would ascertain whether Mexican tuna practices endanger the dolphin population and to what degree the embargo is protectionist.

The public diplomatic sparring between Mexico and the United States on the issue of a toxic waste dump in Texas further reveals how the NAAEC is not equipped to provide conciliatory spaces or to intervene in real trade and environment situations where mechanisms for environmental cooperation are needed. In September 1999, US and Mexican citizens became aware that the state of Texas intended to permit the construction of a toxic waste dump near Sierra Blanca (Paterson 1998), a sparsely populated town just 30 kilometers from the border of the Mexican state of Tamaulipas. The case attracted public attention after a loose bi-national coalition of environmental groups began protesting the pending decision to go ahead with the dump and provided scientific analysis proving the proposal to be unsafe (ibid.). Environmental groups argued that Mexican authorities had not been consulted or considered in the siting, despite the obvious fact that Mexican communities near the dump would share the risks (Boren 1997).

The Mexican government never carried out an independent study of the environmental risks of the dump to its local communities, nor did it present any concerns to the NACEC. The Mexican government ultimately opposed the dump, but it acted weakly and late (Paterson 1998). For its part, the US government informed Mexican officials that the issue fell under state rather than federal jurisdiction (even though, by signing permission for interstate transport of toxic materials to Texas, the federal government had actively participated in the plan to construct the dump). Although Texas finally decided against construction, border citizens were protected not because the NACEC came to the aid of the Mexican government or helped the countries devise a mutually acceptable response but because Texas was forced to respond to citizen pressure and because the politically ambitious governor, George W. Bush, was anxious to minimize public tensions with Mexico and with environmental constituents (LaFranchi 1999; Paterson 1998).

A third example of inadequacy of the NACEC for the task of environmental cooperation is the case of the Agreement on Trans-Border Environmental Impact Evaluation.[26] The environmental side agreement states that projects with trans-border environmental impacts require notification, consultation, and mitigation efforts between the affected parties (article 10.7). An Inter-Governmental Trans-Border Environmental Team began work in early 1995. However, as of April 2002 governments had still not reached a conclusive, final agreement. This generated considerable disappointment and frustration (JPAC 1999; Kelly 1999).[27] One significant problem was that US representatives announced that they could make commitments only for federal projects, which constitute a very small proportion of potential projects (like Sierra Blanca) under consideration in the border area. Mexico insisted that the US government, as a NAFTA signatory, should take responsibility for non-federal projects as well (Kelly 1999). But the US team has rejected each of their proposals. A climate of frustration and cynicism has resulted.

Our fourth example concerns disputes related to foreign investment. In 1994, Metalclad, a US toxic waste disposal corporation, initiated construction of a toxic waste dump on the site of a previous dump in Guadalcazar, a township of 28,000 people in the Mexican state of San Luis Potosi. Despite local community opposition to the dump and the failure of Metalclad to obtain a municipal building permit, construction proceeded. When opposition from citizen's organizations, from the state government of San Luis Potosi, and from the local government of Guadalcazar succeeded in getting operations shut down, Metalclad sued the Mexican government under the investor protection clauses of the NAFTA 's chapter 11.[28] In August 2000, Metalclad won a decision granting them over US$16 million in compensation from the Mexican government.[29] Rewarded for ignoring environmental concerns and challenging local citizens' rights, Metalclad will now apparently receive a large financial payment from the Mexican government that could have been spent on building much-needed safe and legal dumps in Mexico.

Where was the NAFTA's environmental side agreement, not to mention its institutions, during each of these conflicts? Why were the institutions associated with the NAFTA not able to at least inform the processes, if not formally resolve the conflicts?

Optimists call for patience. They remind critics that the NAFTA's environmental side agreement represents a novel approach to regional environmental cooperation, so progress inevitably takes longer than anticipated. While the argument is understandable (anyone familiar with trade and environment issues can attest to their complexity), a serious evaluation must compare the *realized* potential for the NAAEC and its institutions to respond to concrete environmental challenges against the rate of environmental degradation due to trade and investment trends generated by the NAFTA. For an integrated evaluation, the ability of the NAFTA and the NAAEC must also contrast the ability to solve existing environmental problems with tendencies to generate new ones.

When a trade issue is based on real or supposed environmental concerns, the NAFTA's instruments for environmental cooperation should have been able to provide, at the very least, scientific expertise in evaluating the claims, if not specific rules and guidelines for arriving at agreement. Instead, the procedures for resolving controversies have not proved applicable to or useful for solving the actual problems that arise.

The performance of the NAFTA's cooperative mechanisms is a particularly pressing issue because Mexico entered into the NAFTA with a huge backlog of environmental problems and because Mexico continues to lack the public-sector and private-sector funds and the institutions and policies needed to address these problems. Mexicans cannot afford—socially or economically—to allow rapid economic integration to generate more negative environmental impacts.[30] Expanding international trade and investment can stimulate profound changes in patterns of land use, natural resource exploitation, industrial activity, and pollution. In Mexico, for example, foreign investment encouraged by the rush to economic integration has accelerated clear-cutting logging operations at great cost to Mexico's forests and the sustainability of local communities (Carlsen and Salazar 2001). In 1996, the NACEC noted that "the direct and immediate economic and environmental adjustments flowing uniquely from NAFTA will be experienced most heavily and directly by Mexico" (CEC 1996, p. 91). Time has borne out the prognosis, as a visit to any Mexican border community will show.

In sum, the environmental side agreement falls short on the "cooperation" front by failing to provide adequate financial support and cooperative mechanisms to (1) compensate for the disproportionate burden Mexico faces to improve its environmental performance and (2) mitigate the negative environmental effects of increased trade flows and investment patterns of the trade agreement.

Looking forward, the NAFTA governments need to establish guidelines and mechanisms for stronger environmental planning, accountability, public participation and scrutiny, and environmental cooperation, and for resolution of trade-environment tensions. To achieve these ends, several improvements need to be made to strengthen the NAFTA's environmental institutions:

• Increase the budget and scope of the action of BECC and NADBank financing in both loan and grant categories for border environmental protection.
• Consider differentiated financing commitments based on the GDPs of the participating countries. (Commitments are currently equally divided between the United States and Mexico.)
• Organize public consultation in all countries to define main priorities for the NACEC in the coming years.
• Increase the budget of the NACEC and its consultative organs.
• Review existing channels for citizen participation and consultation in the NAFTA process, with the goals of maximizing the range of interest groups represented and defining new mechanisms through which citizens' proposals can be considered in the NAFTA context and in future trade arrangements.
• Expand the scope and the financing of the NAFEC program.

The NAFTA and Beyond: Proposals for Sustainable Economic Integration

It is clear from the NAFTA experience that, if sustainable economic integration is to be promoted in the western hemisphere, each country must develop a national trade and environment agenda that recognizes and responds to the specific requirements for sustainable development. Working together, governments must infuse environmental concerns in all chapters of proposed trade agreements, especially in the areas of agriculture and fisheries, forestry, investment, intellectual property, energy, standards, and conflict resolution.

Governments also must explicitly recognize and invoke the priority of international environmental agreements, both for the resolution of environmental controversies and for the regulation of production, consumption, and capital (particularly international capital) within and between countries.[31] By clearly establishing the principle that countries can enter into specific international environmental accords to strengthen or specify environmental obligations contained in trade agreements (e.g., on specific cross-border, trans-national, or regional environmental issues), governments could reinforce the notion of national responsibility for environmental improvements while averting any tendency for environmental discussions to disintegrate into a cumbersome series of state-by-state environmental agreements.

As they engage in FTAA negotiations, the governments of the western hemisphere should do the following:

• Recognize the relationship between trade and the environment. The first step must be explicit acknowledgement by governments of the need to balance economic and socio-environmental priorities in processes of economic integration. A second step should be to broaden research on the relationship between trade and the environment to include a range of trends stimulated by economic integration in each country. Studies could explore specific issues such as the relationship between natural resource use and environmental degradation, changing patterns of land ownership, and wage and consumption levels. They should involve NGOs, academics and affected communities and their results should be publicly available. Such studies will be crucial to efforts to develop compensatory funds to prevent, mitigate, and reverse the environmental effects of economic integration in vulnerable sectors and ecological regions. A third and related step should be to establish mandatory mechanisms for permanent coordination between trade and environmental officials in the organizations and committees already provided for by the trade agreement as well as in ministerial meetings.

• Dedicate greater political commitment and resources to enhancing environmental protection measures. All governments in the region must work to improve environmental protection programs, to increase environmental budgets (as a percentage of GDP) in accordance with the capacities of each country, and to establish mechanisms that guarantee the real participation of social, business, non governmental, and academic sectors in these programs. This work should invest in initiatives to promote social development alongside respect for natural resources, particularly in poor communities and sectors; to establish protected environmental regions and develop environmental safeguards for ecosystems that are particularly sensitive to investment flows (e.g., forests, coastal zones, communal and indigenous lands, and highly concentrated urban areas); and to recognize Indian peoples' collective intellectual property rights over their traditional knowledge of alternative medicine, plant cultivation, and local and regional management of biological diversity. Safeguards should also be put in place to limit economic activities (such as clear cutting, large aquaculture projects, and tourist development) that displace local communities and flora and fauna.

• Promote improved environmental regulation and compliance. One major challenge for governments across the hemisphere is to create a more effective legal framework for environmental protection within each nation that includes mechanisms to guarantee environmental compliance, such as specific standards and sanctions. Environmental education targeting small and medium-size industries (such as technical consulting and financing packages for environmental projects) will be a tremendously important prerequisite to improving environmental compliance. Efforts to improve both environmental regulation and compliance will depend on data banks and studies that provide reliable information on the environmental situation in the region and on democratic mechanisms that allow for public participation, oversight, and complaints (particularly regarding non-compliance with environmental legislation) within each country and in international initiatives. Two specific programs that could be explored are a prohibition on the importation of pesticides and toxic substances that are prohibited in their country of origin and a hemispheric initiative to promote the development and transfer of "clean" and "cleanup" technologies (e.g., by offering economic incentives for the use of ecological equipment).

• Improve public consultation and accountability. Consultations with the general public and with relevant grassroots organizations, NGOs, and academics will be critical to the prospects of linking trade policy and environment policy in the FTAA context. As in the NAFTA case, governments must revise and enhance existing mechanisms for citizen participation so that a plurality of the communities and sectors most affected (or excluded) from the processes of economic integration are represented (especially Indian peoples). Guaranteed public access to information and broad-based, pluralistic citizen participation will be critical. While working under the rubric of a coordinated global sustainable development, civil-society organizations must work to maximize the decentralization of environmental decision making to countries, states, municipalities, and communities, granting them financial and political support. Among the prerequisites for engaging citizens in the policy dialogue will be raising awareness of the effects of trade on the environment and on particular communities and setting up mechanisms to support collaboration between local organizations, academics, and NGOs on environmental priorities.

Conclusion

Despite the significant shortcomings of the NAFTA's environmental side agreement and related institutions, they constitute an important first step.[32] The NAFTA's environmental side commitments should be considered the *minimum* for future trade arrangements. Unfortunately, however, the Mexican government has abandoned even this modest start in subsequent trade and investment treaties (Barkin 1999).[33] Mexico's recent trade agreements with the European Union, with Costa Rica, and with Israel have not included complementary environmental agreements at all, and the Mexican government continues to express opposition to such linkages (Diaz 2001).[34]

Moving beyond the bare minimum, our analysis of the social and environmental effects of economic integration under the NAFTA reaffirms the urgent need to renegotiate the NAFTA and improve on it in future trade negotiations and to ensure that any new treaties are based on broad public consultation and prioritize citizens' interests and concerns. To advance trade and environment priorities in the context of the FTAA, we have offered a series of policy proposals that emphasize that a sustainable model for hemispheric economic integration must incorporate both environmental and social dimensions.

Notes

1. For an overview of the activities of each of these institutions, see appendix A. For complete descriptions of these institutions, see http://www.cec.org, http://www.becc.org and http://www.nadbank.org. Each web site also contains extensive bibliographies from which to select additional readings. Note that while the NACEC was created by the NAFTA environmental side agreement, the NADBank and BECC were provided for by a second bilateral agreement signed between the US and Mexico—the 1993 Agreement Between the Government of the United States of America and the Government of the United Mexican States Concerning the Establishment of a Border Environmental Cooperation Commission and a North American Development Bank. For the text of this agreement, see Magraw 1995.

2. For an overview of the environmental provisions incorporated into the NAFTA text, see appendix A.

3. For a compilation of many statements and policy proposals advanced by environmental NGOs throughout the NAFTA negotiation process, see Magraw 1995. Many Mexican environmental and development NGOs were also actively engaged in the NAFTA debate, calling for an alternative treaty that would take into account the asymmetries among the three countries, including a strong social agenda. Some groups joined the effort to draft environmental and labor side agreements (Lujan 1998, p. 4). See e.g. Grupo de los Cien 1993 and Arídjis et al. 1992. Mexican analyses of the environment issue in the NAFTA include Land 1993, Peña 1993, RMALC 1992, 1997, 2002, and Sánchez 1991. For an excellent critical exploration of the interaction between Mexican NGOs and their Canadian and US counterparts (1998 and 1996), see Hogenboom 1996.

4. Many civil-society groups have long considered environmental and labor issues to be at the center of a regional social agenda rapidly being defined by an unprecedented coalition of citizen groups, unions, environmentalists and non-governmental

organizations from Mexico, US and Canada (Lujan 1998). For critical assessments of the challenges of North American cross-border organizing at the local and policy levels, see Barry and Sims 1994 and Thorup 1991.

5. Public Citizen (1993), Anderson et al. (1993), and Cameron and Tomlin (2000) offer perspectives on the influence of business interests on the NAFTA negotiations and outcomes. Poitras and Robinson (1994) explore the politics of the NAFTA within Mexico. Alanis Ortega and González-Lutzenkirchen (2002) reveal how little concrete attention trade and environment issues received from Mexican trade officials throughout the NAFTA negotiations.

6. The NAFTA's dispute settlement proceedings, particularly in the investment arena, have attracted particular concern. Environmental NGOs across the region share unanimous concern about the threats to environmental legislation posed by the NAFTA's investor-state arbitration provisions (Mann and Von Moltke 1999; CIEL et al. 1999; Public Citizen 2001). While the investment chapter of the NAFTA prohibits, in very general terms, efforts to attract investment by lowering environmental standards, it fails to incorporate provisions to fulfill the NAFTA's preambular commitment to sustainable development (Seligman 1998).

7. An ongoing series of reviews of the NACEC's performance are available. A range of perspectives is available. For the most recent of these reviews, see Torres 2002, Hufbauer et al. 2001, JPAC 2000, and CEC 1998.

8. The failure to integrate, theoretically and practically, environmental issues into the trade agenda has plagued post-NAFTA environmental performance. A clear illustration of this is the fact that despite the fact that article 10 (6) of the environmental side agreement explicitly calls for close coordination between the two bodies, a meeting between the NAFTA's Free Trade Commission (FTC) and NACEC officials did not take place until June 2000. Even then, the meeting between the NACEC article 10 (6) Environment and Trade Group and the NAFTA Working Group on Standards-Related Measures only focused on labeling and certification with limited discussion on potential areas for information exchange (Hufbauer et al. 2001). The report of the meeting was not made public (CEC 2000).

9. For details on the interaction between the NACEC and the FTC, see Soloway 1999 and Independent Review Committee 1998.

10. While the NACEC has spawned some 50 commissions, groups or committees for work on international environmental issues, an official independent evaluation condemned their failure in even the most basic of the environmental objectives: "With some notable exceptions, the NAFTA Committee work to date has not put into practice the parties' preambular commitment to promote sustainable development. . . ." (CEC 1997)

11. A summary of the meeting can be found at http://www.cec.org/files/PDF/ECONOMY/follow-e_EN.pdf. A summary of all of the projects supported by the NACEC's Fund for Environmental Cooperation can be found on the NACEC web site under "Grants for Environmental Cooperation."

12. At the time of printing a long-awaited meeting with Environment and Trade Officials on NAAEC article 10 (6) was scheduled for December 2001 and provides an excellent opportunity for officials to address this shortcoming.

13. See article X of the NAAEC for a full exposition of these tasks.

14. JPAC is a 15-member, independent, volunteer body that provides advice and public input to Council on any matter within the scope of the NAAEC.

15. Article 14 of the environmental side agreement enables the NACEC to investigate citizen complaints about national enforcement of environmental laws.

16. Governments have delayed and obstructed the citizen submission process in several ways. Most notably, they have failed to impose any kind of limit on how long the Secretariat's may take to review a submission or to respond to comments. The submitter also has to rely on the Secretariat to pursue the claim because they have little opportunity to participate in the review process once the review process is in train. Moreover, there is a lack of guaranteed remedy after the report is completed. Governments have further frustrated the citizen submission process by working, without opportunity for public review, on the revision of the guidelines related to citizen submission and the preparation of factual records (CEMDA 2000).

17. The NACEC's governing council, composed of the federal environment ministers (or equivalents) of the three countries, meets once a year.

18. These examples are from RMALC/Public Citizen 1996.

19. The NAFEC was established by a resolution of NACEC's Council in 1995. See "Creation of the North American Fund for Environmental Cooperation," Council Resolution 95-09, http://www.cec.org.

20. It is important to stress the enormous environmental, economic and social challenges facing Mexico. Poverty rates in Mexico remain depressingly high (Coote 1995). Similarly, Mexico hosts a full range of environmental challenges-from deforestation to rising air pollution and loss of biological diversity (Carlsen and Salazar 2000; Stern 1993). For pre-NAFTA assessments of the state of environmental policy and government responses to Mexico's environmental problems, see Godau Shucking 1985, Quadri de la Torre 1989, and Villegas 1994. For a more general discussion of systemic environmental problems in the Americas, see Faber 1992.

21. Contrary to the frequent assertion and assumption that the environmental movement is largely a Northern phenomena and that it is Northern environmental NGOs that push the need to incorporate environmental considerations into trade policy making, many Mexican NGOs are strongly engaged in the trade and environment discussion (Alfie 1991, Grupo de los Cien 1992; CEMDA 2000, Picard 2001, Sándoval 1991, UGAM 1993). Environmental consciousness is not a "Northern" phenomenon. A broad network of Mexican NGOs is active engaged in efforts to improve Mexican environmental performance and policy (Demmers and Hogenboom 1992).

22. Through March 2000, NADBank's paid-in capital was $348.75 million, of which less than $11 million had been tapped for loans (Kouros 2000). The total paid-in capital is expected to increase to $450 million by 2002 (Kelly et al. 2001). The NADBank itself has recognized the need to change its lending terms, including the possibility of lowering interest rates, and creating revolving loan funds and high-risk pools. In 2000, NADBank announced a $50 million "value lending" program to offer lower interest rates on loans from paid-in capital (NADBank 2000).

23. An overview of grants that have been supported can be found on the NADBank's web site, http://www.nadbank.org.

24. The dispute settlement body of the General Agreement on Tariffs and Trade handed down a decision on the infamous tuna-dolphin dispute in 1991. The panelists found that a US ban on imported tuna from Mexico caught using purse seine nets (alleged to result in unnecessary accidental kill of dolphins) violated GATT rules. For further information about the GATT Tuna-dolphin dispute see, http://www.wto.org.

25. Impacts include a decline of the national fishing fleet; lower rates of tuna fishing; reduced tuna exports (tuna exports declined from 80,000 metric tons in 1988 to 20,000 metric tons in 1993); loss of US$300 million dollars in exports to the US; and the loss of jobs due to the reduction of the tuna fishing fleet (González 1999a; SEMARNAP 1998).

26. The NAAEC recognizes that there is a significant bilateral nature to many transboundary issues and calls upon the three countries to develop an agreement to assess the environmental impacts of proposed projects in any of the three countries party to the NAFTA which would be likely to cause significant adverse trans-boundary effects within the jurisdiction of any of the other parties, to develop a system of notification, consultation, and sharing of relevant information between countries with respect to such projects, and to give consideration to mitigating measures to address the potential adverse effects of such projects.

27. See the April 2002 update on the web site of the US Department of Interior, http://www.mms.gov/eppd/compliance/nepa/procedures/trans/agreement.htm.

28. In 1995, with authorization from the Mexican federal government to build and operate a hazardous waste transfer facility, Metalclad completed construction of the dump. However, the Municipal council of Guadalcazar, which had refused for several years to act on Metalclad's application for a local construction permit, voted to deny the long-pending municipal construction permit (Gaines 2000). Instead, the governor of San Luis Potosi precluded Metalclad from operating the facility by declaring a 600,000-acre zone for the protection of rare cactus species in the area surrounding the site. Metalclad submitted a chapter 11 claim for $90 million in lost profits because the local and state governments had acted to prevent the operation of the landfill it had constructed.

29. In February 2001, Mexico launched a court action launched to set aside the Metalclad Arbitration ruling and was only partially successful in this effort. For the legal details of the final decision, see http://www.dfait-maeci.gc.ca/tna-nac/NAFTA-e.asp#Metalclad.

30. The economic costs (particularly in terms of the costs of environmental clean up, natural disaster and health impacts) are very high. See e.g. Margulis 1991.

31. Although NAFTA's article 104 establishes the preeminence of international environmental agreements on endangered species, the ozone layer, and toxic wastes, several major agreements such as the 1992 Convention of Biological Diversity and the Kyoto Protocol to the 1992 Framework Convention on Climate Change have not been signed by all the parties. Furthermore, no international environmental

agreements have thus far been explicitly used to inform policy making on specific environmental issues relevant to the NAFTA.

32. For an earlier statement of this position, see RMALC 1994a. At a 1994 international meeting of citizen's action organizations on Integration, Democracy and Development, participants concluded that "in spite of the insufficiency of the parallel agreements, they constitute a first draft of the kind of supranational institutionality that must be generated in processes of regional integration" (RMALCa 1994, p. 6).

33. Barkin (1999) notes that the Mexican Trade Ministry "has energetically opposed any effort to examine the environmental and social impact of economic integration" and that "often it has directly intervened in the new tri-national institutions to block these efforts."

34. The trade agreement signed with the European Union on March 23 1999 culminated a long, drawn out process of negotiations that began in 1995. It was approved by the Mexican Senate on March 20 2000 and went into effect in July 2000. Like Mexico's Agreement with the EFTA countries of Iceland, Liechtenstein, Norway, and Switzerland, the EU Agreement contains some NAFTA-like environmental provisions within the main text (Diaz 2001). Labor and environmental considerations are omitted from all of Mexico's FTA's with Latin America (Diaz 2001).

References

Abel, A. 2000. "NAFTA's Institutional Framework: Lessons Learned and Relevance to the FTAA: Some Issues to Consider Regarding the BECC and NADBank Models." Paper presented at Washington conference on "The FTAA and the Environment: What Can We Learn from the NAFTA Model?"

Alanis-Ortega, G., and A. González-Lutzenkirchen. 2000. "No Room for the Environment: The NAFTA Negotiations and the Mexican Perspective on Trade and the Environment." In this volume.

Alfie, M. 1991. "Ecología y Tratado de Libre Comercio." *El Cotidiano* 43.

Alfie, M. 1995. "La realidad del movimiento ecologista en México." *El Cotidiano* 70.

Anderson, S., J. Cavanagh, and S. Gross. 1993. *NAFTA's Corporate Cadre. An Analysis of the USA-NAFTA State Captains.* Washington: Institute for Policy Studies.

Anderson, S. 2001. *Seven Years Under NAFTA.* Washington: Institute for Policy Studies.

Associated Press. 2001. "Mexican Fishermen Call US Tuna Ruling Unfair, Unproductive." July 24.

Audley, J. 1997. *Green Politics and Global Trade: NAFTA and the Future of Environmental Politics.* Georgetown University Press.

Audley, J. 1995. "What Environmentalists Got in NAFTA: Environmental Participation in Trade Policy." Paper presented at 36th annual International Studies Association Conference.

Audley, J. 1993. "Why environmentalists are angry about the North American Free Trade Agreement." In *Trade and Environment*, ed. D. Zaelke et al. Island.

Audley, J., and E. Uslaner. 1994. "NAFTA, the Environment, and American Domestic Politics." *North American Outlook* 4, no. 3: 23–58.

Barba Pírez, R. 1993. "La Unión de Grupos Ambientalistas en el proceso de negociación del Tratado de Libre Comercio." *Frontera Norte 5*, no. 10.

Barkin, D. 1999. "Free Trade and Environmental Policymaking in Mexico." *Borderlines* 7, no. 9: 14–15.

Barry, T., and B. Sims. 1994. *The Challenge of Cross-Border Environmentalism: The US Mexico Case*. Resource Center Press and Border Ecology Project.

Boren, R. 1997. "Waste on the Way: West Texas Town Targeted for Nuclear Dump." *Borderlines* 5, no. 7: 1–3.

Carlsen, L., and H. Salazar. 2001. *The Social and Environmental Impacts of NAFTA in Mexico*. RMALC-Sierra Club.

CEC. 1996. "NAFTA effects: Potential NAFTA Effects—Claims and Arguments 1991–1994." *Environment and Trade Series 2*.

CEC. 1997. Las instituciones del TLC: Rendimiento y potencial ambiental de la comisión de Libre Comercio y otros organismos relacionados.

CEC. 1998. Four Year Review of the North American Agreement on Environmental Cooperation.

CEC. 1999. Assessing Environmental Effects of the North American Free Trade Agreement (NAFTA): An Analytic Framework (phase II) and Issue Studies. Environment and Trade Series 6.

CEMDA 2000. Comentarios del Centro Mexicano de Derecho Ambiental, A.C. (CEMDA) Sobre las Lecciones Aprendidas en Relación a las Peticiones Ciudadanas Contenidas en los Artículos 14 y 15 del Acuerdo de Cooperación Ambiental de América del Norte. CEMDA.

CIEL, Instituto del Tercer Mundo, Preamble Center. 1999. Investment Agreement of the Americas: Environmental, Economic and Social Perspectives. Center for International Environmental Law (CIEL), El Instituto del Tercer Mundo and Preamble Center.

Coote, B. 1995. *NAFTA: Poverty and Free Trade in Mexico*. Oxfam Publications.

Defenders of Wildlife. 2001. "Keeping Americas Tuna Dolphin Safe: The Facts." Available at http://www.defenders.org/wildlife/new/dolphins.html.

Demmers, J., and B. Hogenboom. 1992. Popular Organization and Party Dominance. The Political Role of Environmental NGOs in Mexico. Master's thesis, University of Amsterdam.

Diaz, M. 2001. "Mexico: Free Trade Agreements Anyone?" Available at http://www.usmcoc.org/nafta_MexFTAs.html.

DiMenta, J., and P. Doughman. 1998. "Soft Teeth in the Back of the Mouth: The NAFTA Environmental Side Agreement Implemented." *Georgetown International Environmental Law Review* 10, no. 3: 651–752.

Esty, D. 1993. "Integrating Trade and Environment Policy Making: First Steps in the North American Free Trade Agreement." In *Trade and Environment*, ed. D. Zaelke et al. Island.

Esty, D. 1994. "Making Trade and Environmental Policies Work Together: Lessons from NAFTA." *Aussenwirtschaft* 49: 59–79.

Faber, D. 1992. "The Ecological Crisis of Latin America, A Theoretical Introduction." *Latin American Perspectives* 19, no. 1: 3.

Faulkner, T. 1999. "BECC Technical Assistance to Small Communities." *Borderlines* 7, no. 2: 11.

Friends of the Earth and Public Citizen. 1999. *NAFTA's Corporate Lawsuits.* Friends of the Earth.

Gaines, S. 2000. "NAFTA Chapter 11 As a Challenge to Environmental Law Making: One View from the United States." Presented at EnviReform First Annual Conference: Civil Society Participation in NAFTA, University of Toronto. Text available at http://www.library.utoronto.ca/envireform/publications/Conference/Gaines.htm.

Gitli, E., and C. Murillo. 1999. "Factores que desalientan la inclusión del tema ambiental: El ALCA y una posible agenda positiva." In *Comercio y ambiente: Temas para promover el debate*, ed. M. Araya. Organization of American States.

González, A. 1999. *Visión preliminar sobre el estado del debate comercio y medio ambiente en México, primera parte: La postura gubernamental.* Centro Mexicano de Derecho Ambiental.

Grupo de los Cien. 1993. Mexico's Environmental Movement and Its Influence on the NAFTA Negotiations.

Hemispheric Social Alliance. 1999. *Social Exclusion, Jobs, Poverty, and the Proposed Free Trade Area of the Americas.* Development Gap.

Hogenboom, B. 1996. "Cooperation and polarization beyond borders: The transnationalisation of Mexican environmental issues during the NAFTA negotiations." *Third World Quarterly* 17, no. 4: 989–1006.

Hogenboom, B. 1998. *Mexico and the NAFTA Environment Debate: The Transnational Politics of Economic Integration.* International Books.

Hufbauer, G., D. Esty, D. Orejas, L. Rubio, and J. Schott. 2000. *NAFTA and the Environment: Seven Years Later.* Institute for International Economics.

Independent Review Committee. 1998. Four-Year Review of the North American Agreement on Environmental Cooperation: Report of the Independent Review Committee. Commission for Environmental Cooperation.

JPAC. 1999. Advice to Council, no. 99-07 Regarding The Negotiations Toward a North American Agreement on Transboundary Environmental Impact Assessment. CEC.

JPAC. 2000. JPAC Report to Council. Montreal: CEC.

Kelly, M. 1998. "Discussion of Recent BECC/NADBANK Issues." Paper presented at Environmental Law/Border Conference CLE International, Austin, Texas.

Kelly, M. 1993. "Review of the NAFTA Environmental Side Agreement: Does it Meet Expectations?" *Frontera Norte 5*, no. 10.

Kelly, M. 1999. *Report of the Texas Center for Policy Studies on CEC JPAC and Council Meetings in Banff, Canada, June 27–29th 1999*. Texas Center for Policy Studies.

Kelly, M., C. Reed, and L. Taylor. 2001. "A Call for the NAFTA-Created Environmental Institutions to Be Strengthened—Not Weakened—by the US and Mexican Governments." Draft discussion paper, Texas Center for Policy Studies, Austin.

Kouros, G. 2000. "NADBank Update: Interview with Raúl Rodriguez." *Borderlines* Updater, November 10.

La Franchi, H. 1999. "No Environmental Borders." *Christian Science Monitor*. February 2.

Land, G. 1993. "North American Free Trade and the Environment: Border Environmental Groups and the NAFTA." *Frontera Norte 5*, no. 10: 99–115.

Luján, B. 1998. La Acción de las redes ciudadanas en la región del TLCAN: 1990–1997. Red Mexicana de Acción Frente al Libre Comercio.

Magraw, D. 1995. *NAFTA and the Environment: Substance and Procedure*. American Bar Association.

Mann, H., and K. von Moltke. 1999. *NAFTA's Chapter 11 and the Environment: Addressing the Impacts of the Investor-state Process on the Environment*. International Institute for Sustainable Development.

Margulis, S. 1991. Back of the Envelope Estimates of Environmental Damage Costs in Mexico. Internal discussion paper, World Bank.

Mayer, F. 1998. *Interpreting NAFTA: The Science and Art of Political Analysis*. Columbia University Press.

Mayer, F. 2002. "Negotiating the NAFTA: Political Lessons for the FTAA." In this volume.

McFayden, J. 1998. NAFTA Supplemental Agreements: Four Year Review. Working paper 98-4, Institute for Environmental Economics.

Milner, H. 1988. *Resisting Protectionism: Global Industries and the Politics of International Trade*. Princeton University Press.

Mumme, S. 1999. "NAFTA and the Environment." *Foreign Policy in Focus* 4, no. 28: 1–8.

NADBank. 2000. Expansion of NADBank Sectors of Activity. Board Resolution 2000-10.

National Advisory Committee. 2000. National Advisory Committee to the US Representative to the Commission for Environmental Cooperation. Available at www.cec.org.

Nauman, T. 2001. "Hopes Rise for Pollutant Register in Mexico." *Hispanic Vista*, April 6.

OECD. 1996. *Trade Liberalisation Policies in Mexico*.

OECD. 1998. Mexico: Environmental Performance Review.

Osorio, V. 1998. Reinventando el Continente: Agenda Social y Libre Comercio en las Américas. RMALC.

Paterson, K. 1998. "A Movement Blossoms: Cross-Border Activism Picks Up Speed." *Borderlines*, October 20.

Peña, D. 1993. "Letter From Mexico: Mexico's Struggle Against NAFTA." *Capitalism, Nature, Socialism* 4, no. 4.

Picard, A. 2001. El TALCA: Objectivos y Resultados 7 Años Despues. RMALC.

Poitras, G., and R. Robinson. 1994. "The Politics of NAFTA in Mexico." *Journal of InterAmerican Studies and World Affairs* 36, no. 1: 1–35.

Public Citizen. 1993. *NAFTA's Bizarre Bazaar. The Deal Making that Bought Congressional Votes on the North American Free Trade Agreement.*

Public Citizen. 2001. *Down on the Farm: NAFTA's Seven-Years War on Farmers and Ranchers in the US, Canada and Mexico: Dwindling Incomes For Small Farmers in the US, Mexico, and Canada, Lost Farms, and Rural Crisis is NAFTA's Legacy.*

Public Citizen, Sierra Club, and US Business and Industry Council Education Foundation. 1997. *The Failed Experiment: NAFTA at Three Years.* Economic Policy Institute.

Quadri de la Torre, G., and E. Durazo. 1994. *Partidos politicos y medio ambiente: Experiences internacionales y perspectives para México.* Colegio de México.

Red Mexicana de Acción Frente al Libre Comercio. 1991. Libre Comercio o Explotación Libre. RMALC.

Red Mexicana de Acción Frente al Libre Comercio. 1992. *Memoria de Zacatecas. La opinion pública y las negociaciones del Tratado de Libre Comercio: Alternativas ciudadanas. 26 y 27 de octubre de 1991.* RMALC.

Red Mexicana de Acción Frente al Libre Comercio. 1994. *Memoria del Encuentro Internacional—Integración, Democracia y Desarrollo: Hacia una Agenda Social Continental.* RMALC.

Red Mexicana de Acción Frente al Libre Comercio. 1997. *Espejismo y realidad: El TLCAN tres anos despues: Analisis y propuetsa desde la sociedad civil.* RMALC.

Red Mexicana de Acción Frente al Libre Comercio. 1998. *Alternativas para las Américas.* RMALC.

Red Mexicana de Acción Frente al Libre Comercio. 2000. "El Tratado de Libre Comercio de Norteamérica (TLCAN) y el medio ambiente en México: Tendencias de integración económica y sus impactos ambientales." RMALC.

RMALC/Public Citizen. 1996. "Broken Promises." RMALC/Public Citizen.

Rodriguez, R. 2000. "The North American Development Bank: Lessons for the FTAA." Paper presented at Washington conference on "The FTAA and the Environment: What Can We Learn from the NAFTA Model?"

Salazar, H., and C. Laura. eds. 2001. *Impactos socioambientales del TLCAN/The Social and Environmental Impacts of NAFTA: Grassroots Responses to Economic Integration.* RMALC.

Sanchez, R. 1991a. "Environment: Mexican Perspective." In *US Mexican Industrial Integration: The Road to Free Trade*, ed. S. Weintraub et al. Westview.

Sanchez, R. 1991b. "El Tratado de Libre Comercio en América del Norte y el Medio Ambiente de la Frontera Norte." *Frontera Norte* 3, no. 6.

Sanchez, R. 1995. "El TLC y el medio ambiente." In *México frente al TLC: Costos y beneficios*, ed. V. Bulmer-Thomas et al. Colegio de México.

Sandoval, J. 1991. "Los nuevos movimientos socials y el medio ambiente en México." In *Servicios Urbanos, Gestíon Local y Medio Ambiente*, ed. M. Schteingart and L. d'Andrea. Colegio de México.

Seligman, D. 1998. "Broken Promises." Sierra Club.

SEMARNAP. 1998. "Pesca del Atún y Protección del delfín." *Cuadernos de la SEMARNAP*.

Simon, J. 1997. *Endangered Mexico: An Environment on the Edge*. Sierra Club Books.

Soloway, J. 1999. "Environmental Trade Barriers Under NAFTA: The MMT Fuel Additives Controversy." *Minnesota Journal of Global Trade* 8, no. 1: 55–95.

Spalding, M. 1997. "Lessons of NAFTA for APEC." *Journal of Environment and Development* 6, no. 3: 257–275.

Spalding, M. 1999. "Governance Issues under the Environmental Side Agreement to the NAFTA." In *Sustainable Development in San Diego-Tijuana*, ed. M. Spalding. Center for US-Mexican Studies/University of California, San Diego.

Spalding, M., and J. Audley. 1997. *Promising Potential for the US-Mexico Border and for the Future: An Assessment of the BECC/NADBank Institutions*. National Wildlife Federation.

Stern, M. 1993. "Mexican Environmental Policy Revisited." *Journal of Environment and Development* 2, no. 2.

Thorup, C. 1991. "The Politics of Free Trade and the Dynamics of Cross-Border Coalitions in the US-Mexican Relations." *Columbia Journal of World Business* 26, no. 2: 12–26.

Torres, B. 1993. "La cuestión del medio ambiente en la agenda bilateral (1991–1992)." In *México, Estados Unidos, Canadá*, ed. G. Vega et al. Colegio de México.

Torres, B. 2002. "The North American Agreement on Environmental Cooperation: Rowing Upstream." In this volume.

UGAM. 1993. *NAFTA and the Environment Round II*. Letter, March.

Weintraub, S. 1997. *NAFTA at Three*. Center for Strategic and International Studies.

WWF and Fundacion Futuro Latinamericano. 2000. The International Experts Meeting on Sustainability Assessments of Trade Liberalization, Full Meeting Report, Quito, Ecuador, 6–8 March. WWF.

IV

Linking Trade and Environment in the Americas

13

The Environmental Characteristics of South American Exports

Marianne Schaper

The debate on trade and the environment has been criticized for its high level of abstraction. Discussions, conferences, negotiators, and papers frequently embark on lengthy discourses on trade and environment issues without much data or any real analysis of who is exporting what to whom and with what environmental implications.[1] Ongoing negotiations for a Free Trade Area of the Americas (FTAA) must be informed by a concrete understanding of the environmental implications of the existing export structures of Latin American countries and future changes in trade policy.

By the early 1990s, most Latin American countries had embarked upon a process of deep macroeconomic reforms to stimulate long-term growth, increase economic efficiency and competitiveness, and improve income distribution (Edwards 1995). Nations across the western hemisphere opted for growth and development strategies based on exports, market-driven resource allocation, and a reduced role for the state in production and other activities (Ramos 1997). New efforts at trade liberalization were expected to catalyze shifts in relative prices and the reallocation of productive resources according to comparative advantage (ECLAC 1997; Rodrik 1996).

There can be no doubt that these economic reforms altered Latin American economic growth patterns and export profiles and, with them, the region's environment (ECLAC 1992; Schaper 1999; Reed 1996). For many countries of the region this has meant an incentive to investment, production, and export of natural-resource intensive industries. If economic forces are leading countries to a rate of exploitation that exceeds natural regeneration rate of their environmental resources, their long-term economic prosperity may be jeopardized.

Changes in export structures can be associated with three different types of environmental effects (Grossman and Krueger 1993; Esty 2001):

• The scale effect. If trade and investment grow and economic activities expand accordingly while the nature of these activities remains unchanged, then the total pollution load to be carried by the ecosystem will also grow.

• The composition effect. With trade liberalization and greater outward orientation, countries tend to reallocate their productive resources, specializing in sectors with greater or lesser impact on the environment (Mandeng 1991; Sorsa 1994).

• The technology effect. Changes can occur in production processes as a result of the introduction of—or the failure to introduce—new technologies. The environmental impact of the introduction of new technologies is closely linked to the environmental intensity in terms of the pollution load per unit of output.[2]

This chapter seeks to ground the FTAA discussion of trade and the environment in practical reality by assessing the environmental sensitivity of the evolving trade flows of South American countries. The analysis in the chapter refers to the Mercosur region (Argentina, Brazil, Paraguay, Uruguay) and the Andean Community (Bolivia, Colombia, Ecuador, Peru, Venezuela). It focuses on exports with different potential environmental impacts, including the following:

• exports from environmentally sensitive or "dirty" industries,[3] such as iron and steel, non-ferrous metals, pulp and paper, chemicals and petroleum products[4]

• exports from clean industries, such as textiles, non-electrical machinery, electrical machinery, transport equipment, and instruments[5]

• primary products, including primary agricultural, mining, and energy products.

The analysis covers the period 1990–1998 and presents mainly aggregate results.[6]

Scale Effect: Magnitude and Direction of Trade Flows

Information on the development of exports in terms of volume (tons) can be useful for estimating the environmental (physical) impact of the expansion of export activity within a particular sector. When looking at export volumes, one must take account of the fact that a large number of products are aggregated, each with different environmental impacts. Moreover, a higher volume of exports does not necessarily imply a greater level of environmental degradation; the scale of degradation depends on how natural resources are exploited and processed (in other words, on whether their use is sustainable).

During the 1990s, the export volumes of a number of sectors with obvious environmental impact (such as primary products and environmentally sensitive industries) increased by factors of 2 or 3 in most of the Mercosur and Andean Community countries. Another important finding is that the predominant destination for environmentally sensitive exports from Mercosur and the Andean Community was the NAFTA countries (Canada, Mexico, and the United States). Exports from the Mercosur region (mainly Brazil) increased particularly rapidly. Figures 1 and 2 provide an indication of the pollution loads that the ecosystems and societies of the exporting countries carry. They also raise the question of whether the share of environmentally sensitive exports is being intensified by closer trade links with the NAFTA member countries.

It is interesting, however, that the volume of trade flows in environmentally sensitive industries within the Mercosur and Andean Community regions has been growing at a faster rate than flows to countries outside of

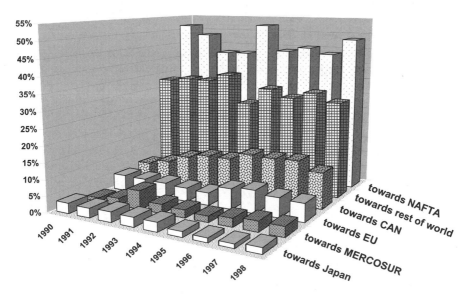

Figure 1
Export volumes of pollution-intensive Andean Community industries by destination.

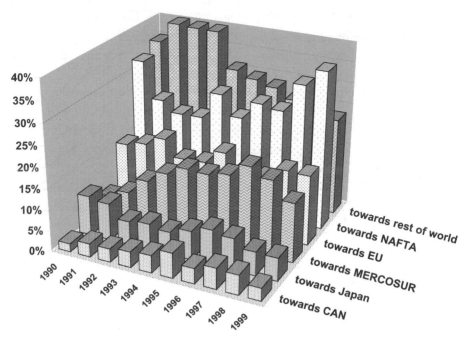

Figure 2
Export volumes of pollution-intensive Mercosur industries by destination. Sources:
Badecel, ECLAC.

Table 1
Annual average growth rates of trade in environmentally sensitive industries
(derived from ECLAC Badecel database).

	From Mercosur (1990–1999)	From Andean Community (1990–1998)
Toward Mercosur	19.6%	34.2%
Toward Andean Community	16.5%	15.9%
Toward NAFTA	4.9%	7.0%
Toward European Union	4.0%	12.6%
Toward Japan	–3.0%	1.8%

the regions over this period. (See table 1.) Although this is certainly a matter of concern, it should be emphasized that this higher growth started at very low levels.

Trade Specialization

Overall, the economic performance of both environmentally sensitive and natural-resource-based exports (such as iron and steel, pulp and paper, petrochemicals, non-ferrous metals, copper, and aluminum) improved during the 1990s for both Mercosur and Andean Community countries. However, as illustrated in the following paragraphs, the relative trade shares and the competitiveness of these sectors show, on average, a downward trend during the second half of the decade.

Composition Effect
The relative trade shares of products in the nine countries during the 1980s and the early 1990s suggest an increasingly natural-resource-intensive export specialization. However, there seems to have been a decline in relative export shares of environmentally sensitive goods and an increase in clean industry products during the 1990s. (See table 2.) Among these very aggregated figures, some countries stand out as the most important providers of these goods. Brazil was responsible for more than half of Mercosur's total agricultural exports in the 1990s, with Argentina the main provider of Mercosur intra-regional trade. Although Colombia is still the main Andean Community agricultural exporter, its share in total agricultural exports within

Table 2
Share in the value of total exports of selected product groups. Source: Badecel, ECLAC.

	Andean Community		Mercosur	
	1990	1998	1990	1998
Primary agricultural products	17.2%	24.9%	43.3%	43.3%
Primary mining products	9.5%	7.9%	9.8%	7.0%
Primary energy products	59.4%	42.5%	3.5%	2.9%
Environmentally sensitive industries	30.7%	24.9%	25.0%	17.5%
Clean industries	2.7%	5.2%	16.9%	23.3%

this regional grouping declined from 48 percent in 1990 to 43 percent in 1998. At the same time, the value of Ecuador's agricultural exports more than doubled. Peru and Brazil were the main exporters of primary mining products; Venezuela was the main exporter of primary energy products.

Both regional blocks show a clear downward trend in the relative share of environmentally sensitive industries as a proportion of total exports, particularly after the mid 1990s. Although the value of exports in this group of industries increased at an average annual rate of 2.6 percent in Mercosur and 1.1 percent in the Andean Community during the 1990s, the total value of exports increased at a higher rate.

The aggregations in table 2 conceal a number of different situations.

First, economic developments in Brazil strongly influence Mercosur's export figures. Brazil accounts for about 80 percent of Mercosur exports from environmentally sensitive industries and for about 75 percent of Mercosur exports from clean industries. In terms of value, exports from environmentally sensitive industries (mainly aluminum, iron, and steel) as a proportion of total Brazilian exports declined from 27 percent in 1995 to 22 percent in 1999. The relative share of clean industries (mainly transport equipment and non-electrical machinery) increased from 22 percent to 26 percent during the same period.

Second, in the case of the Andean Community, Venezuela is responsible for about 70 percent of the value of environmentally sensitive industry exports (mainly petroleum products), followed by Peru (copper) and Colombia (mainly petroleum and chemical products). The declining share of environmentally sensitive products in the total exports each of the countries is less clear than in the case of Mercosur countries. Nonetheless, the proportion of clean industry exports relative to total exports of the Andean Community group increased from 2.7 percent in 1990 to 5.2 percent in 1998, mainly from Colombia and Venezuela.

The Index of Revealed Comparative Advantage

A second way to analyze changes in trade patterns is to calculate indices of revealed comparative advantage (RCA).[7] RCA indices for environmentally sensitive industries and clean industries are presented in tables 3 and 4. I will examine each in turn.

In terms of environmentally sensitive exports, Mercosur and Andean Community countries as a whole have lost markets since 1990, but they

Table 3
Revealed comparative advantage indexes in environmentally sensitive industries (derived from ECLAC Badecel and the UN Comtrade databases).

	1990	1998
Mercosur	1.37	1.11
Argentina	1.19	0.72
Brazil	1.55	1.35
Paraguay	0.11	0.24
Uruguay	0.29	0.41
Andean Community	1.69	1.58
Bolivia	0.89	0.58
Colombia	0.66	0.71
Ecuador	0.36	0.37
Peru	2.14	1.71
Venezuela	2.26	2.48

Table 4
Revealed comparative advantage indexes in clean industries (derived from ECLAC Badecel and UN Comtrade databases).

	1990	1998
Mercosur	0.36	0.45
Argentina	0.16	0.34
Brazil	0.46	0.53
Paraguay	0.01	0.06
Uruguay	0.15	0.23
Andean Community	0.06	0.10
Bolivia	0.0	0.30
Colombia	0.07	0.13
Ecuador	0.01	0.06
Peru	0.16	0.11
Venezuela	0.04	0.08

are still competitive (RCA > 1). However, the aggregate data hide large differences between regions, countries, and products.

While Mercosur is very competitive (RCA between 2 and 5) in pulp and paper, iron and steel, and aluminum, the Andean Community is competitive in copper, aluminum, and petroleum products. Brazil has the highest and most rapidly increasing RCA in pulp and paper and in iron and steel. Between 1990 and 1998, Brazil's RCA in pulp and paper increased from 3.1 to 6.2 and its RCA in iron and steel from 20.4 to 23.5. During the same period, Argentina lost competitiveness not only in environmentally sensitive industries as a whole but also in individual products within the industry. Paraguay, Uruguay, Colombia, and Ecuador gained market share in environmentally sensitive industries during the 1990s but are not yet competitive overall in that grouping. Paraguay and Ecuador are, however, very competitive in veneers and plywood (RCA > 3). Colombia is competitive in cement and in iron (RCA > 15); Uruguay is competitive in cement (RCA > 3).

In terms of clean industry exports, the overall importance of Mercosur and Andean Community countries in the world's clean-industry trade has increased (table 4). However, while all the countries except Peru did better by the end of the decade, none of them are competitive in world markets (RCA < 1). In Mercosur countries, the increase in the RCA index is mainly due to the automotive industry, which is considered to have a clean production process. It is worth mentioning, however, that the consumption of final automotive industry products cannot be considered to be clean, in spite of technological improvements and the decline in emissions per unit.

Conclusions and Recommendations

The 1990s were years of tremendous change in the export sectors of the Andean Community and Mercosur countries. Trade liberalization emerged as a powerful force for resource allocation and the most influential force for the restructuring of productive and export activities in all countries (IDB 1997). Several conclusions can be drawn.

First, problems of pollution and environmental degradation have become more acute in all Latin American countries, but in different ways and with different levels of intensity. Even though the institutional and regulatory framework for the environment has improved in almost all countries and could help mitigate potential negative effects on the environment, the problem of weak enforcement persists. One important promising trend is the increase in some countries of the number of companies with ISO 14000 certification.[8] The number of certified companies is an interesting indicator of environmental progress within the country and is likely to coincide with better practices in the use of clean technologies. The number of companies with ISO 14001 certification in Brazil, for example, increased from only two in 1995 to 300 in June 2001, followed by Argentina with 145. All other countries, however, lag far behind.

Second, there is no simple relationship between trade liberalization policies and environmental impacts. The export structure of Latin American countries varies considerably from country to country. Environmental implications will depend on the particular situation of a country in terms of the environmental institutions in place, the composition of its trade, its income distribution, the enforcement of environmental norms, the pressures of particular interest groups, the geographical density of economic activity, the levels of education of its population, and the income elasticity of demand for environmental quality. The data on environmentally sensitive exports presented in this chapter should be considered as suggestive. The analysis needs to be enriched with more detailed data and information on the environmental performance of specific industries operating within the environmentally sensitive sectors.

Third, South America maintains a specialization in environmentally sensitive and natural-resource-intensive industries but has suffered a relative loss in the ability to compete in knowledge-intensive or high-technology sectors. The resource-allocation patterns and the export growth path in most Mercosur and Andean Community countries are environmentally vulnerable and unsustainable.

South America's specialization in environmentally sensitive and natural-resource-intensive industries has several implications. It places the region in a vulnerable situation in terms of the environmental requirements it will have to face in the markets of developed countries. This specialization also means that countries will face the growing need for (and the costs associated with) investments in environmental services, technology, and equipment and in improved environmental management practices to address the environmental impacts of the environmentally sensitive sectors, to comply with requirements for the international market, and to ensure the economic viability, health, and sustainability of their own productive bases.

Latin American countries do not have much flexibility for adjusting their productive structures to the environmental requirements of the main export markets. The developed countries with the largest share in world trade will most likely continue to define the productive, technological, and environmental patterns and adjustments that will prevail in the rest of the world.

In response, South American countries must balance their trade-opening strategy with appropriate improvements in environmental policy and in institutional structures. In the context of the current environmental sensitivity of exports, it is difficult to justify a continuation of passive attitudes toward the environment. The need for a clear environmental regulatory framework and appropriate environmental enforcement mechanisms is beyond any doubt. Current environmental regulations have been defined mostly in isolation and should be integrated with other policies in related areas.

To contribute to a change of attitudes toward the environment, governments must emphasize the formulation and implementation of converging economic, technological, and environmental policies that contribute to improving the competitiveness, productivity, and exports of sectors with due regard to the environment. To facilitate environmental compliance, particularly by smaller firms, governments need to foster consensus between the public and private sectors, providing incentives to induce companies to improve their environmental performance, combined with adequate financial mechanisms and the design and staged implementation of environmental policies (Lopez 1997).

Taking into account the specific sectoral features in each country, the design of these policies should start with a diagnosis of the current situation, encourage systematic improvements of technological and environmental performance in the relevant sectors and lead to changes that reflect due consideration of short-term, medium-term, and long-term social and economic objectives. Ongoing FTAA negotiations must ensure that the increase of South American exports is accompanied by increased efficiency in the use of natural resources, emphasizing the need for a sustainable development path in the region.

Notes

1. These questions have been explored in detail by Chudnovsky et al. (1996) and by Ferraz and Young (1999). Some of the most interesting work has explored the

environmental implications of trade liberalization as part of the broader package of structural adjustment. See e.g. Panayotou and Hupé 1997 and Reed 1996.

2. The technology effect is not subject of analysis in this chapter given shortages of necessary data.

3. By this criterion, among a selection of 40 three-digit SITC (Rev. 1) industries, the following sectors emerge as leading candidates for dirty industry status in their production processes: pulp and paper (251), Petroleum products (332), organic chemicals (512), inorganic chemicals (513), other inorganic chemicals (514), radioactive material (515), mineral tars and petrochemicals (521), manufactured fertilizers (561), agricultural chemicals (599), veneers, plywood (631), wood manufactures (632), paper and paperboard (641), paper articles (642), cement, etc. (661), iron and steel (67), non-ferrous metals (68), and metal manufactures (69).

4. A conventional approach in the literature (Low and Yeats 1992) has been to identify environmentally sensitive sectors as those incurring in the highest level of pollution abatement and control expenditures/output ratio in the US in 1988. These industries incurred pollution abatement and control expenditures of approximately 1% or more of the value of their total 1988 sales in the US.

5. Using pollution intensity rankings, Wheeler and Mani (1997) identified the following two-digit SITC (Rev. 1) as the cleanest sectors are: textiles (65), non-electrical machinery (71), electrical machinery (72), transport equipment (73), and instruments (86).

6. Data for 1999 are distorted by that year's financial crisis.

7. The measure of revealed comparative advantage was first developed by Balassa (1965). RCA provides a measure of relative export performance by country and industry, defined as a country's share of world exports of a particular good divided its share of total world exports (ibid.). The formula for calculating RCA is $RCA_{ij} = (x_{ji}/X_{jt})/(x_{it}/X_{tw})$, where j is product or industry, i is country, w is world, and t is total. This index provides a useful way to measures changes in the share of a country's exports of a product in world exports of the product compared to changes in the country's total share in world exports, i.e., relative to the size of the country. If the RCA > 1, a country is deemed to have a comparative advantage in a product, i.e. its share in the market of a product was larger than its overall share in world trade.

8. ISO 14000 is a system of voluntary fulfillment of international norms of environmental management and establishes guidelines of good environmental performance of a company.

References

Balassa, B. 1965. "Trade Liberalization and Revealed Comparative Advantage." *Manchester School of Economic and Social Studies* 33, no. 2: 99–123.

Chudnovsky, D., F. Porta, A. López, and M. Chidiak. 1996. *Los límites de la apertura: Liberalización, reestructuración productiva y medio ambiente*. Buenos Aires: Alianza.

ECLAC. 1992. El comercio de manufacturas de América Latina: Evolución y estructura 1962–1989. Estudios e Informes de la CEPAL no. 88, Economic Commission for Latin America and the Caribbean, Santiago de Chile.

ECLAC. 1997. CANPLUS: Competitive Analysis of Nations. Economic Commission for Latin America and the Caribbean.

Edwards, S. 1995. *Crisis and Reform in Latin America, from Despair to Hope.* Oxford University Press and World Bank.

Esty, D. 2001. "Bridging the Trade-Environment Divide." *Journal of Economic Perspectives* 15, no. 3: 113–130.

Ferraz, C., and E. Young. 1999. Trade liberalization and industrial pollution in Brazil. Serie Medio Ambiente y Desarrollo no. 23, Economic Commission for Latin America and the Caribbean, Santiago de Chile.

Grossman, G., and Krueger, A. 1993. "Environmental Impacts of a North American Free Trade Agreement." In *The Mexico-US Free Trade Agreement*, ed. P. Garber. MIT Press.

IDB. 1997. "Latin America After a Decade of Reforms: What Comes Next?" Presented at Thirty-eighth Annual Meeting of the Board of Governors of the IDB and Twelfth Annual Meeting on the Board of Governors of the IIC, Barcelona.

Lopez, R. 1997. *Protecting the "Green" Environment in a Context of Fast Economic Growth: The Case for Demand-based Incentives.* University of Maryland.

Low, P., and A. Yeats. 1992. "Do 'Dirty' Industries Migrate?" In *International Trade and the Environment*, ed. P. Low (Discussion Paper 159, World Bank).

Mandeng, O. 1991. "Competitividad Internacional y Especialización." *Revista de la CEPAL* no. 45: 25–42.

Panayotou, T., and Hupé, K. 1996. Environmental Impacts of Structural Adjustment Programmes: Synthesis and Recommendations. Harvard Institute for International Development .

Ramos, J. 1997. "Un balance de las reformas estructurales neoliberales en América Latina." *ECLAC Review* no. 62: 15–38.

Reed, D. 1996. *Structural Adjustment, the Environment and Sustainable Development.* Earthscan.

Rodrik, D. 1996. "Understanding Economic Policy Reform." *Journal of Economic Literature* 34, March: 1–41.

Schaper, M. 1999. Impactos ambientales de los cambios en la estructura exportadora en nueve países de América Latina y el Caribe: 1980–1995. Serie Medio Ambiente y Desarrollo. no. 19, Economic Commission for Latin America and the Caribbean, Santiago de Chile.

Sorsa, P. 1994. Competitiveness and Environmental Standards: Some Exploratory Results. Policy Research working paper 1249, World Bank.

Wheeler, D., and M. Mani. 1997. In Search of Pollution Havens? Dirty Industry in the World Economy 1960–1995. Washington: World Bank.

14

Trade and the Environment in the FTAA: A Chilean Perspective

Mario Matus and Edda Rossi

In the past several years, Chile's international economic policy has focused on advancing the nation's economic development by expanding and deepening Chile's trade relations with individual third countries, or groups of countries, and with the multilateral organizations most important to Chilean trade. This approach ensures a balanced and organized process for participation in regional integration processes and in the globalization of the international economy. The core of this strategy is to build a framework for an open economy oriented to foreign markets. Owing to Chile's economic openness, current trends of globalization and regionalization significantly influence Chile's development. They can also pose significant challenges.

This chapter provides an overview of Chile's position on trade and the environment. We present Chile's approach in the context of the Canada-Chile Free Trade Agreement. We also advance a Chilean government perspective on possible approaches to environmental issues in the Free Trade Area of the Americas (FTAA).

Chile's Environmental Policy

Despite important precedents, it was only in the early 1990s, with the reinstatement of democracy in Chile, that environmental issues began to be fully included as a major aspect of government policy. Based on the agreements and commitments undertaken at the 1992 United Nations Conference on Environment and Development (the "Earth Summit"), Chile adopted the objective of ensuring that economic growth proceeds in the context of environmental protection, the sustainable use of natural resources and increasing social equity.

Chile's environmental achievements of the 1990s include advances in environmental legislation and in the institutional framework, the culture, and the management of environmental issues. The overarching goals of these initiatives are

• to recover and improve the quality of the environment,
• to prevent environmental deterioration,
• to foster protection of the environmental heritage and the sustainable use of natural resources,
• to involve citizens in environmental management,
• to strengthen the institutional framework for environmental protection at the national and regional level,
• to improve environmental management and develop new management tools,

and

• to introduce environmental considerations into productive economic activities.

There is a specific focus on environmentally sustainable public policies, holding polluters responsible for their emissions, and also contributing to international environmental cooperation. Tools for achieving these goals include policies favoring clean production practices, promotion of internationally accepted environmental standards, design and adoption of environmental certification systems, and efforts to make environmental and trade policies mutually supportive.

Chile's Institutional Framework for the Environment

An important step in the legal history of Chile was the inclusion of environmental protection provisions in the 1980 constitution,[1] which guarantees Chileans the right to live in a pollution-free environment and makes the state responsible for "ensuring that this right is not affected and the preservation of the environment is protected" (article 19, no. 8).[2] The constitution further acknowledges that the social function of property includes everything required to "preserve the environmental heritage" (article 19, no. 24).

Long before the concept of environment and environmental protection became more widely discussed, and before the enactment of the constitution, many regulations in Chile addressed environmental matters. About 1,300 provisions exist that, in one way or another, are important in environmental terms, some dating back to as early as 1916. However, the full translation into action of Chile's constitutional commitment to environmental protection did not occur until the 1993 enactment of Chile's broad-based General Environmental Law. This law established the main management principles for Chile's environment.[3] Based on this law, new pollution-control and natural-resource-management regulations were drafted, including laws related to Environmental Impact Assessment, Environmental Quality, and Emission Standards as well as Pollution Prevention and Abatement Plans. A National Environmental Management System was developed, and in 1998 an Environmental Policy for Sustainable Development was drafted and approved (CONAMA 1998).

The General Environmental Law established a National Commission on the Environment (CONAMA) charged with the horizontal coordination of the different ministries and government agencies with pre-existing institutional environmental responsibilities and functions.[4] While CONAMA is subject to the supervision of the Chilean president through the Ministry of the Secretary-General to the Presidency, it operates as a public agency endowed with its own legal status and funds. CONAMA's actions are decentralized via Regional Commissions for the Environment (COREMA) located in each of Chile's thirteen administrative regions.

The General Environmental Law also provided procedures for the operation of Consultative Councils—one national and one in each of the 13 regions of the country. Established in 1996, the Consultative Councils are the formal and permanent channels for citizens to participate in public environmental management and for CONAMA to obtain different opinions. The Consultative Councils consist of two scientists from academia, two representatives of non-governmental organizations concerned with environmental issues, two representatives of the business sector, two labor representatives, and one representative of the President of the Republic. Together, CONAMA and the Consultative Councils are responsible for devising how to operationalize a National Environmental Management System that emphasizes the need to incorporate environmental sustainability considerations into public policy making as a whole.

Citizen Involvement in Chile's Environmental Policy

The environmental arena is the first arena in which citizen involvement was formally included in Chilean public management. Citizens can participate through the Environmental Impact Assessment System (SEIA) and by making submissions regarding Environmental Quality, Emission Standards, and Pollution Prevention and Abatement Plans. They also participate through the Consultative Councils and through the Sustainable Development Council (which was established in 1999 to provide advice directly to the President of the Republic). The Sustainable Development Council has 92 members, who represent the main lines of public and private activities (business, academia, non-governmental organizations, trade unions, etc.). Their duty is to propose actions to promote sustainable development in Chile.

Toward Sustainable Development

Chile's economic future will depend on the government's ability to integrate sustainability considerations into development policy. The principles adopted and the commitments undertaken at the 1992 "Earth Summit" require the political will to internalize the concept of sustainability in all spheres of public and private activities. In recent years, Chile's environmental problems have become more evident as a result of a combination of enhanced awareness, economic progress that has made investments in environmental protection possible, advances in scientific research, and the full operation of Chile's institutional framework for the environment (which now encourages free expression and citizen involvement). Chile's present environmental situation is the outcome of the accumulation of economic activities of past decades and even centuries (Gligo 1995; O'Ryan 1994; Simonetti et al. 1995). Chile is now focused on the task of reducing the environmental costs associated with its economic growth of the last 20 years.[5]

To address these challenges, national efforts in favor of sustainable development seek to combine economic growth and expansion with poverty alleviation and social justice, a more equitable distribution of the benefits of growth, and the fulfillment of basic needs (such as housing, health, education, and energy), while preserving the environment and improving the quality of life for all Chileans. As noted above, new environmental regulations are gradually being phased in, environmental management systems are being implemented and there are initiatives to modernize government administration and the state that will pay environmental dividends. Chile aims to turn environmental protection into a central component of development as opposed to a concept divorced from other factors critical to economic progress. The task has not been an easy one. Many obstacles and challenges remain.

Trade and the Environment

Over the past few decades, the international community has become increasingly aware of the dangers to survival posed by growing environmental deterioration. There is now widespread recognition of the link between the health of the environment and other issues of international politics, including tensions related to economic regulation and liberalization, the future of North-South relationships, and territorial sovereignty.

The relationship between international trade, the environment, and sustainable development is becoming increasingly important on the international agenda and within Chile. The debate on trade and the environment is a vigorous one, complicated by the different approaches adopted both by countries and by citizens' organizations.[6] The Chilean government's view is that we must be prepared to address these issues—in the most constructive way possible—in the negotiating processes underway in international, subregional, and bilateral forums and also to fulfill the commitments undertaken in such negotiations. The World Trade Organization has an important role to play in this regard, as do multilateral agreements on environmental matters.

The Chilean government believes that the goal must be to advance both the principles of sustainable development and the growth of trade. To achieve these dual goals, we need to address the trade and environment challenge in a flexible manner, developing policies on scientific grounds and looking for the political mechanisms and tools that will allow us to make progress on both fronts. We accept that international trade and the environment are increasingly connected. On the one hand, economic development produces environmental impacts (Esty 1994). On the other hand, environmental measures sometimes impose restrictions on trade (ibid.). It is important to promote the values of sustainable development, but also to remove barriers that prevent the expansion of Chile's international exports. We need a multilateral, open, non-discriminatory, equitable trade system, and we need to increase our knowledge on how to overcome trade barriers and future threats. To this end, there are several principles on which the Chilean government is clear:

• We are in favor of free, non-discriminatory international trade. Trade policy measures seeking to advance environmental goals should not operate as disguised restrictions on trade or be used as arbitrary or unjustified barriers to international trade.

• We are completely opposed to restrictive trade measures or commercial penalties to enforce compliance with environmental goals. In particular, we are against the adoption of unilateral measures aiming to protect the environment, natural resources, or living species beyond national borders.

• Multilateral environmental agreements are the best way to address global environmental issues.

• International cooperation, through the transfer of technology, is very important. Aid from developed countries should be targeted to environmental capacity building. Each country should be helped to adopt the measures it considers most appropriate for achieving economic development while also protecting the environment.

• We are in favor of "win-win" initiatives that involve advantages for trade and also for the environment. An important example is the elimination of environmentally damaging export subsidies (Fisher and Toman 2000).

• We are fully convinced that there is no contradiction between developing international trade and protecting the environment. Environmental protection systems and the multilateral trade system can and should mutually reinforce each other.

At the international level, on the basis of the above principles, Chile has maintained active participation in bilateral, regional, and international forums, symposia, and seminars and meetings where these issues are discussed. These include the World Trade Organization's Committee on Trade and Environment,[7] the Mercosur Working Group 6 on the Environment (in which Chile participates as an observer),[8] and the FTAA Committee of Government Representatives on Civil Society Participation (on which Chile attempts to maintain a middle position to bridge the different approaches).[9] Chile also considers it important to follow up on trade and environment issues in multilateral environmental agreements, many of which include trade-related clauses that need to be fleshed out.[10]

At the domestic level, Chile favors making its export structure compatible with environmental standards so as to position Chile in the world as a reliable supplier of high-quality goods and services produced sustainably in compliance with environmental standards. Chile is also making increasing efforts to exchange points of view with business associations, academicians, and NGOs, to obtain their views, and to help define Chile's environmental approach to international trade.

In early 1999, Chile's General Directorate for International Economic Affairs (DIRECON), attached to the Ministry of Foreign Affairs, established a Trade and Sustainable Development Department. That department is charged with coordinating and supporting discussions about trade and the environment at meetings of the World Trade Organization and in regional or bilateral international forums. It is also responsible for supporting negotiations in multilateral agreements involving environmental matters when they include trade-related clauses. These duties are carried out in close collaboration with the National Commission for the Environment and other appropriate state agencies.

The Chilean government stands prepared and willing to discuss the highly sensitive matter of trade and environment issues whenever necessary, in a manner consistent with its principles, in the forums listed above and in negotiations already underway or that will be undertaken in future (for example, with the European Union and Korea).

The Canada-Chile Agreement on Environmental Cooperation
The Canada-Chile Agreement on Environmental Cooperation (CCAEC) is Chile's most significant achievement in terms of international negotiations on trade and environment issues.[11] Signed together with a Free Trade Agreement and a Labor Cooperation Agreement, this environmental side agreement offers a constructive approach to issues that are increasingly prominent on the international agenda.[12] The goal of the CCAEC is to generate the conditions needed to ensure the social and economic development of both countries while ensuring protection and improvement of environmental and labor conditions. While the main concepts, format, and content of the CCAEC were modeled after the NAFTA's environmental side agreement (the North American Agreement on Environmental Cooperation (NAAEC)), some major differences—reflecting Chile's unique circumstances and priorities—exist.[13] The agreement emphasizes environmental cooperation and the sharing of experiences between the parties,

but its institutional framework and procedures are much simpler than the NAAEC's. The CCAEC:

• Affirms the right of each party to establish its own environmental priorities, policies and levels of protection. Each party undertakes to establish high levels of environmental protection in its domestic laws and strive for their continuous improvement, and to avoid encouraging investment by decreasing environmental protection measures.

• Does not impose trade penalties for alleged non-compliance with environmental regulations. In the event that one party deems that the other party is persistently failing to comply with environmental legislation, the parties may conduct consultations and make use of conflict-resolution mechanisms. Fines may be levied, but they may not exceed US$10 million or the equivalent in the currency of the party subject to the fine. Funds so collected must be used to improve or enhance the environment or to improve enforcement of environmental legislation in the country of the offending party (in compliance with its own legislation). In short, the penalties are intended to benefit the offending party and not the party that initiated the conflict-solving procedure. This is quite a novel approach, since the monetary contribution never leaves the country against which the complaint was filed.

• Allows citizens and NGOs of the two countries to make submissions alleging a party's failure to effectively enforce its environmental laws. Such submissions may not include complaints affecting a private individual or a specific productive activity, although they may be filed against the parties if they fail to enforce their own environmental legislation.

• Provides for the establishment of a Canada-Chile Commission for Environmental Cooperation. The commission is responsible for ensuring compliance with the agreement. In general, efforts were made to use the institutional framework already in place in each country and thus avoid bureaucratic channels or methods that involved high maintenance costs. This is one of the main features that distinguish the CCAEC from the corresponding NAFTA agreement. The Canada-Chile Commission is governed by a council headed by the Canadian Minister for the Environment and the Executive Director of CONAMA. The council's main duties include serving as a forum to discuss environmental matters and draft recommendations in this respect, supervising the duties assigned to the national secretariats, acting as mediator in potential conflicts arising from the interpretation of the agreement, and helping to prevent or solve conflicts.

• Establishes two national secretariats. The secretariats are not independent entities, as in the case of NAFTA. Instead, they form part of the government structure of each country. The secretariats are responsible for managing and handling the agreement. The Joint Submissions Committee, consisting of two members, one from each party, is the channel for managing citizen submissions. It also recommends the preparation of a factual record exploring alleged instances of environmental non-compliance, if applicable. There is also a Joint Public Advisory Committee comprising six members of civil society, three members from each of the parties.

The CCAEC is an important policy tool. It produces benefits for Chile as a whole and particularly for the government, for producers, and for citizens' organizations. The Canada-Chile environmental negotiations prompted a deep process of review and gradual systematization of Chilean environmental legislation. Thanks to the environmental agreement, the Chilean government has put itself at the forefront of the global trend toward the inclusion of environmental variables and considerations in international trade. By acting in advance and adopting a proactive approach, Chile's experience shows that it is possible for developing countries to influence and anticipate the integration of environment and trade concerns. In so doing, developing countries can ensure that the process is sensitive to their circumstances and avoid suffering the unpredictable consequences of a reactive approach.

The CCAEC will also play a preventive role regarding potential environmental conflicts. It includes several tools to anticipate potential disputes, facilitate negotiation and conflict-resolution efforts, and allow the parties to compromise where necessary. In sum, the commitments made in the CCAEC reaffirm the political will of the government of Chile to the principles of sustainable development—harmonizing economic growth with environmental protection and social equity.

Trade and the Environment in the Free Trade Area of the Americas

During the preliminary stages of the FTAA, many commitments to the environment were made by the members of the process. (See box.) Despite these commitments, trade and environment issues are currently excluded from the FTAA process. Since the preparatory phase, only the United States and (to some extent) Canada have encouraged discussion of the environment within the FTAA. In October 1997, the United States made a formal proposal to create a Study Group on the Environment with a mandate to examine the work of the World Trade Organization Committee on Trade and the Environment (and in particular to review its work on the environmental benefits of removing trade restrictions and distortions), to discuss how to inform and foster a dialogue with members of civil society on these issues, and to

Statements and Actions Reflecting Commitment to the Environment in FTAA-Related Discussions

Summit of the Americas, December 1994. Miami, Florida:

Free trade and increased economic integration are key factors for . . . better protecting the environment.

Free trade and increased economic integration are key factors for sustainable development. This will be furthered as we strive to make our trade liberalization and environmental policies mutually supportive, taking into account efforts undertaken by the GATT/WTO and other international organizations.

We will avoid disguised restrictions on trade, in accordance with the GATT/WTO and other international obligations.

Joint Declaration of the First Trade Ministerial, June 1996, Denver, Colorado:

We are committed to transparency in the FTAA process. As economic integration in the Hemisphere proceeds, we welcome the contribution of the private sector and appropriate processes to address the protection of the environment . . . through our respective Governments.

Joint Declaration of Second Trade Ministerial, March 1996, Cartagena, Colombia:

We reaffirm our commitment to transparency in the FTAA process. We direct our Vice-Ministers to consider appropriate processes to address the protection of the environment. After having received the report of the committee that will be presented at the WTO Ministerial Meeting in Singapore (Note: The committee refered to here is the WTO Committee on Trade and the Environment), we will consider creating a study group on this issue based upon recommendations from our Vice-Ministers. To this end, we and our Vice-Ministers will consider how to proceed in the construction of the FTAA in this area.

Joint Declaration of Third Trade Ministerial, May 1997, Belo Horizonte, Brazil:

The issue of the environment and its relation to trade has been considered by our Vice-Ministers since the Cartagena meeting and is the subject of ongoing discussions within the WTO and within the FTAA process. We will keep this issue under consideration, in light of further developments in the work of the WTO Committee on Trade and the Environment.

We reiterate our commitment to transparency in the FTAA process. In this sense, we consider the inputs from stakeholders or our civil societies to be important to our countries to take them into account through mechanisms of dialogue and consultation.

At Belo Horizonte there was a compromise on the more limited issue of how to incorporate inputs from non-governmental groups in the process. Brazil, as the host country, supported the U.S. idea to create a mechanism for incorporating non-governmental views other than business views.

Joint Declaration of Fourth Trade Ministerial, March 1998, San José, Costa Rica:

A Committee of Government Representatives on Civil Society Participation was established. Its purpose is to receive recommendations from civil society addressing issues directly related with the negotiations. The environment was one of them but, by no means, the only one. Vice-Ministers and Ministers are to decide the treatment and a response to be given to these contributions.

Source: http://www.alca-ftaa.org/English Version/view.htm

compile information on environment-related provisions in the existing sub-regional and bilateral arrangements in the hemisphere. This proposal was not accepted. With different levels of resistance, the majority of Latin American countries and some Caribbean countries opposed this initiative.

Chile has tried to raise the level of the debate on trade and the environment. Chile submitted, for example, an informal proposal for a FTAA Committee of Government Representatives to discuss trade and environment issues that would be presided over by a high-level political authority. Like the US initiatives, the proposal was not well received.

We must not, however, be confused. Every Latin American country considers the environment to be a very important issue. Each Latin American country attended the 1992 Rio "Earth Summit." They all participate in the WTO's Committee on Trade and the Environment and have signed most of the Multilateral Environmental Agreements, such as the UN Basel Convention on the Control of Transboundary Movements of Hazardous Wastes and their Disposal, the UN Montreal Protocol on Substances That Deplete the Ozone Layer, and the UN Convention on Biological Diversity's Cartagena Protocol on Biosafety.[14]

Nevertheless, many Latin American countries do not want to deal with environmental issues in the context of hemispheric trade generally or the FTAA specifically. Their rationale is simple: they fear that developed countries will use the environment, or the assumption of non-compliance with environmental regulations, as a non-tariff barrier blocking market access for developing country products. The common view is that the creation of policy links between trade and the environment would facilitate forms of disguised protectionism aimed at undermining an important competitive advantage of developing countries: less burdensome regulatory regimes.

As we see it, the pressure for an environmental linkage stems from some very well organized civil-society groups—in the United States and Canada—that are concerned about the environment and quality of life (Mayer 1998). And there also are pressures from business and labor organizations that fear unfair competition, due to the fact that the other countries do not internalize in the price of their goods the potential harm to the environment caused by production processes.

The Chilean government understands the various positions and shares some of them, but it also understands clearly that trade and environment issues are becoming increasingly important in national and international agendas and that it is very difficult to ignore. As a small country, Chile fears the absence of clear and objective rules. It is time therefore to set aside fears and misconceptions and to look at the trade and environment debate from a more scientific and technical perspective. We believe that it is better to address the issue in a way that takes seriously both the very real environmental concerns and the fears of disguised protectionism.

Finding appropriate ways to build environmental sensitivities into the FTAA will not be easy. But the issue must be tackled head on. It makes sense to move cautiously, to build understanding, and to take our time to examine the options carefully. We need to find actively creative ways to address environment issues that provide assurance to all the parties involved. While one cannot go against the will of the majority, several approaches should be explored. Chile's Agreement for Environmental Cooperation with Canada represents an excellent starting point from which to continue developing these policies. Chile and Canada have committed to an ongoing dialogue among trade and environment officials. To advance the conceptual debate and propose actions, a round table was established—to which experts from different sectors are invited (the first session was held at the beginning of October 2000). Future activities related to the WTO are also important, particularly after the Seattle ministerial meeting, where the issue of trade and the environment emerged as a very sensitive one.

We believe that it would be useful to deepen the work carried out by the WTO Committee on Trade and Environment. Subregional analyses of these issues, such as in Mercosur (where there are already some ongoing activities) and the Andean Community, would also be helpful. In sum, Chile stands willing to continue the dialogue and to search for constructive proposals to address—on bilateral, regional, and global levels—the challenges associated with trade and the environment.

Notes

1. The full text of the Political Constitution of Chile can be found at http://www.congreso.cl/biblioteca/leyes/aa1.htm.

2. This guarantee is specifically established in article 20, subparagraph 2 of the Chilean constitution.

3. The General Environmental Law is no. 19.300. From now on we shall refer to it as "the Law" or "the Environmental Law." The CONAMA documents referred to in this chapter are available at http://www.conama.cl.

4. CONAMA's Board of Directors, its highest authority, consists of thirteen Cabinet Ministers. It is presided over by the Minister Secretary-General to the Presidency. CONAMA's Executive Office is responsible for implementing and carrying out the duties entrusted to CONAMA. The Executive Office is also responsible for complying with the agreements adopted by the Board of Directors and the guidelines it establishes.

5. CONAMA's web site provides documentation of Chile's environmental challenges in many areas, including climate change, biodiversity, eco-labeling, air pollution, and ozone depletion.

6. The NAFTA debate provides an excellent case in point. For a compilation of the various policy positions and recommendations of civil-society organizations from across the North American region, see Magraw 1995. For histories of the environmental debate including a discussion of the different viewpoints and political strategies of environmental NGOs, see Hogenboom 1998 and Mayer 1998.

7. The WTO's Committee on Trade and Environment was created at the WTO's 1994 Ministerial Meeting in Marrakech. The CTE has both analytical and prescriptive functions: to identify the relationships between trade and environmental measures in order to promote sustainable development, and to make recommendations on whether any modifications to the provisions of the multilateral trading system are required. See http://www.wto.org for further information.

8. Mercosur (Mercado Comun del Sur, meaning Southern Common Market) is a regional trade agreement between Argentina, Brazil, Paraguay, and Uruguay, with Chile as an associate member. An Additional Protocol to the Treaty of Asunción Regarding the Environment was proposed in 1997 and revised in 1999. A new Framework Agreement on the Environment has since been replaced this initiative. At the time of printing, the Framework Agreement was pending ratification by Member states. For more information about Mercosur, see http://www.mercosur.org.

9. At the 1994 Summit of the Americas in Miami, 34 countries in the western hemisphere committed to negotiating a free trade agreement (the Free Trade Area of the Americas) among them by 2005. Discussion of environmental issues is currently minimal and is de-linked from trade discussions.

10. One example of an MEA with trade provisions is the 1987 Protocol on Substances That Deplete the Ozone Layer (the Montreal Protocol). It restricts trade in ozone-depleting chemicals and provides for the exchange of information and technology relating to substitutes for ozone-depleting substances.

11. The Agreement on Environmental Cooperation between the Government of Canada and the Government of the Republic of Chile can be found at http://www.sice.oas.org/trade/chican_e.

12. More information about the trade and labor agreements can be found at http://aceis.agr.ca/itpd-dpci/tal.html. The bilateral Canada-Chile Free Trade Agreement, which came into force on July 5, 1997, covers trade in goods and services, investment, and dispute settlement mechanisms.

13. The full text of the North American Agreement on Environmental Cooperation can be found at http://www.cec.org.

14. For further information about these agreements and their signatories, see http://www.unep.ch/basel/, http://www.unep.ch/ozone, and http://www.biodiv.org/biosafe/Protocol/Protocol.html.

References

CONAMA. 1998. *Una politica ambiental para el desarollo sustentable*. Santiago: CONAMA.

Esty, D. 1994. *Greening the GATT: Trade, environment, and the future*. Washington: Institute for International Economics.

Fisher, C., and M. Toman. 2000. *Environmentally and Economically Damaging Subsidies, Concepts and Illustrations*. Washington: Resources for the Future.

Gligo, N. 1995. "The Present State and Future Prospects of the Environment in Latin America and the Caribbean." *CEPAL Review* no. 55: 109–126.

Hogenboom, B. 1998. *Mexico and the NAFTA Environment Debate, the Transnational Politics of Enconomic Integration*. International Books.

Magraw, D. 1995. *NAFTA and the Environment: Substance and Process*. American Bar Association.

Mayer, F. 1998. *Interpreting NAFTA: The Art and Science of Political Analysis*. Columbia University Press.

O'Ryan, R. 1994. *Sustainable Development and the Environment in Chile: A Review of the Issues*. Universidad de Chile.

Simonetti, J., A. Arroyo, A. Spotorno, and E. Lozada. 1995. *Diversidad Biologica en Chile*. Santiago: CONICYT.

15

A Latin American Agenda for a Trade and Environment Link in the FTAA

Eduardo Gitli and Carlos Murillo

At present, environmental issues do not feature as a subject for negotiation in the Free Trade Area of the Americas (FTAA) process. Most Latin American governments explicitly object to any discussion of the combination of trade and environment issues. This opposition is expressed primarily through their trade negotiators, who wield the dominant influence in the process of making trade policy. As a general rule, moreover, most developing countries have not designed their own environmental agendas. Participation in international environmental discussions is almost entirely reactive. Developing countries see themselves as constantly on the defensive, and as more likely to be defendants than plaintiffs in trade and environment disputes.[1]

Similarly, most developing countries have failed to develop proactive, substantive positions on trade and the environment. This shortcoming undermines the capacity of developing countries to take advantage of the many opportunities to leverage resources for environmental protection, to trade environmental commitments for concessions on other trade issues, or to challenge developed countries' environmental performance. As a result, Latin American countries are giving their northern partners the upper hand in trade negotiations by implicitly allowing them to define the "real" issues and agenda.

In this chapter we present a Latin American viewpoint on the NAFTA trade and environment model and provides some recommendations for the FTAA negotiations. Our objective is to make the case that governments can and should develop a trade-environment agenda for the FTAA that simultaneously advances their environmental, economic, and political priorities.

General Objectives for a Regional Trade Agreement

Any trade and environment agenda in the FTAA context must be built on a recognition that, from a Latin American perspective, the primary objective of trade diplomacy and ensuing institutional arrangements in Latin America must be to promote open trade and investment. The underlying assumption of this chapter is that open markets, especially greater market access for developing country goods and services and open markets for investment, are beneficial for economic development, particularly in developing countries.[2] Important objectives of Latin American countries in regional trade arrangements should therefore be to promote market access for Latin American goods (this will require not only duty-free access to the relevant markets, but also predictable and transparent export-import systems in all countries to ensure that exports are not encumbered by "export harassment measures"), to expand investment in their countries, to ensure that expanding trade and investment contributes to a better quality of life for Latin American people,[3] and to provide a flexible dispute resolution mechanism that is as *inexpensive* as possible for participating countries. In pursuit of these objectives, regional agreements must also guarantee respect for national sovereignty, constitutions, and laws. They should also facilitate cooperation among countries to reduce transportation costs (through public and private investment in infrastructure), expand trade, solve migration problems, and reduce general transaction costs associated with trade.

The challenge for countries in the Americas is to promote trade while ensuring the responsible and sustainable use of natural resources. However, several challenges confront efforts to integrate environmental considerations into this trade agenda.

First, in considering trade and environment questions in the Americas, we must first acknowledge the following:

• Economic development in Latin America is currently based on the use of natural resources and the function and services of ecosystems throughout the hemisphere.

• Environmental problems in Latin America transcend political boundaries (UNEP 2000). Long-term solutions demand shared efforts among countries and regions (IDB 1996).

• Few environmental problems arise from trade per se. Rather, they emerge from the production methods and consumption patterns of economies that involve not only domestic production and consumption but also exports and imports.

Second, countries in Latin America and the Caribbean have little experience in the area of trade and the environment. Beyond the NAFTA environmental side agreement and the Canada-Chile Agreement on Environmental Cooperation, Latin American efforts are largely restricted to general statements with insufficient focus on implementation. The Central America region's "Alianza para el Desarrollo Sostenible" (Alliance for Sustainable Development) and Mercosur's new Framework Agreement on the Environment are two recent and promising innovations (Segger and Borregaard 2002).

Third, the trade and environment issue is complicated by the fact that the FTAA region comprises 34 countries with distinct economic, social, ecological, and geographic circumstances. In this context, any effort to integrate environmental considerations into the FTAA must focus on cooperation and on providing for a long-term process of implementation.

Applying the NAFTA Model to the Americas

Countries in the western hemisphere have at least three options regarding the issue of environment and the FTAA. Plan A is to avoid all discussions and negotiations on this issue. Plan B is to adopt the NAFTA model of parallel environment negotiations and a side agreement for the environment. Plan C is for countries to work to insert the environmental negotiations directly into the FTAA through an environmental chapter in the FTAA.

Plan A is not a viable option. Environmental issues cannot be ignored or avoided. If not addressed explicitly, they will emerge implicitly and perhaps chaotically. Plan C creates other risks, especially unnecessary and excessive entanglement of trade and environment policies and the possibility of scaring some Latin American countries out of the FTAA process.

Plan B, the NAFTA model of addressing environmental issues through a parallel track of negotiation along with incorporation of selected environmental principles in the main trade text, offers the most compelling strategy for the Americas for several reasons. It provides a mechanism for addressing trade and environmental issues in the trade context where appropriate and on separate tracks otherwise. It ensures a thoughtful and careful approach to trade liberalization's effects on the environment and the environment's effects on trade.

In addition, we see two approaches within Plan B. The first is the "pure NAFTA" model, in which an environmental agreement (the North American Agreement on Environmental Cooperation (NAAEC)) is negotiated in parallel to the trade agreement. The second is the "Canadian-Chilean interpretation" used in the Canada-Chile Agreement on Environmental Cooperation (CCAEC). The CCAEC improves on the NAAEC by providing several important refinements, including provisions for fines rather than sanctions as the last resort in dispute settlement proceedings.

The benefits of the side agreement model include respect for national sovereignty, obligations on parties to establish and maintain open and transparent environmental management systems, the establishment of institutions to foster cooperation, and provisions to enhance the participation of civil society in decision making. The NAFTA model illustrates that countries can address important environmental issues that arise in the context of trade, and provides examples of several important provisions that could feasibly and usefully be applied to the Latin American context.

First, the most important benefit of the side agreement approach is that it moves countries beyond a focus on discrete trade and environment issues to provide for a broad range of environmental cooperation initiatives and solutions. The NAFTA model obliges parties to establish and maintain comprehensive, accountable, open, and transparent systems for environmental management. It has had considerable success in promoting stronger environmental performance in each of the countries (Hufbauer et al. 2000). The CCAEC takes up this model, focusing on preventing and anticipating environmental conflicts, strengthening environmental cooperation, enforcing environmental legislation, and stimulating gradual modernization, revision, and adaptation of environmental legislation.

Second, the NAFTA allocates responsibility for environmental cooperation to appropriate agencies. With the North American Commission for Environmental Cooperation (NACEC), the North American Development Bank (NADBank), and the Border Environmental Cooperation Commission (BECC), the NAFTA model creates an institutional and organizational framework for environmental cooperation that relies primarily on technical environmental expertise.[4] The CCAEC similarly establishes a Commission for Environmental Cooperation to oversee the implementation of its environmental provisions. In so doing, these two side agreements entrench the principle that environmental issues require expertise that is often beyond the scope of trade negotiators and allows environment authorities as well as trade authorities to intervene in discussions of trade-environment linkage.

Third, the NAFTA creates important and useful avenues for the participation of civil society in the environmental aspects of economic integration. Articles 14 and 15 of the NAFTA environmental side agreement, for example, provide that any citizen or non-governmental organization from the parties may send to the secretariat a submission asserting that a party is failing to effectively enforce its environmental law in order to promote exports or investment.[5] In response, the NACEC's secretariat may be obliged to prepare a factual record. The factual record does not have legal value and cannot trigger the imposition of trade sanctions. The CCAEC similarly encourages, and provides mechanisms for, the participation of civil society in the enforcement of environmental laws. It also grants recognition to the efforts made by the private sector to improve environmental performance.

Despite these positive features, the FTAA context is different than that in which the NAFTA was negotiated. Moreover, an effort should be made to learn from the NAFTA's weaknesses. Three crucial components of the NAFTA's environmental side agreement might be improved upon or adjusted in the broader context of the FTAA: cooperation between national environmental agencies and civil society, enforcement of the environmental laws in each country, and trade sanctions.

The main issues addressed by the NAFTA's efforts at environmental cooperation stem from the unique geographical, political, and economic ties of Mexico, Canada, and the United States—for example, territorial jurisdiction, waste management, and the quality of international rivers. Environmental cooperation in the western hemisphere will require a much broader framework. The most important environmental priorities in the Americas are likely to be transferring cleaner technologies, building local capacity for environmental management, and exchanging information and experience, and training.

Second, by obliging each country to implement its own environmental laws, the NAFTA model addresses the important task of ensuring that countries do not weaken their attention to environmental issues in order to attract investment or promote trade. At the same time, this provision prevents one country from imposing its foreign environmental standards on other parties to the agreement. To properly understand the implications of such a provision in the FTAA context, further research is needed to assess the internal coherence of the environmental legislation of each Latin American country as well as its record of, and capacity for, enforcement (Gitli and Murillo 1999). In all likelihood, each Latin American country will find that it does not have the capacity to properly enforce even its existing environmental legislation. Research will also help countries identify their needs for technical and financial assistance to achieve effective environmental enforcement.

Latin American countries also have the opportunity to improve upon the NAFTA's provisions related to citizen submissions regarding domestic noncompliance with environmental legislation (articles 14 and 15). While important, the NAFTA's provisions for citizen submissions have proved difficult to implement effectively.

As of August 2001, the NACEC had received 31 citizen submissions. Twelve of these submissions were under review and nineteen files had been closed. Ten of the nineteen closed submissions were terminated because they did not meet the established criteria and one submission was withdrawn.[6] Two Factual Records have been prepared. The first Factual Record concerned allegations that the Canadian government was failing both to enforce provisions of its Fisheries Act and to utilize its powers pursuant to the National Energy Board Act to ensure the protection of fish and fish habitat in British Columbia's rivers from ongoing and repeated environmental damage caused by hydroelectric dams.[7] The second Factual Record concerned a pier off the island of Cozumel in Mexico whose construction had allegedly involved destruction of an ecologically critical and delicate coral reef.[8] The environmental community was unified in its dissatisfaction with both Factual Records. In the Cozumel case, the NACEC concurred that Mexico had failed to enforce its environmental laws in order to attract the investment in Cozumel but had failed to make one recommendation for action or to censure the Mexican government. In any similar future provisions, greater efforts should be made to guarantee that any investigating body attracts the necessary political support so that it can properly respond to environmental concerns that arise. At the same time, it is important to note that in their responses to the Factual Records all three NAFTA parties expressed strong reservations about the NACEC's reports arguing that the NACEC had proceeded beyond the scope of investigation and recommendation antici-pated by articles 14 and 15.[9]

Finally, we believe that the NAFTA's provisions for the use of trade sanc-tions to promote compliance with its environmental provisions are inap-propriate for the Americas. Part Five of the NAFTA environmental side agreement, "Consultation and Resolution of Disputes," details the proce-dures in cases of an alleged persistent pattern of failure by one NAFTA party to effectively enforce its environmental law (to promote exports or invest-ment) and includes provisions for trade sanctions. To date, no sanctions have been imposed under the auspices of the NAFTA and the process lead-ing up to the imposition of sanctions could take more than a year and a half.[10] Nonetheless, opposition to, and fear of, the potential use of trade sanctions is the issue that most provokes Latin American opposition to the inclusion of environmental considerations in the FTAA negotiations. Whereas the environmental community in North America tends to favor the identification and punishment of environmental offenders, the approach favored by many within Latin America's environmental community is coop-eration to avoid and repair environmental damage.

Even if the real likelihood of trade sanctions is remote, provisions for sanctions should be rejected by developing countries in any hemispheric investment or trade agreements for three reasons.

First, the history of international relations teaches developing countries not to underestimate the potential for international rules to be exercised and twisted by stronger economic or political forces.

Second, the "sanctions" issue provides developing countries a perfect opportunity to stand up to frequent efforts by developed countries to "over-reach" on purportedly trade-related issues such as intellectual property rights. No developing country should be forced to suffer a loss of national sovereignty, dignity or welfare as a result of external efforts that force it to change internal policies or practices. Developing countries must defend the principle that every country should be in control of its own destiny and the enforcement of any regulation should only take place nationally. This prin-ciple does not preclude the commitment to binding rules established in mul-tilateral forums, such as multilateral environmental agreements.

Third, there are other more effective means for providing developing countries economic incentives for "good environmental behavior." The threat of loss of prestige has, for example, proved effective for many international environmental issues in several instances (e.g., whaling). Economic fines and compensation measures, such as those contained in the CCAEC, could also be used in cases of persistent patterns of violations of domestic laws to encourage better performance. Provisions for financial and technical assistance to developing countries could be one of the most effective ways to improve environmental performance, particularly in countries that simply cannot afford to dedicate more of their limited national budget to environmental concerns.

The CCAEC incorporates several improvements on the NAFTA model, particularly in the area of dispute resolution. Most importantly, the CCAEC has no provisions for the suspension of trading benefits through trade sanctions. Instead, after efforts to resolve disputes through consultations and dialogue, a monetary fine can be imposed, as a last resort, on the country that continues to persistently fail to effectively enforce its environmental law. The fine can be no greater than US$10 million. To determine the amount of the assessment, the dispute panel is required to take into account, among other considerations: the pervasiveness and duration of the party's persistent pattern of failure to effectively enforce its environmental law; and/or the level of enforcement that could reasonably be expected of a party given its resource constraints (Durbin 2000). The CCAEC also reproduces the NAFTA's important provisions for the preparation of factual reports on environmental concerns at the request of the public.

Despite the Canada-Chile model's improvements on the NAFTA model, both models are flawed. On the NAFTA front, a recurring complaint is that the institutions created by the NAFTA's environmental side agreements have failed to ameliorate the infamous environmental degradation along the US-Mexico border (Hufbauer et al. 2000; Public Citizen 1999). Even though the NAFTA has encouraged investment away from the border area to other parts of Mexico, the number and scale of border factories continues to grow despite inadequate investment infrastructure.[11] This trend has resulted in increasing rather than decreasing environmental impacts on the border. Mexican and US NGOs working in the border region argue that not only has the NAFTA exacerbated environmental and social problems on the border, the NAFTA's environmental institutions, the Border Environmental Cooperation Commission (BECC) and the North American Development Bank (NADBank), have been relatively impotent in the face of mounting problems (Public Citizen 1999; Public Citizen and RMALC 1996; Spalding and Audley 1997). They argue that the growth of the maquila industry at the border and the intensification of industrial activities has lead to an increase in the amount of hazardous waste and of illegal imports of hazardous waste into the United States, to deficient sewage treatment infrastructure, and to border air pollution.

Design flaws have also hampered the work of each of the NAFTA's environmental institutions. The NACEC, for example, is troubled by a lack of independence, a limited scope for work and weak enforcement power (Carlsen and Salazar 2002; Hufbauer et al. 2000). Problems encountered by the BECC and the NADBank reflect a similar lack of political will. Both institutions suffer inadequate funding due to a combination of Mexican economic depression and US government budget cuts. NADBank funding is largely limited to projects certified by the BECC, which are mainly focused on water issues, excluding air, natural resources and species conservation projects (Carlsen and Salazar 2002; Kelly et al. 2001). In addition, the NADBank provides predominantly loans (not grants), which shrinks the pool of successful applicants to a limited number that can demonstrate the capacity to repay NADBank loans (Carlsen and Salazar 2002; Kelly et al. 2001).

Both the NAFTA and the Canada-Chile Agreements on Environmental Cooperation fail to address the intensification of environmental problems due to export expansion. For example, several Chilean NGOs argue that increased Canadian investment in the mining sector, without adoption and enforcement of domestic environmental laws, will generate tremendous environmental problems. They also complain that increased pressure to favor foreign investment will contribute to weakening of institutions charged with protecting the environment. Even though general environmental cooperation is a legitimate central focus for environmental side agreements, governments should be careful not to ignore the crucial trade-related issues.

Recommendations for an FTAA Trade and Environment Agenda

Addressing environmental concerns in the FTAA context is unavoidable. This is good news for countries in Latin America! Each country in the Latin American region stands to benefit from taking advantage of opportunities to promote simultaneously environmental protection and trade. Rather than trying to deny the trade-environment linkage, it is time for Latin American countries to develop their own *positive* agenda for incorporating environmental considerations into the FTAA negotiations in ways that will benefit their own environments and economies. As governments and negotiators pursue this task, we offer the following recommendations.

Environmental Principles in the FTAA Trade Agreement

The FTAA should include express commitments to do the following:

• Maintain environmental standards. Environmental standards should not be relaxed in order to attract investment or promote free trade. Rather, special incentives should be offered for trade and investment that improve environmental performance.

• Develop and enforce regulations at the national level. This would not exclude commitments to minimum standards or provisions in multilateral environmental agreements (MEAs). However, the centerpiece of environmental activity must be national environmental programs.

• Ensure that trade-related environmental measures included in MEAs prevail over any contradictory provisions of trade agreements. The dispute settlement mechanisms of MEAs, rather than that of the FTAA, should be the primary forum for resolutions of any conflicts that arise. Every FTAA member should also ratify, at minimum, the following MEAs: the Montreal Protocol on Substances That Deplete the Ozone Layer; the Basel Convention on the Control of Transboundary Movements of Hazardous Wastes and Their Disposal; the Convention on International Trade in Endangered Species (CITES); the Kyoto Protocol to the Framework Convention on Climate Change; and the Convention on Biological Diversity (CBD) and its Biosafety Protocol.[12]

Environmental Side Agreement

An environmental side agreement would provide FTAA negotiators an opportunity to promote cooperation which leads to solutions rather than disputes, to discuss environmental issues with trade repercussions, and to secure the participation of civil society in environment and trade-related decision making. Governments have two choices: they can negotiate the environmental cooperation efforts either within or outside of the context of the FTAA negotiations.

In our view, negotiations for an environmental side agreement should be an integral part of the FTAA process. This approach would enhance the credibility of the FTAA, promote greater consistency between international trade and environmental commitments, and ensure that environmental strategies advanced benefit from the expertise of trade negotiators. The experience of weak follow-up after the 1996 Summit of the Americas on Sustainable Development (held in Santa Cruz, Bolivia) highlights the important role that credibility and political will play in ensuring proper implementation and enforcement of any environmental commitments.[13] Relegating discussion of trade and environment issues to separate environmental forums outside the context of the FTAA are unlikely to be perceived as credible by the public. Nor are they likely to attract the political support necessary for effective responses to the trade and environment issues that might arise.

We therefore recommend that governments do the following:

• integrate environmental considerations into the FTAA process through the negotiation of a side agreement—learning from the existing hemispheric experiences (e.g., NAAEC and the CCAEC)
• ensure the FTAA's environmental cooperation agenda includes efforts to: assist developing countries and their private sectors to meet the costs of "green certification"; address the trade in products that are forbidden in countries of origin (e.g., hazardous waste); clarify the relationship of trade rules with international environmental initiatives, such as the Climate Change Convention's Clean Development Mechanism; promote transfer of clean technologies; promote efficient and renewable energy; and promote common environmental principles for all bilateral and regional trade agreements[14]
• exclude provisions for trade sanction provisions from the FTAA an environmental side agreement. The FTAA should rely on cooperative measures, not penalties, to improve environmental performance and to avoid and repair environmental damage. However, provisions for fines could be included in the case of persistent patterns of domestic environmental legislation violations. The fines could be used to fund efforts to resolve the problem that caused the controversy. Public exposure, "shaming," and technical and financial assistance could also be valuable tools for promoting environmental enforcement.

Negotiating Process
We see a number of ways to improve the negotiating process to build greater environmental sensitivity into the FTAA. Specifically, we suggest that a commitment be made to do the following:

• increase public access to information including: official and provisional negotiating agendas, official reports, provisional and final agreements,

working documents from the negotiating groups, and formal and informal contributions from members during the negotiating process
• prepare non-technical summaries of policy issues, workshops, and seminars to strengthen the understanding and participation of civil-society representatives and enhance the possibility of acceptable agreements
• incorporate environmental considerations into each of the FTAA's nine negotiating groups (see box below)
• create a regional environmental advisory committee (REAC) under the Trade Negotiations Committee. The REAC would be comprised of designated experts and one representative from each of the region's existing free trade groupings: the Andean Community (CAN), the Central America Common Market (CACM), Caribbean Community and Common Market (CARICOM), Mercosur and the NAFTA.[15]

At the National Level
As we have stressed, improved *national* environmental performance will be a critical element of any effort to address environmental issues in the FTAA context. We thus recommend that the governments of countries in the western hemisphere commit to studying the internal effectiveness and coherence of domestic environmental legislation and enforcement regimes, designing programs that address the negative social and environmental costs of growing specialization in international trade, and developing proactive stances on global environmental issues.

Important environmental issues will inevitably arise as trade expands in the Americas. Without a concerted policy effort, the deterioration of the environment could result in significant short- and long-term burdens for the developing countries of the hemisphere.

Potential Environmental Considerations for the FTAA Negotiating Groups

While an FTAA trade and environment agenda could become a market access issue for Latin American countries, it could also be used constructively. Developing countries could use each of the FTAA's negotiating groups as an opportunity to negotiate for the inclusion of mutually beneficial environmental considerations in exchange for trade and investment concessions. We see constructive opportunities that developing country governments might consider advancing within each of the FTAA's Negotiating Groups.[1] Examples follow.

Market Access
• Advance incentives for ecologically friendly products for trade.
• Develop multilateral environmental management systems or performance related standards for certification and eco-labels. Eco-labels and certification efforts could enhance market access opportunities that arise due to increased environmental consciousness in many markets, particularly in North America.
• Explore opportunities and mechanisms for taking advantage of the growing demand for environmental goods and services in many countries.

Investment
• Agree that environmental standards should not be waived to attract foreign investment.
• Prevent requirements for financial compensation to companies for expenses incurred due to the legitimate development and enforcement of environmental legislation.
• Provide better access to information about ethical investment criteria.

Services
• Evaluate potential environmental impacts of: higher levels of inter-hemispheric transportation; increased shipping and port services; and other large infrastructure projects resulting from increased international trade.
• Develop low-impact service sectors (e.g. eco-tourism).

Government Procurement
• Improve transparency in government procurement choices.
• Establish incentives for environmentally sound goods and services.

Dispute Settlement
• Negotiate provisions to establish the binding precedence of multilateral environmental agreements in cases of conflict between environmental objectives and trade agreements.
• Ensure inclusion of provisions allowing dispute settlement bodies to access environmental expertise when necessary.
• Establish procedures for multilateral monitoring of, and accountability for, environmental misdemeanors.

Agriculture
• Take into account the impacts of intensive use of pesticides and agro-chemicals (such as soil exhaustion).
• Address new environmental challenges for phytso-sanitary institutions due to increases in the scale and scope of inspections and other activities they must undertake (possible measures could include greater information sharing and capacity building).
• Coordinate with the relevant Secretariats of multilateral environmental agreement (MEA) on biosafety issues related to genetically modified organisms.

Intellectual Property Rights
• Investigate protocols for access to genetic resources, bio-prospecting and traditional collective resource rights.
• Promote the development of, access to, and protection of clean technologies.

Subsidies, Anti-Dumping Issues, and Countervailing Duties
• Identify and eliminate subsidies that disrupt trade flows and create incentives for unsustainable levels of resource exploitation (e.g., agricultural subsidies which support chemical-intensive practices, and energy subsidies).
• Develop a position on the issues of antidumping and ecological dumping that will likely arise in the FTAA discussions.[2]

Competition Policy
• Improve information about products and production processes to ensure that consumers have better knowledge of what they are buying.
• Increase environment-related information on products and production processes to promote the differentiation of products, competition and the creation of new market niches resulting from a friendlier relationship with the environment.
• Harmonize domestic environmental policies, including environmental impact assessment systems, standards, market-based or voluntary instruments, etc.

sources: Murillo 2000; Segger et al. 1999

1. Most of these recommendations are based on a Submission of the Permanent Group on Trade and Environment from Costa Rica to the FTAA's Committee of Government Representatives on Civil Society (2000). The Permanent Group consists of representatives from academia, non-governmental organizations, and the private sector. This submission was signed by representatives from IUCN-Meso America, the Development Observatory of the University of Costa Rica, the Research Foundation of the University of Costa Rica, International Centre of Political Economy for Sustainable Development (CINPE), the Center for Environmental Law and Natural Resources (CEDARENA), the Fundacion Ambio, the Fundación de la Universidad de Costa Rica para la Investigación (FUNDEVI), and Cámara de Exportadores de Costa Rica (CADEXCO).

2. Ecological dumping occurs when negative environmental externalities are not included in the prices of traded products resulting in lower prices for the products.

Notes

1. The environmental debate is systematically unbalanced due to this phenomenon. For example, most US complaints concern environmental damage caused in the production process (e.g., where damage is caused by developing countries looking for unfair shortcuts to competition). There are few, if any, complaints about environmental damage provoked by consumption patterns in developed countries.

2. The importance of open markets for investment should not be underestimated. In the NAFTA context, for example, Mexico's primary interest was to attract investment while market access was a subsidiary issue. See Gitli and Ryd 1992.

3. Countries should be conscious of potential negative social and environmental impacts of international trade and investment. For example, a by-product of the growth of tourism can be an increase in prostitution. Increased investment in the mining, cattle and manufacturing sectors can lead to heavy environmental deterioration. While these impacts do not necessarily arise from free trade agreements, trade negotiations can provide an opportunity to anticipate and address them. See Gitli and Murillo 1999.

4. For further information about these institutions, see http://www.cec.org, http://www.becc.org and http://www.nadbank.org.

5. Where the secretariat determines that specific criteria outlined in NAAEC article 14(1) criteria are met, it can determine whether the submission merits requesting a response from the party named in the submission. In light of any response provided by that party (during the next 30–60 days to the notification by the secretariat), the secretariat can recommend to the Council that a factual record be prepared, in accordance to article 15. The Council, comprised of the environmental ministers (or their equivalent) of Canada, Mexico, and the US, may then instruct (by two-thirds vote of the Council) the secretariat to prepare a factual record on the submission. The final factual record is made publicly available upon a two-thirds vote of the Council. See the NACEC web site, http://www.cec.org/citizen.

6. For a complete summary of the submissions and their status, see http://www.cec.org.

7. The citizen submission was by the B.C. Aboriginal Fisheries Commission in April 1997. For a summary of the BC Hydro Submission and Factual Record (released in June 2000), see http://www.cec.org.

8. For a summary of the Cozumel Submission and Factual Record, see http://www.cec.org.

9. See e.g. formal comments to the Executive Director of the NACEC from the US, Mexico, and Canada regarding the BC Hydro submission. Copies of these letters can be located on http://www.cec.org.

10. The process begins with a petition presented by one the parties. The first step to solve the conflict is that the parties involved shall make every attempt to arrive at a mutually satisfactory resolution of the matter. If it fails, then the requesting party should request a special session of the Council. If the matter has not been resolved within 60 days, "the Council shall, on the written request of any consulting party

and by two-thirds vote, convene an arbitral panel to consider the matter." The panel has 180 days to release an Initial Report, which determines if there has been a persistent pattern of failure in enforcing environmental legislation. An action plan is also proposed, giving 30 days to the parties to make observations and prepare the Final Report. After that comes a process of implementation of the Final Report and review of its implementation. If the failure persists, the panel can impose a monetary enforcement assessment. If the party fails to pay this, countries have the option of applying trade sanctions to suspend benefits derived from the NAFTA. For further information, see http://www.nafta-sec-alena.org.

11. According to a study by Public Citizen (1999), the number of maquiladoras in Mexico increased by 37% during the first five years of NAFTA implementation.

12. Information about and links to each of these conventions can be located at http://www.unep.org/SEC/env3.htm.

13. For background on the 1996 Summit, see http://www.summit-americas.org/ Boliviadec.htm.

14. For a basic overview of the Clean Development Mechanism, see http://www.wri.org/cdm/.

15. For further information about these agreements, see http://www.mercosur.org, http://www.caricom.org, http://www.communidadandina.org, http://www.cacm.org.

References

Carlsen, L., and H. Salazar. 2002. "Limits to Cooperation: A Mexican Perspective on the NAFTA's Environmental Side Agreement and Institutions." In this volume.

Comision Nacional del Medio Ambiente. 1998. Comercio Internacional y Medio Ambiente. Contexto Normativo e Institucional. Documento de Trabajo no. 5, CONAMA, Costa Rica.

Durbin, W. 2000. A Comparison of the Environmental Provisions of the NAFTA, the Canada-Chile Trade Agreement and the Mexican-European Community Trade Agreement. Yale Center for Environmental Law and Policy.

Gitli, E., and C. Murillo. 1999. Factors Hindering the Inclusion of environmental Issues into Trade Negotiations: FTAA and a Positive Agenda. INCA Project, CINPE-UNA, Costa Rica.

Gitli, E., and G. Ryd. 1992. "Latin American Integration and the Enterprise for the Americas Initiative." *Journal of World Trade* 26, no. 4: 25–45.

Hufbauer, G., D. Esty, D. Orejas, L. Rubio, and J. Schott. 2000. *NAFTA and the Environment: Seven Years Later*. Washington: Institute for International Economics.

IDB. 1996. Environmental Management in the Southern Cone: A Study on the Legal and Institutional Framework. Background Studies. Report ATN/II-5109-96, Interamerican Development Bank.

Kelly, M., C. Reed, and L. Taylor. 2001. The Border Environmental Cooperation Commission (BECC) and the North American Development Bank (NADBank): Achieving Their Environmental Mandate. Texas Center for Policy Studies.

Murillo, C. 2000. "Environmental Opportunities in the FTAA Negotiating Groups." In *Environmentally Sound Trade Expansion in the Americas*, ed. R. Rosenberg. Miami: North-South Centre.

Permanent Group on Trade and Environment from Costa Rica. 2000. Letter to the FTAA Committee of Government Representatives on Civil Society. September 21.

Public Citizen and Mexican Action Network on Free Trade (RMALC). 1996. *NAFTA's Broken Promises: The Border Betrayed*. Public Citizen.

Public Citizen. 1999. *NAFTA at Five*. Public Citizen.

Schaper, M. 2002. "The Environmental Characteristics of South American Exports." In this volume.

Segger, M., and N. Borregaard. 2002. "Sustainability and Hemispheric Integration: A Review of Existing Approaches." In this volume.

Segger, M., M. Muñoz, P. Meirles, J. Taurel, and V. Paul. 1999. *Trade Rules and Sustainability in the Americas*. Winnipeg: International Institute for Sustainable Development.

Spalding, M., and J. Audley. 1997. *Promising Potential for the US-Mexico Border and for the Future: An Assessment of the BECC/NADBAnk Institutions*. Washington: National Wildlife Federation.

UNEP. 2000. *Global Environmental Outlook 2000*. Nairobi: United Nations Environment Program.

16

Civil Society and the Environment: An Environmentalist's Agenda for Sustainable Trade and Investment in the Americas

Douglas Jake Caldwell

Trade and the environment are inexorably linked. Trade has the potential to contribute in a significant manner to improving the quality of life of all peoples as we collectively strive to build a sustainable future. However, neither trade nor trade liberalization alone can meet the ambitious goals of sustainable development in the Americas (Konz 2000; UNEP et al. 2001). Environmental groups and scholars have consistently advocated improvement and progress in the Free Trade Area of the Americas (FTAA) process (Sierra Club 2001; Hemispheric Social Alliance 2001; Segger et al. 2000). The FTAA negotiations must strive to balance the pursuit of trade and market liberalization with the need to address environmental concerns and the potential environmental impacts of increased market integration (Rosenberg 2000; Runge et al. 1997). Such an approach is not protectionism. It is the cornerstone of sustainable development.

The FTAA partners have formally recognized the importance of trade-environment linkage. The 1998 Ministerial Declaration of the Summit of the America's Fourth Trade Ministerial in San José states as a general objective "to strive to make our trade liberalization and environmental policies mutually supportive, taking into account work undertaken by the WTO and other international organizations (San José Declaration)."[1] In light of this objective, environmental observers note with serious concern the relative lack of progress in addressing environmental issues within the FTAA process (Gitli and Murillo 2000; Sierra Club 2001). This concern is compounded by inadequate opportunities for public input into negotiations, particularly on the environment front. While many environmental NGOs welcome the creation of the FTAA's Committee of Government Representatives for the Participation of Civil Society (CGR) (sometimes known simply as the Committee on Civil Society), the absence of a specific work agenda or precisely defined role for the CGR within the FTAA process raises serious questions regarding the current and future effect of the CGR as an effective vehicle for public input in the FTAA negotiations (Caldwell 1999).

This chapter identifies a series of issues that merit priority attention in the FTAA context. The views expressed are not intended to be a comprehensive review of all concerns within the environmental community.[2] Indeed, any such effort would be extremely difficult given the diversity of views among environmental groups. Rather, this chapter represents an effort to contribute to the advancement of a constructive agenda for sustainable trade and investment in the FTAA negotiations. After a brief reflection on the current state of play, a number of areas for action are identified.

Trade Liberalization and Environmental Protection Must Go Hand in Hand

Improve FTAA Deference to National Environmental Standards and Multilateral Environmental Agreements

Trade rules must be crafted so they do not diminish the environmental protections that governments have provided for their citizens and resources. Each FTAA member country should retain the right to develop and enforce high conservation measures through trade measures "even if they exceed the international norm" without running afoul of FTAA rules. Further, the FTAA must allow for appropriate deference to national regulatory authorities in the development of high environmental standards that treat foreign and domestic producers alike—even if they exceed relevant international standards. The North American Free Trade Agreement made limited and certainly not complete progress in certain areas, but the NAFTA language provides a useful starting point in a number of contexts. The FTAA, however, will have to improve on the NAFTA's environmental components to earn the support of the environmental community (and the broader public) (Hemispheric Social Alliance 2001; Sierra Club 2001). Specifically, in matters addressing Sanitary and Phytosanitary measures, Technical Barriers to Trade, and other FTAA negotiations:

• The burden of proof should be explicitly placed on the challenging party in disputes involving health or environmental measures. (See, e.g., NAFTA, art. 723.6.)

• Environmental standards higher than the international norm can be maintained as long as they have a scientific basis. "Scientific basis means a reason based on data or information derived using scientific methods" and not "scientific justification." (See, e.g., NAFTA art. 712.3.) In the WTO context, the term "scientific justification" has, unfortunately, allowed dispute settlement panels to sit in judgment of the *quality* of the science utilized by domestic regulatory authorities in risk management decisions (Hughes 1998). Environmental standards higher than the international norm should comport with the obligations of the FTAA as long as there is a rational relationship between a risk assessment and the restriction imposed by the regulation. One option would be for negotiators to adopt the following NAFTA language: "A Party may adopt, maintain or apply a . . . measure that is more stringent than the relevant international standard, guideline or recommendation." (See, e.g., NAFTA, Art. 713(3).)

• Any harmonization of standards should not "reduc[e] the level of protection of human, animal, plant life or health." (See, e.g., NAFTA, Art. 713.1.)

• The parties should provide a procedure for citizen submissions and independent investigations relating to the environmental provisions of the FTAA, similar to that under articles 14 and 15 of the North American Agreement on Environmental Cooperation (NAAEC) between Canada, Mexico, and the United States. Citizens should be able to make submissions regarding lack of enforcement of environmental laws, adequacy of environmental regulatory framework, and investor responsibility provisions of the free trade agreement. The independence of the investigative body should be clearly established, as should its authority to issue findings of fact, conclusions, and recommendations.

• The FTAA should explicitly recognize the Precautionary Principle and allow FTAA countries to adopt environmental, health, and safety measures in instances where the scientific evidence is insufficient to determine the actual risk posed by a given product or service. (See, e.g., NAFTA, Art. 715.4, 907.3.)

Allow Countries to Distinguish Products on the Basis of How They Are Produced

The long-standing distinction between environmental product standards (acceptable under trade rules) and regulations aimed at production processes or methods (PPMs), which are considered to be uniformly unacceptable in the trade world, cannot be maintained (Esty 1994). The FTAA should therefore permit each party to make distinctions concerning market access based partially on the environmental impacts of production, as long as there is no clear and convincing violation of any FTAA provisions on national treatment. Laws that address the environmental impact of how products are made (such as the US import ban on shrimp harvested by trawl nets that kill endangered sea turtles) must be accepted under trade rules. For example, trade rules should allow countries to label products or restrict the importation of products that are produced or brought to market in a way that harms endangered species and/or the global commons.

At a minimum, the FTAA should explicitly affirm the permissibility of eco-labeling programs that take into account the environmental impact of production (Deere 2000). Such a policy on eco-labeling would reflect the position that multilateral trade rules should provide sufficient flexibility to permit all forms of eco-labeling.

Make Environmental Impact Assessments Integral to Trade Negotiations

Trade negotiators must look before they leap. The environmental ramifications of any trade agreement must be carefully evaluated before the agreement is concluded or put into effect. The goal of the environmental assessments and their open public review-and-comment process should be to provide accurate information on the environmental impact of the proposed trade agreement and to suggest alternatives to mitigate the effect of trade on the environment (OECD 2000)

A broad and comprehensive assessment of trade-related social and environmental effects, initiated well in advance of any future FTAA negotiations, is necessary to assess the positive and negative environmental implications of trade liberalization. The appropriate use of environmental assessments early on in the negotiations can help identify relevant considerations, establish preventive and mitigative measures, and proffer reasonable alternative actions (UNEP 2001). Environmental assessments could also strengthen public participation in FTAA negotiations by making the best use of NGO and other inputs and analyses from civil society (WWF and CIPMA 2001).

Although many nationally based and project specific environmental impact assessments (EIAs) traditionally undertaken in the hemisphere share similar features, in order to maximize the utility of EIAs to trade negotiators, FTAA partners should explore common experiences with EIAs to develop a set of shared EIA principles. For the FTAA, an effective environmental assessment would include the following:

• A new approach to trade negotiations, fully integrating environmental concerns in trade liberalization through an environmental review process initiated as early as possible in the "pre-negotiation" stage of trade negotiations.

• Environmental reviews of all major trade agreements, identifying environmentally preferable outcomes from trade liberalization, assessing global effects, recommending corrective actions, and recognizing the need for specific capacity-building assistance in developing countries.

• Broader participation, including involvement of the public, legislatures, and executive branch agencies with environmental responsibilities, in the development of a more balanced hemispheric trade policy through a timely, open, transparent, and responsive environmental review process.

Eliminate Environmentally Perverse Subsidies and Promote Trade in Environmental Technologies

Renewed attention and energy must be devoted to delivering eminently achievable "win-win" solutions relating to trade and the environment. For example, the elimination of perverse and environmentally damaging subsidies in natural resource sectors such as fisheries and forest products may result in positive gains for both the environment and trade (Nordstrom and Vaughan 1999). Nonetheless, the timing and transition efforts must be carefully managed so as to avoid short-term dislocations (WWF 1998). In addition, efforts should be made in the FTAA to facilitate the trade in environmental technologies. While not a cure-all to the resolution of all trade-environment conflicts, eliminating environmentally destructive subsidies and fostering trade in clean technology would represent a positive step forward.

Environmentally Responsible Investment Agreements

FTAA negotiators should develop specific policy guidelines regarding the relationship between investment negotiations and domestic environmental regulation (CIEL et al. 1999). Future investment negotiations should, at minimum, do the following:

• Seek mandatory, enforceable measures in the agreement to prohibit the lowering of environmental standards to attract investment and an active monitoring system to ensure compliance.

• Review the traditional "investor-to-state" principle found in numerous bilateral investment agreements with an emphasis on ensuring its compatibility with procedural openness, transparency and environmental protection efforts. Recently, in the NAFTA context, several private investors have attempted to use the investor-to-state provisions to challenge domestic regulations with detrimental consequences for local and national environmental laws. FTAA negotiators must ensure that laws and regulations protecting the environment will not be eroded in new trade and investment agreements and the NAFTA parties should undertake efforts to reinterpret or renegotiate the NAFTA to conform to this approach.

• Investment negotiations should include obligations allowing legitimate measures designed to conserve the environment, natural resources and the promotion of cooperative environmental programs to be maintained and strengthened.

Openness and Accountability

A significant step toward a comprehensive trade and environment policy in FTAA negotiations is recognition that the old, exclusive, and secretive deal-making process of trade negotiations must give way to an inclusive, transparent, and democratic process. The negotiating strategies pursued by the FTAA trading partners must reflect this new reality. It is past time to follow through with a process that takes fully into account the views of developing as well as developed countries and of citizens and citizen groups as well as those of industry and government officials from all countries. The era of international trade negotiations being insulated from public concerns, including respect for the environment, is over.

Institutions that govern public conduct must be accountable to the public. The FTAA must adopt modern, democratic principles of due process, including the right of the public to review and comment on the written record of future trade disputes, access to working documents and a permanent role for non-governmental organizations in future FTAA activities. Public participation should be integral to any trade or investment negotiations. Such a linkage confirms the relationship between open markets and democratic principles, and provides citizens with the information they need to make sound and informed choices about policies that affect their future. Despite the clear mandates in the 1994 Miami Summit Plan of Action, and in the Joint Declarations of the Third Trade Ministerial in Belo Horizonte (1997) and of the Fourth Trade Ministerial in San José (1998) strongly supporting the full integration of civil society in the decision-making process, few steps have been taken to date by the CGR or through any other mechanism in the FTAA. The time has come for FTAA countries to make public participation a cornerstone of the FTAA process.

Citizen participation is essential to the success of the FTAA. It is necessary to the development of sound laws and policies because the trust of citizens in the FTAA negotiating process will only be gained when there is real consideration of views through processes that allow for active engagement. Citizen participation in the FTAA process will allow the FTAA negotiators to gather more complete information on issues and set priorities. In addition, citizen participation in monitoring and enforcement efforts can complement efforts by governments because national, state and local governments, as well as regional bodies, do not have the financial or human resources to identify and prosecute all violations of law. Citizens can also serve a watchdog function assuring that the various branches of government properly and fully carry out their respective responsibilities.

In this regard, it is notable that the WTO is taking steps to establish a constructive relationship with civil society. Healthy discussions among stakeholders in trade policy development are essential if governments expect to secure the support of people for trade and investment negotiations.

As noted earlier, environmentalists have serious reservations regarding the FTAA public participation experience to date under the auspices of the CGR. The CGR still has the potential to play a key role in defining the relationship between the western hemisphere trading system and the environment. Clearly, with or without a CGR, the challenge remains before the respective FTAA governments to establish the connection between sustainable development goals and the trade negotiations themselves.

With an eye toward the reintegration of sustainable development priorities in trade agreements via the CGR and/or other mechanisms, the FTAA negotiators should adopt and implement the following recommendations as essential components of the trade negotiations:

Designate Public Participation as a Specific Overall Trade Negotiation Objective

Making public participation a general objective of the FTAA negotiations would send a strong signal to participating countries and to business interests that democratic decision making is integral to good trade and investment

policy. In order for the CGR process to earn the confidence of people in the
hemisphere, it must give priority attention to social and environmental concerns as part of its primary agenda. In addition, the CGR must ensure that
civil-society comments are being properly balanced with business interests.
Most importantly, the CGR must establish a mechanism in the FTAA that
adequately incorporates the views of civil society directly into the individual negotiating groups. The performance of this Committee should be given
as high a priority as the FTAA's other committees (such as the Trade
Negotiations Committee (TNC), the Committee on Small Economies and
the nine sectoral negotiating groups).[3]

Provide a Specific Work Plan Designed to Overcome the Obstacles That Restrict Citizen Participation

While several members of the business community enjoy the financial capacity, technical skills, and personal and professional relationships required to
engage government officials in useful policy dialogue, citizen groups—especially those working in emerging economy countries—frequently do not pos-
sess such resources. To overcome these obstacles to effective participation,
we have recommended in the past the creation of a work plan that includes
establishment of an information clearing house, establishment of national
advisory committees, the promotion of research, training and capacity building, and funding for civil-society participation in the FTAA negotiations.

Define Information Disclosure Policies and Facilitate the De-restriction of Relevant Documents

The results of the first several rounds of negotiations show little concrete
progress in promoting transparency in the FTAA negotiations. The FTAA
negotiators must define information disclosure policies as a means to ensure
public participation in the negotiations. Specifically, we believe the following "classes" or types of documents might be candidates for de-restriction
as an important first step in empowering civil society in a process to encourage "constructive" comments during FTAA negotiations:

• work plans and meeting schedules of the negotiating groups
• an outline of preliminary and official agendas of the CGR and the negotiating group meetings
• working papers and draft texts of the negotiating groups, CGR and other
committees

- minutes, records of discussions, or factual summary of the main elements stemming from the negotiating groups, Trade Negotiations Committee (TNC), and Trade Ministers' meetings
- country position papers on topics addressed by the negotiating groups
- the annotated outline produced by the negotiating groups
- formal reports of the negotiating groups, CGR and other committees to the trade discussion
- a schedule of the next meetings of each of the FTAA negotiating groups, CGR, and other committees
- names and contact information of the national representatives involved with the CGR, committees and negotiating groups and contact information for the lead negotiators from each country in each of the negotiating groups.

We strongly urge governments to release a document describing the information disclosure policies to be utilized under the FTAA and to consider developing it into a comprehensive public document which would also integrate public consultation guidelines.

Establish a Verifiable Avenue for Public Comments on the Activities of Each FTAA Negotiating Group

The establishment of the CGR as a forum to provide civil society with an opportunity to comment on the FTAA process is a step in the right direction. And, we must acknowledge and appreciate the challenge that negotiators faced, and will continue to face, in the creation of the CGR and in maintaining the Committee as a viable and meaningful public participation vehicle.

Nevertheless, for the CGR to successfully function as one of several potentially appropriate vehicles for ensuring that public participation and environmental protection are integrated into the context of trade negotiations, the Committee must adopt several important components into the core of its work program. These components include an opportunity for public dialogue including proactive efforts on behalf of the CGR to actively encourage public involvement in its activities, the development of a comprehensive information disclosure policy and related communications policy, providing creative funding mechanisms for citizen groups throughout the hemisphere which currently lack the financial and technical resources to engage fully in consultations, and clear and measurable operational procedures which indicate how civil-society concerns will be addressed within the context of the negotiations.

Building Hemispheric Cooperation for Trade and the Environment in the FTAA

A comprehensive work plan to advance a western hemisphere agenda on trade and sustainable development must include the following cooperative elements.

Capacity Building

Governments in the western hemisphere need to enhance their capacities on trade and the environment.

• Strong participation of environment ministries in the FTAA negotiations should be encouraged. Their participation should be on equal footing with trade ministries. Countries in the western hemisphere should promote and coordinate regular meetings between trade and environment ministers and their technical advisors in the hemisphere. As the interactions between trade policy and environment policies and outcomes intensify, environment ministries must assume a larger role in the development of trade policy.

• Governments should aim to develop concrete steps to enhance the capacity of all national governments for environmental protection, constructive domestic dialogue and creating their own programs on trade and the environment. A systematic capacity-building program is needed—one which includes an evaluation of needs, significant incentives and financial assistance, milestones, and reporting and evaluation of results. The program should be grounded in a cooperative process involving non-governmental organizations (NGOs) as well as foreign governments.

A systematic plan to assess and improve international environmental performance should also be adopted. Such a plan might include initiatives to achieve and report on results through diplomacy and through trade law remedies. It should also include capacity building to improve enforcement and environmentally favorable technology transfer. What is needed is not retail help for the environment, but wholesale efforts that are at least as systematic as international efforts on trade barriers and on human rights.

To ensure negotiators are better prepared for an FTAA that promotes ecologically sustainable trade, there should be at least one expert government representative in each country who has enough knowledge and understanding of the connections between trade and the environment. If this is not the case, the Tripartite Committee (a joint initiative of the Inter-American Development Bank (IADB), the Organization of American States (OAS), and the Economic Commission for Latin America and the Caribbean (ECLAC)) should provide the necessary support to develop that technical capacity.[4]

Seek Support from the Tripartite Committee for a Technical Analysis of Trade and Environment Issues in the Western Hemisphere

The governments of the western hemisphere should seek support from the Tripartite Committee for the following:

• A technical analysis of country-specific carrying capacities as a means to help develop strategies for sustainable growth and development. Many FTAA negotiators continue to view integration of the environment into trade negotiations exclusively in terms of obstacles to greater market access, and infringements on a country's right to use their own resources. Further analysis might assist in expanding the scope of this view. Studies should be undertaken to identify the degree and intensity of causal relationships between trade barriers and environmental degradation in specific sectors and for particular natural resources in countries in the western hemisphere. For example, Tripartite Committee studies could seek to identify and eliminate tariff escalation efforts that contribute significantly to poor sustainable management of natural resources in the hemisphere.

• As part of the technical cooperation in customs operations discussions, FTAA negotiators should seek to identify each country's interagency cooperation mechanisms for linking environment and trade policy and current information-sharing mechanisms in detecting and recommending measures to correct noncompliance with international and national environmental regulations along borders. We recommend that negotiators discussing customs procedures identify training needs for customs officials in the hemisphere to appropriately enforce international and national environmental regulations along the borders.

• Convene regional dialogues, involving NGOs, government, and business, on substantive issues that characterize trade and sustainable development. To date, political differences over environmental postures and development priorities have deterred the development of a trade and sustainable development agenda under the FTAA process. The complexity and political sensitivity of many regional environmental problems call for solid trans-national collaboration. More information is required, for example, to identify habitats and resources in need of protection, to assess the human and environmental impact of environmental threats, to identify environmentally sound policies, to adjudicate disputes over boundaries, on the environmental impacts of trade, and concerns over sovereignty, in the context of the western hemisphere.

• Draw on lessons learned from examining the different approaches and effectiveness of existing multinational institutions that have already started addressing the linkages between trade and the environment. While more analysis needs to be done on the different approaches used by the institutions such as the International Joint Commission (IJC),[5] the Organization of American States (OAS),[6] and the NAFTA environmental institutions (in making, monitoring and enforcing international policies as well as in national capacity building),[7] the FTAA negotiators would undoubtedly benefit from evaluating the results of these institutions and in addressing environmental concerns.

Conclusion

The FTAA negotiations have the potential to support a hemispheric integration process consistent with the vision articulated at the Summit of the Americas held in Miami in 1994. In Miami, governments committed to linking the advancement of human prosperity to three fundamental principles: social progress, economic prosperity, and a healthy environment. The Miami Declaration states: "Social progress and economic prosperity can be sustained only if our people live in a healthy environment and our ecosystems and natural resources are managed carefully and responsibly. . . . We will advance our social well-being and economic prosperity in ways that are fully cognizant of our impact on the environment."

The increased market integration accompanying the FTAA negotiations create important opportunities to develop parallel mechanisms for cooperative hemispheric mechanisms to address the myriad environmental challenges of the hemisphere. To date, few concrete steps have been taken to ensure that environmental issues are addressed in the FTAA. The principles and negotiating objectives articulated since the 1994 Miami Summit (i.e., in the 1998 San José Declaration) fail to encourage the kind of trading relationships likely to promote healthy economies and cleaner environments. It appears, in fact, that resistance to integrating trade and sustainable development in the FTAA has grown. For example, a decision to consider creating a Study Group on Trade and the Environment (included in the 1996 Joint Declaration of the Summit of the Americas Second Ministerial Trade

Meeting in Cartagena Colombia was rejected at the Fourth Trade Ministerial (San José, Costa Rica, March 1998). In addition, specific opportunities for raising environmental concerns *directly* in each of the FTAA's nine negotiating groups have yet to be identified. As a result, the only remaining official avenue for consideration of environmental implications in the FTAA appears to be through the CGR.

This chapter has set out a series of detailed recommendations to governments indicating how they can advance the trade and environment agendas simultaneously in the FTAA context. Governments must ensure trade liberalization and environmental protection go hand-in-hand; promote openness and accountability in trade negotiations; and assist in the development of hemispheric cooperation and capacity building for trade and the environment. In the interest of building essential broad-based public support for the FTAA negotiations, the FTAA negotiators must take these concrete actions toward assigning meaningful value to environmental concerns by fully integrating environmental protection goals in the FTAA negotiating agenda.

Notes

1. For the complete text of this declaration, see http://www.ftaa-alca.org.

2. A version of this chapter was originally presented by the National Wildlife Federation (NWF) to the FTAA's Committee of Government Representatives for the Participation of Civil Society (CGR) in response to their Open Invitation to Civil Society in FTAA Participating Countries (September 2000) for a presentation of views on trade matters related to the FTAA.

3. At the Fourth Trade Ministerial in March 1998 in San José, the ministers of trade from the hemisphere established a structure for negotiations based on nine negotiating groups among which all commercial issues subjected to negotiation were distributed. The groups are: Market Access, Investment, Government Procurement, Dispute Settlement, Intellectual Property Rights, Subsidies, Antidumping and Countervailing Duties, Agriculture and Competition Policy.

4. The Tripartite Committee has been charged with providing technical assistance to the FTAA negotiations process, developing compendia and databases on a variety of trade policy issues, and technical and analytical support, particularly for smaller economies of the Hemisphere.

5. The International Joint Commission was established by the 1909 Boundary Waters Treaty. The IJC assists the governments of the US and Canada to find solutions to problems in the many rivers and lakes that like along, or flow across, the border between the US and Canada. The IJC has six members. Three are appointed by the president of the US, with the advice and approval of the US Senate and the

remaining three are appointed by the Governor in Canada, on the advice of the Prime Minister. The Commissioners must follow the Treaty as they try to prevent or resolve disputes. They must act impartially, in reviewing problems and deciding on issues, rather than representing the views of their respective governments. The Commission has set up more that 20 boards, made up of experts from the US and Canada, to help it carry out its responsibilities. For more information, see http://www.ijc.org.

6. For more information on the Organization of American States, see http://www.oas.org.

7. In particular, they should examine the experience of the North American Development Bank (NADBank), the Border Environmental Cooperation Commission (BECC) and NAFTA's Commission for Environmental Cooperation (CEC).

References

Caldwell, J. 1999. NWF's Communication to the Committee of Government Representatives for the Participation of Civil Society (CGR). Washington: National Wildlife Federation.

CIEL, Instituto del Tercer Mundo, Preamble Center. 1999. *Investment Agreement of the Americas: Environmental, Economic and Social Perspectives*. Washington: Center for International Environmental Law, El Instituto del Tercer Mundo, and Preamble Center.

Deere, C. 2000. *Ecolabelling and Sustainable Fisheries*. World Conservation Union and Food and Agriculture Organization.

Esty, D. 1994. *Greening the GATT: Trade, Environment and the Future*. Washington: Institute for International Economics.

Gitli, E., and C. Murillo. 2000. Factors Hindering the Inclusion of Environmental Issues into Trade Negotiations: FTAA and a Positive Agenda. Costa Rica: INCA Project, CINPE.

Hemispheric Social Alliance. 2001. Alternatives for the Americas. Washington: Development Gap.

Hughes, L. 1998. "Limiting the Jurisdiction of Dispute Settlement Panels: The WTO Appellate Body Beef Hormone Decision." *Georgetown International Environmental Law Review* 10, no. 3: 915–942.

Konz, P. 2000. Trade, Environment and Sustainable Development: Views from Sub-Saharan Africa and Latin America. ICTSD and UNU-IAS.

Nordstrom, H., and S. Vaughan. 1999. Trade and Environment. Special Study 4, World Trade Organization.

OECD. 2000. *Assessing the Environmental Effects of Trade Liberalisation Agreements*. Paris: OECD

Rosenberg, R., ed. 2000. *Environmentally Sound Trade Expansion in the Americas: A Hemipsheric Dialogue*. North-South Center Press.

Runge, C., E. Cap, P. Faeth, P. McGinnis, D. Papageorgiou, J. Tobey, and R. Houseman. 1997. *Sustainable Trade Expansion in Latin America and the Caribbean*. Washington: World Resources Institute

Segger, M., M. Munos, P. Meirles, J. Taurel, and V. Paul. 2000. *Trade Rules and Sustainability in the Americas*. Winnipeg: International Institute for Sustainable Development.

Sierra Club. 2001. "Five Environmental Reasons to Oppose the FTAA: From Swordfish to Services." In Materials Prepared for the People's Summit of the Americas, Quebec City, April 2001. Sierra Club Canada.

UNEP. 2001. Reference Manual for the Integrated Assessment of Trade-related Policies. United Nations Environment Program, Division of Technology, Industry and Economics (Economics and Trade Unit).

UNEP, IISD, UNAM, COMEDES, and UNCTAD. 2001. *Conferencia Internaciónal sobre Comercio, Medio Ambiente y Desarrollo Sostenible: Perspectivas de América Latina y el Caribe*. Mexico: UNEP ROLAC.

WWF. 1998. Developing a Methodology for the Environmental Assessment of Trade Liberalisation Agreements: Discussion Paper. WWF International.

WWF and CIPMA. 2001. Sustainability Assessments in the Americas: Workshop Report. Gland: WWF International.

17

Sustainability and Hemispheric Integration: A Review of Existing Approaches

Marie-Claire Cordonier Segger and Nicola Borregaard

The negotiations for a Free Trade Area of the Americas (FTAA) are but one component of a broader Summit of Americas agenda in which sustainable development has attracted considerable attention. The 1994 declaration of the First Summit of the Americas in Miami, for example, proposed a partnership between the countries of the Americas for environmental protection, economic growth, democratic reform and social justice, thus opening the doors to mutually supportive environmental, social and trade policies, and to the participation of civil society in their development.[1] In 1996, at the Second Summit of the Americas in Santa Cruz, Bolivia, the 34 governments devised a hemispheric agenda for sustainable development modeled on the 1992 Earth Summit's Agenda 21, stating that "development strategies need to include sustainability as an essential requirement for the balanced, interdependent, and integrated attainment of economic, social, and environmental goals (article 2)."[2]

While no legally binding sustainability accords were opened for signature in Santa Cruz, several sectoral environmental initiatives were launched.[3] In 2001, we heard more of the same. The 2001 Declaration of the Third Summit of the Americas in Quebec City incorporated environment and sustainable development issues as a critical component of the quest for greater prosperity in the western hemisphere.

To respond to increasingly serious environmental challenges across the Americas, an active Forum of Environment Ministers of Latin America and the Caribbean has been meeting for several years independent of the Summit of the Americas process (UNEP 2000).[4] At their meeting in April 2001, the Environment Ministers of the Americas emphasized the need to maximize the potential for mutually supportive policies on economic integration and environmental protection.[5] Forging a way forward, they began to consider and address economic issues at the October 2001 session in Rio de Janeiro.

Despite political acknowledgement of the importance of sustainability considerations and of the interdependence of environmental and economic progress in the context of the Summit of Americas process, in practice, environmental cooperation and economic integration efforts proceed along alarmingly separate tracks. While the governments of the Americas incorporated a package of useful environmental initiatives into their 2001 Quebec City Agenda for Action, political commitment, funding and concrete initiatives to forge real linkages between the environmental and trade agendas have been elusive.

How can governments make real their declared commitments to promote sustainability alongside economic integration? What kind of hemispheric legal and policy framework would best advance sustainability? To date, what little attention has been devoted to these questions has focused on whether the North American Free Trade Agreement's environmental package offers a possible model for the FTAA. The NAFTA and its environmental negotiations engaged just three parties. The process for economic integration in the Americas proposes to join the 34. Moreover, the countries of the Americas are already linked by a complex web of hemispheric and subregional inter-governmental organizations and legal instruments.

The central argument of this chapter is that as policy makers grapple with the challenge of devising "ways forward" on the trade and sustainability front, they would do well to consider the whole range of existing subregional trade agreements and environment accords as potential models.[6]

Traditional international relations theory sharply divides the western hemisphere between North and Latin America (with the sotto voce addition of "the Caribbean"). Reality is, however, more complicated. The greatest advances in regional trade discussions have been made through five subregional trade agreements: Mercosur, the Andean Community (ANCOM), the Caribbean Community (CARICOM), the Central American Common Market (CACM), and the NAFTA.

In their efforts to ensure that the FTAA builds upon subregional economic agreements, governments should also learn from subregional approaches to questions of trade and sustainability. At the same time, policy makers ought to take stock of strategies for environmental cooperation and sustainable development that are evolving in five broad eco-regions:

the Insular Caribbean, the Andean Zone, the Amazon and Southern Cone, Central (or Meso) America, and North America (Segger et al. 2001).[7] Existing and proposed subregional environmental agreements cover everything from the coordination of environmental laws (including standards and certification practices) to capacity building and public awareness programs, the management of forest fires, trans-boundary air and water pollution, desertification, and flood management as well as efforts for joint ecosystem management, wildlife protection and the biodiversity conservation (Segger et al. 2001). The progress underway in the environmental arena, albeit underfinanced, highlights that environmental issues are very much on the political agendas of countries in Latin America and the Caribbean. The challenge is to build stronger relationships between the economic and environmental arenas.

While the NAFTA environmental package is a model worthy of consideration, more systematic study of other subregional models for trade and sustainability policy linkages also promises practical and political benefits. The current demandeurs on the trade and environment front—namely the United States and Canada—may find far greater political traction for this agenda if they work to build on the existing, if nascent, architecture for international economic, development and environmental cooperation in the Americas.

Openness must also be a cornerstone of any effort to address trade-environment linkage. Public participation, consultation, and transparency benefit governments and societies by enabling them to pool expertise, analysis and creative policy ideas. Growing public political interest in the FTAA negotiations increases the political urgency of progress in this area. Here again, subregional models offer valuable precedents. The NAFTA stands out as the most advanced model for openness, but other subregional agreements and processes also provide useful lessons.

Sub-Regional Approaches to Trade and the Environment

Subregional integration initiatives in the Americas approach environmental issues in a number of different ways. Some agreements delegate environmental cooperation to a completely independent parallel track, while others make links to environmental issues through the main text of trade agreements. Other subregional strategies work to integrate social and environmental cooperation mechanisms through a broader project of economic and political cooperation. The combinations vary, as do the institutional arrangements. This section examines each subregion in turn.

Mercosur

Established in 1991 by the Treaty of Asunción, the Common Market of the Southern Cone (Mercosur) integrates Brazil, Argentina, Uruguay, and Paraguay, with Bolivia and Chile as associate members. Mercosur's central focus is economic integration. It combines more than 200 million people with a collective GDP of more than a trillion US dollars in 1999.[8] Intra-regional exports accounted for 20.5 percent of the region's total exports in 1999. A common external tariff scheme provides for tariffs of between 0 and 20 percent (depending on the product) and a timetable for implementation.[9]

The reality of linkages between trade and the environment were recognized early in the Mercosur process. The Mercosur structure, though still evolving, incorporates several innovative environment and trade mechanisms. Efforts originally focused on integrating environmental provisions into the structure of the Mercosur. In more recent years, the new Mercosur Framework Agreement on the Environment represents a shift toward a parallel or separate regime for environmental cooperation (Mercosur 2001a).

Environmental considerations began in 1992, when the Canela Declaration created an informal working group, the Reunion Especializada en Medio Ambiente (REMA), to study environmental laws, standards and practices in the four countries. This forum evolved into "Sub-Grupo no. 6" on the environment that was subsequently established as one of Mercosur's technical working groups. Sub-Grupo no. 6 examines issues such as environment and competitiveness, non-tariff barriers to trade, and common systems of environmental information. In addition, several resolutions of Mercosur's highest decision-making organs—the Grupo Mercado Comun and the Consejo de Mercado Comun—have focused on issues of environmental protection, including rules to regulate pesticide residues in food products, contaminants in food packaging, eco-labeling and transportation of dangerous goods (IADB 1996).[10] A protocol focused on the environmental impacts of new physical infrastructure has been under negotiation for several years. Concurrent meetings of the four Environment Ministers bolstered this cooperation on trade and environment challenges in the subregion.

The most exciting aspect of Mercosur's efforts is the new Framework Agreement on the Environment. This Framework Agreement sets out the parties' shared objective of "sustainable development and environmental protection" (chapter 2, article 4). A draft of this Framework Agreement was negotiated, and later approved, by Mercosur's Sub-Grupo no. 6.[11] Upon ratification by member States, it will be appended to Mercosur's Treaty of Asunción (Mercosur 2001b). The Framework Agreement provides for upward harmonization of environmental management systems, increased cooperation on shared ecosystems, and mechanisms for social participation and the protection of human health. Upon ratification, it will commit member states to cooperation on the development of various instruments for environmental management including quality standards, environmental impact assessment methods, environmental monitoring and costs, environmental information systems and certification processes (chapter 3). For the settlement of environment-related disputes, the draft Agreement defers to Mercosur's existing dispute settlement process (chapter 4, articles 8–11). The annex to the agreement provides a framework for the negotiation of future protocols in three areas:

• sustainable management of natural resources (such as protected areas, biological diversity, wildlife, forests, and hydrological resources);
• quality of life and environmental management (such as hazardous waste management, urban planning, renewable energy, and improvement of soil and atmosphere/air quality)
• environmental policy (such as environmental impact assessment, economic instruments, environmental information exchange, environmental awareness programs).

Though governments have much to do to ensure that the promise of the Framework Agreement on the Environment is realized, the elements are there, and a number of environmental advocacy groups have expressed cautious optimism about this subregional effort to link the trade agenda to the environmental agenda.[12] Particularly noteworthy is the fact that the stimulus for Mercosur's Framework Agreement on the Environment came from within the Mercosur's trade structures. Governments worked explicitly to account for environmental priorities—creating a place for environmental cooperation with the general subregional economic integration process to ensure it attracts ongoing political commitment.

The Caribbean Community and Common Market

The Caribbean Community and Common Market (CARICOM) was formally established in 1973 by the Treaty of Chaguaramas.[13] It comprises 14 small countries: Antigua and Barbuda, the Bahamas, Barbados, Belize, Dominica, Grenada, Guyana, Jamaica, Montserrat, St. Kitts and Nevis, St. Lucia, St. Vincent and the Grenadines, Suriname, and Trinidad and Tobago.[14] From its inception, CARICOM has concentrated not only on economic integration, but also on cooperation with regard to foreign and a range of specific social issues.[15] However, policies for economic integration and environmental cooperation have evolved largely independently of each other.

In 1999, the CARICOM region had a population of approximately 13.4 million people and a combined GDP of US$23.6 billion (CARICOM 2000a). Between 1990 and 1998, intra-regional imports accounted for between 8 and 10 percent of CARICOM's total imports, while intra-regional exports accounted for between 12 and 23 percent of the region's exports over the same period (ibid.). It is noteworthy that between 1990 and 1998 CARICOM's total imports grew by 55 percent while total exports expanded by only 4 percent (ibid.). This large difference in the growth of the Region's overall exports and imports points to the significant imbalance in the growth of trade with external markets. Imports from non-CARICOM sources expanded much faster than exports to those destinations.

CARICOM has largely eliminated intra-regional barriers to trade—over 95 percent of intra-regional trade is free of all restrictions (CARICOM 2000b). CARICOM also provides for a Common External Tariff and a common trade policy (uniform customs legislation and documentation is also under consideration). The region is protected by non-reciprocal trade preferences (Inter-American Dialogue 1997) under the US Caribbean Basin Initiative and other agreements. In the 1989 Grand Anse Declaration, CARICOM Heads of Government articulated their goal of strengthening the Caribbean Community to form a Caribbean Single Market and Economy (CSME).[16]

Caribbean structures for environmental cooperation are proceeding separately from the CARICOM agenda. The legal foundation for environmental cooperation in the Caribbean was laid in 1983 with the adoption of the Convention for the Protection and Development of the Marine Environment of the Wider Caribbean Region (Cartagena Convention). The centerpiece of Caribbean environmental cooperation is now the Caribbean Environment Program (CEP) established in 1986.[17] The CEP is administered by UNEP's Regional Seas Program and coordinated by a secretariat in the Caribbean.[18]

Since 1983, the Cartagena Convention has been supplemented by protocols to foster Cooperation in Combating Oil Spills in the Wider Caribbean Region and Specially Protected Areas and Wildlife (SPAW) Protocol, and to reduce Land Based Sources of Air Pollution. At a 1997 meeting on the implementation of the Small Island Developing States Program of Action, delegations recommended the formation of the Caribbean Sea as a Special Area under the framework of the Regional Seas Program of United Nations Environment Program (UNEP). Other notable environmental initiatives include a Caribbean Action Plan in support of the International Coral Reef Initiative and a Caribbean Regional Energy Action Plan.

While a new unit has been developed at the CARICOM secretariat to promote increased cooperation on environmental issues, the Caribbean has traditionally kept trade and economic negotiations fully separate from environmental discussions and legal frameworks. One strong reason for this is that Caribbean environmental cooperation is largely externally stimulated and financed. That said, growing concerns about economic vulnerability due to natural disasters and the impacts on tourism revenue of environmental degradation have prompted one regional grouping—the Association of Caribbean States—to combine its focus on regional cooperation in trade, integration, and transportation with cooperation on issues of sustainable tourism and natural disasters.

The Andean Community

The Andean Community is both a political and economic endeavor. It comprises Bolivia, Colombia, Ecuador, Peru and Venezuela. Established in 1969, the Andean Community had a total population of 115 million and a GDP of about US$ 226 billion in 1999 (Andean Community 2001). Intra-regional trade accounted for 16 percent of total non oil exports in 1999 (Andean Community 2001). The common external tariffs of this customs union range from 5 to 35 percent in five tiers that each country has implemented to varying degrees (Rodriguez-Mendoza 1997, p. 10).

There is little available information about the embryonic environmental cooperation efforts in the Andean Community. In 1998, the Andean Community created a Committee of Andean Environmental Authorities

(CAAAM). This committee offers considerable potential for linking environmental cooperation measures with the Andean integration process. The functions of CAAAM include the elaboration of strategies for the management of natural resources and support to the Andean Community's General Secretariat regarding the design of an Environmental Action Plan for the Community. As of beginning of 2001, the CAAAM had met twice. Together with the General Secretariat, the CAAAM is working to develop a biodiversity strategy for the Andean Community.[19]

The Central American Common Market and the Sistema de la Integración Centroamericana

Established in 1961, the Central American Common Market (CACM) is a customs union comprising Guatemala, El Salvador, Nicaragua, Costa Rica, Honduras, and Panama. In 1999, CACM had a combined population of about 42 million people and a GDP of about US$54 billion (CACM 2000). Inter-CACM trade accounts for roughly 20 percent of total exports. Common external tariffs average 15 percent (tiered from 5 to 20 percent) (CACM 2000). Mexico and Belize are negotiating with CACM for the creation of a single treaty for the whole region by the year 2002 (Iglesias 1997).

The CACM clearly adopted a separate, parallel course in the development of subregional environmental laws. Although environmental progress by individual countries has been uneven, harmonization and coordination of national activities is increasing. The environment became a significant issue in 1989, following the signature of the Central American Convention for the Protection of the Environment (CPC) and the subsequent creation of the Central American Commission for the Environment and Development (CCAD). Three tangible successes of the CCAD deserve mention.

First, the CCAD coordinated the development of a joint Central American position ("Agenda 2000") for the 1992 United Nations Conference on Environment and Development (UNCED). Second, the CCAD supported the creation of the Central American Inter-Parliamentary Commission on the Environment, consisting of members of parliament from its seven member countries. This Commission was instrumental in spurring member countries to sign a regional Forests Convention, create a Central American Forest Council charged with its implementation and develop a Tropical Forestry Action Plan (which included efforts to devise common guidelines for forestry concessions). Third, the CCAD coordinated the development of a 1992 Central American Convention for the Conservation of Biodiversity and the Protection of Priority Natural Areas. This Agreement led to the creation of the Central American Council for Protected Areas (CCAP) charged with coordinating regional efforts toward the development of Regional Protected Area System.[20] In 1997, at the Nineteenth Central American Summit, the presidents of the region approved CCAP's proposal for implementation of a regional protected area referred to as the Mesoamerican Biological Corridor (MBC).

CCAD's considerable achievements stem in part from its transparent and participatory decision-making process: civil-society organizations, representatives of indigenous peoples, and businesses all participate in CCAD's quarterly meetings and other sponsored events. Another significant element is its regional rather than global approach. Because only a small number of member countries with clear common interests are involved, progress on sensitive issues is possible.

Perhaps most important, Central America boasts an inter-governmental Alliance for Sustainable Development (ALIDES). Established in 1994, ALIDES was conceived to provide a comprehensive subregional conceptual and operational framework through which to strengthen environmental protection and economic development priorities. National Councils on Sustainable Development act as instruments for its implementation. ALIDES served as the starting point for the 1994 CONCAUSA (CONvenio CentroAmerica-USA), a partnership for sustainable development that provided US funding to the region for a list of concrete environmental measures such as the conservation of biodiversity, development of renewable energy, environmental legislation standards and eco-friendly industrial processes.

In the immediate future, it appears likely that the CACM will continue to develop environmental protection as a separate agenda to the trade integration process. The subregional environmental institutions—such as ALIDES—are not strongly linked to the regional common market agenda. One hopeful area for greater integration of environmental and economic issues is the Sistema de la Integración Centroamericana (SICA)—this initiative includes all of the CACM countries with the addition of Belize and with the Dominican Republic as an observer.[21] Established in 1991 by the treaty of Tegucigalpa, the SICA mandate extends beyond CACM's focus on trade to include social, cultural, and ecological objectives. The challenge will be to transform into practice the political commitment to addressing an integrated set of economic, environmental, and social issues.

Bilateral Accords within the Americas

Many bi-lateral trade and investment accords exist in the Americas. A 1998 estimate counted over 106 bilateral trade agreements in force or under negotiation within the region (Morton 1998).[22] This represents a rapid proliferation—in 1994, only 26 bilateral or trilateral free trade agreements or customs unions were in place. Some of these agreements focus on simple tariff elimination or selective strategic objectives, while others are geared toward bilateral common markets.

On the environmental front, two bilateral trade agreements present examples of innovative mechanisms that work to integrate economic and environmental objectives. The Canada-Chile Free Trade Agreement was accompanied by a Canada-Chile environmental side agreement—developed largely to ensure Chile's compatibility with the provisions of the NAFTA in anticipation of its accession.[23] The side agreement includes several innovative improvements on the NAFTA's environmental side agreement—the North American Agreement on Environmental Cooperation (NAAEC) upon which other countries can reflect and build, including: exclusion of the possibility of trade sanctions; an agenda for gradual improvements in compliance and enforcement; and a simplified institutional framework. In terms of bilateral accords, the technical assistance program established under the framework of the NAFTA deserves mention. In the context of the NAFTA negotiations, Mexican authorities and the US Environmental Protection Agency (EPA) worked to establish the 1991 Integrated Border Environmental Plan to improve the environment, protect public health and natural resources, and encourage sustainable development along the US-Mexico border (Esty 1994).[24] This initiative later evolved into the US-Mexico Border XXI Program. This program includes collaborative projects with strong social and environmental components to improve health, working conditions and polluted areas on the border.[25] Alongside the NAFTA, Mexico and the United States negotiated an additional bilateral environmental agreement to complement the NAFTA side agreement that created the Border Environmental Cooperation Commission (BECC) and the North American Development Bank (NADBank).[26] Similarly, to accompany their 2001 bilateral free trade agreement, Costa Rica and Canada negotiated an environmental accord focused on access to environmental information and capacity building for environmental policymakers and lawmakers.[27] Bilateral agreements—with only two parties to coordinate—provide considerable opportunity for flexibility and innovation in the development of integrated social, environmental and economic legal instruments. They tend to be more amenable than regional agreements to sharing information, resources, and technical capacity. Their evolution warrants close observation for models that could be useful for larger processes.

This brief survey of subregional economic and environmental processes in the Americas prompts several observations. First, while none of subregional accords in the Americas have attained the full level of supra-national integration present in the European Union, several initiatives have strong levels of supra-nationality as their ultimate goals. The future prospects for such subregional supra-nationality warrant detailed appraisal as does the question of the extent to which Hemispheric cooperative efforts ought to build from subregional agreements.

Second, the diversity of economic, social and environmental realities in the Americas has prompted a series of quite distinct policy responses to issues of sustainability. Each of the five subregional models integrates environment and trade priorities to a different degree—indicating varying levels of recognition of the interface of trade and environmental policy and its potential effects.

Finally, while each of the subregional trade processes have accorded relatively low priority to environmental (and social) cooperation, several trade and environment initiatives do provide useful precedents. In particular, Mercosur's Framework Agreement on the Environment, the Central American Alliance for Sustainable Development, and the Andean Community Council of Environmental Authorities warrant further exploration.[28] Overall, recognition of the need to accompany economic integration efforts with parallel structures for environmental and social policy coordination is growing. The challenge for hemispheric trade discussions is to quickly correct course—incorporating environmental consideration throughout the negotiation process, not as an afterthought.

Moving the Trade and Environment Agenda Forward: Models for Openness

The western hemisphere boasts a host of initiatives to promote openness, the lessons from which governments can import into the trade and sustainability context. The Aarhus Convention has become a reference point on issues of openness. It proposes three essential dimensions of openness: access to information (including transparency), access to mechanisms for civil-society participation, and access to justice.[29] Achievement of each of these relies on ongoing processes of democratization at the national level.

Civil society participation is not a novel or revolutionary idea in the Americas. Inter-American government events were among the first to officially include certain sectors of civil society, such as private enterprise, in multilateral conferences (Charnovitz 1997, p. 211). For example, the first Pan-American conference, held in Washington in 1889, saw manufacturers, merchants, and several lawyers on the US delegation (Inman 1965). Some current regional agreements include provisions for the direct participation of the business community, but not civil society. For example, the Andean Group allows companies, through member countries, to request measures to prevent or correct damage to production or exports caused by business practices that restrict free competition within the subregion (UNCTAD 1997, p. 222).

Transparent and participatory negotiation processes will help governments respond effectively to trade and sustainability linkages in the Americas. Public involvement can stimulate higher quality and more diverse exchanges of expertise, data and ideas which in turn can lead to more informed decisions, more effective domestic implementation, and greater legitimacy for subsequent agreements and activities (IISD 1994).

Some of the most innovative models for increased openness can be found in existing efforts to link subregional trade and environmental agendas. Beyond the NAFTA, none of the subregional agreements in the Americas provide explicitly for citizen access to justice, but they do each contain some provisions on public participation and several contain guarantees of access to information.

First, in the Preamble to the NAFTA's environmental side agreement—the North American Agreement on Environmental Cooperation (NAAEC)—the parties acknowledge the importance of civil-society participation to the conservation, protection and improvement of the environment. The Agreement sets out a series of provisions related to public access to information at all levels, obliging all parties to produce periodically, reports about the state of the environment and to ensure their availability to the public (article 2). The NAAEC also promotes public participation in the activities of the North American Commission for Environmental Cooperation (NACEC), mandating its Council to hold public meetings alongside each of its ordinary sessions and consult with non-governmental organizations, as well as independent experts, in decision making processes. Regarding access to justice, the NAAEC promotes opportunities for individuals as well as organizations to access mechanisms to resolve environmental controversies at the national level. The NAAEC also includes provisions for individuals and citizens organizations to challenge countries for environmental non-compliance (articles 14 and 15). (In addition, chapter 11 of the NAFTA's main text contains controversial provisions to facilitate the access of foreign investors to legal remedies in particular situations. See Mann and von Moltke 1999.)

Second, the Mercosur 2001 Framework Agreement on the Environment incorporates provisions to promote openness. The preamble to the agreement includes specific commitments to ensure public access to information and participation of civil society in efforts to address environmental issues. The Framework Agreement also foresees a role for civil-society organizations and other national organizations in the implementation of the accord, particularly through increased information exchange regarding environmental laws, regulations, procedures, policies and practice, including their social, cultural, economic and health aspects, particularly those which might affect trade or competitiveness (article 6a). Mercosur's Economic and Social Council provides mechanisms for direct participation by civil-society organizations and technical experts through, for example, informal consultations held before every meeting of its Working Group focused on the environment (Sub-Grupo no. 6).

Third, CARICOM appears to be the most advanced subregional arrangement in terms of *formal* mechanisms for participation of civil society. Recognizing the need to strengthen existing mechanisms of consultation between government and civil society, and the need for participation by a wide range of actions, CARICOM ratified a Civil Society Charter in 1997.[30] Leaders are currently working to identify new mechanisms and seek a commitment to ongoing collaboration at national and regional levels. In their 1999 Consensus of Chaguaramas, Caribbean leaders proposed the creation of a Caribbean Community Forum to bring together representatives of the region's private sector, NGOs, labor, and government.[31]

Recommendations

To date, FTAA negotiators have viewed environment and social issues only through the prism of their potential disruptive effects on trade relations. Concurrently, despite considerable commitments to openness and inclusiveness at the subregional level, hemispheric trade discussions remain tightly closed. To achieve broad-based support for any new trade agreement, environmental concerns will have to be addressed in a sensitive step-by-step policy discussion that emphasizes Latin American and Caribbean priorities and addresses current fears and concerns. To conclude, we offer three sets of recommendations for a constructive approach to the trade-sustainability-openness interface in the Americas:[32]

Encourage Greater Coordination, Collaboration, and Information Exchange

For the proposed FTAA to foster rather than frustrate sustainability objectives, countries with extremely diverse development trajectories and economic conditions must be satisfied. Last minute environmental negotiations will not be productive. Instead, governments need to prepare themselves substantively for environmental discussions. To begin, governments will have to work to address the lack of communication between trade and environment specialists in intergovernmental organizations and government departments at all levels (national, subregional, hemispheric). To address the resulting problems of poor policy coherence and misunderstanding, governments must invest in mechanisms for inter-ministerial and inter-organizational communication (including joint events and dialogues) and systematically work to address differences in the technical and administrative capacity of governments in the region.

Consider the Long-Term Sustainability Aspects of a Potential FTAA

One basis for synergy between trade and sustainable-development objectives in the FTAA context is a preamble asserting that the long-term goal of the agreement is sustainable development. Taking the lead from Mercosur, the environment must not be perceived as a separate "non-trade" issue, but rather as an integrated element of both social and economic issues. In preparation for inevitable discussion of how to incorporate appropriate environment provisions and measures into the FTAA, governments should explore the models that have already been developed and tested at the subregional level.

To inform the FTAA negotiations, governments should launch coordinated processes to conduct comprehensive, participatory sustainability reviews of the proposed FTAA in each subregion. Such efforts ought to include partners such as the Inter-American Development Bank (IADB), the Economic Commission for Latin America and the Caribbean (ECLAC) and the Organization of American States (OAS) each of which have provided in-depth analysis of the region's trade structures. Civil society organizations should also participate as partners in all aspects of the sustainability reviews.

In particular, the research should examine the specific environmental and related social issues relevant to each of the nine FTAA negotiating groups and help develop options for the sequencing of liberalization efforts to mitigate or lower any negative environmental effects. The reviews could benefit from a range of recent work on environmental assessments undertaken by the United Nations Environment Program (UNEP), the Organization for Economic Cooperation and Development (OECD), various national governments and the North American Commission for Environmental Cooperation, as well as work by NGOs and research organizations (such as the World Wildlife Fund).[33]

Create a Space for Continuous Trade and Sustainability Dialogue

In the FTAA context, efforts to create space for continuous dialogue on trade and sustainability—including mechanisms for greater openness and public participation—have been frustrated by concerns that: a) environmental voices are stalking horses for protectionist interests; and b) the uneven capacity of civil-society organizations across the Americas might lead to unbalanced participation (Murillo 1998). Disparities in regional and subregional representation could hinder the development of effective processes on the hemispheric level. There are also fears that civil-society participation might simply be used to advance distinct environmental agendas of some developed country environmental organizations instead of a broader sustainable development agenda.

Many civil-society groups and marginalized communities in the Americas lack the capacity, analysis and resources to take advantage of existing spaces for dialogue let alone new opportunities for participation that arise. Civil society organizations need to work to build strong Hemispheric networks of organizations with the capacity to articulate their own interests and participate effectively in shaping trade policy.

Several promising initiatives are already underway. A Hemispheric Trade and Sustainability Working Group has been created to link some environmental NGOs in the different subregions, generate joint analysis, and build capacity.[34] Support is also needed to establish civil-society-led mechanisms to build bridges between the trade and sustainable-development communities. Civil-society-led "Peoples Summits" that parallel the FTAA Trade Ministerial meetings are one option. Another possibility is the creation of a non-advocacy center to undertake capacity building, increase information analysis and flow, provide technical support on sustainable development issues for the FTAA and to facilitate comprehensive policy dialogues among the different interests, subregional perspectives and sectors.[35] In 2000, for example, the "Grupo Zapallar" was created to facilitate information exchange, participation and cooperation among different regional actors on trade and sustainable development issues.[36]

At the official level, spaces for dialogue between civil society and government need to be created. Many civil-society groups feel that the existing FTAA Civil Society Committee initiative must be strengthened or transformed (Caldwell 1999). Their recommendations include calls for more proactive public dialogue or consultation, comprehensive information disclosure and communication policies, funding mechanisms for underrepresented groups, and measurable operational procedures to indicate how civil-society concerns will be addressed in negotiations (Caldwell 1999; Lucas 2000). They have also recommended that: environmental and social concerns must be a specific agenda item for the Committee; that verifiable links be established to the work of the FTAA Negotiating Groups; and all reports must be made publicly available. Another potential model that governments and civil society could look to in this regard is the mechanisms created by the OAS to facilitate civil-society input into the Summit of Americas process more broadly.[37]

A final possibility for stimulating more informed dialogue on the intersection of trade and sustainability issues is the formation of a High Level Hemispheric Trade and Sustainability Forum comprising senior officials from governments, regional and hemispheric institutions, NGOs, academic institutions and the private sector. Building on the groundwork laid by the existing Forum of Environment Ministers of Latin America and the Caribbean, the Trade and Sustainability Forum would work to build consensus on a focused and integrated trade and sustainability agenda that could be integrated into both trade policies and environmental policies.

Notes

1. The full text of each of the Summit of the Americas declarations and action plans can be found at http://www.ftaa-alca.org.

2. A complete text of the declaration can be found at http://www.oas.org.

3. These included the Red Inter-Americana de Recursos Hidricos (RIRH), Iniciativa Energia Renovable en las Americas (EREA), and the Inter-American Biodiversity Information Network (IABIN). For more information about these initiatives and the full text of the Santa Cruz Declaration, see the Summit of the Americas sections on the web site of the Organization of American States (http://www.oas.org).

4. This forum is organized under the auspices of the United Nations Environment Program's Regional Office for Latin America and the Caribbean, with Canadian and American observers.

5. For information on the outcomes of that meeting, see http://www.ec.gc.ca/press/2001/010330_f_e.htm.

6. The focus of this chapter is on the environmental aspects of sustainable development. Complementary information about social, cultural, political, and other dimensions of subregional agreements would be extremely useful.

7. Segger et al. (2000) divides current international environmental regimes in the Americas into species-specific accords which protect particular migratory or transboundary species and populations; natural resources accords which work to ensure that the productive capacity of certain natural resources is respected, restored or managed; ecosystem areas accords that encourage conservation of a particular ecosystem, habitat, or heritage area; and comprehensive ecological cooperation accords formed as part of a broader integration package. Many of these agreements include provisions for the use of "TMEPs" (trade measures for environmental purposes).

8. See the link to statistics at http://www.mercosur.org.

9. In January 1995, Mercosur members agreed on a list of more than 8,700 products to be exempted from import duties. See the Mercosur web site for further information.

10. The Consejo Mercado Commún is the highest decision-making body of Mercosur, comprising ministers of foreign relations and economic ministers. The Grupo Mercado Commún is the executive body of Mercosur. Coordinated by the ministers of foreign relations of each of the member countries, it oversees the implementation of the treaty by elaborating work plans and concrete activities.

11. The Framework Agreement replaces an earlier, less comprehensive effort to develop an environmental protocol for Mercosur (Protocol Adicional Al Tratado De Asunción Sobre Medio Ambiente, draft, Montevideo, Uruguay).

12. In February 2001, leading NGOs engaged in trade and environment discussions in the Mercosur context issued a statement welcoming the new Framework Agreement and setting out a series of recommendations.

13. The origins of the Caribbean Community lie in the West Indies Federation. That federation, which came to an end in 1962, was followed by a series of efforts to strengthen cooperation in the region. A 1963 Heads of Government Conference proved to be the first in a series of conferences among the leaders of Commonwealth Caribbean Countries. These culminated in 1968 in the creation of the Caribbean Free Trade Association (CARIFTA) and the Commonwealth Caribbean Regional Secretariat. (The Caribbean Development Bank was created in 1969.) In October 1972, the Caribbean Leaders decided to transform CARIFTA into a common market and to establish the Caribbean Community, of which the common market would be an integral part.

14. The British Virgin Islands and the Turks and Caicos Islands are associate members. Haiti and the Dominican Republic are negotiating to join.

15. The Caribbean Community has three objectives: economic cooperation through the Caribbean Single Market and Economy, coordination of foreign policy, and common services and cooperation in functional matters such as health, education and culture, communications, and industrial relations. The main organs of the community are the Conference of Heads of Government and the Community Council of Ministers (commonly called "The Council"). Major issues currently on the regional agenda include analysis of the impact of the NAFTA on existing arrangements such as the Caribbean-Canada Trade Agreement (CARIBCAN) and the Caribbean Basin Initiative; resolution of the Haitian crisis; and strengthening of relations with the wider Caribbean through the establishment of trade and economic agreements with Venezuela, Columbia, and the wider Caribbean. For more information, see http://www.caricom.org.

16. Governments completed negotiations of the CSME treaty in 2000. Nine protocols have been opened for signature and ratification. Whereas the common market's focus was inward and on goods, the CSME's focus is both inward and outward and extends beyond goods to services, labor, capital and access to land.

17. UNEP's Regional Seas Program for the control of marine pollution and the protection and management of the marine and coastal environment was launched in 1974. In 1976, UNEP launched a Regional Program for the Wider Caribbean Region and the Caribbean Environment Program. A draft Action Plan for the Caribbean Environment Program was prepared in 1979. For further information, see http://www.cep.unep.org.

18. The work of the Caribbean Environment Program is discussed at intergovernmental meetings every 2 years. A nine-nation Monitoring Committee is responsible for supervising the development of the program and providing policy direction. The secretariat provides assistance to all countries to strengthen national and subregional institutions, coordinate international assistance, and stimulate technical cooperation among countries.

19. The biodiversity strategy incorporates a joint Andean Declaration on phytosanitary measures (which includes provisions on biosafety). See http://www.comunidadandina.org and ICTSD 1999.

20. The council was instructed to work with the World Commission on Protected Areas with financial assistance from the Regional Fund for Environment and Development. The concept was endorsed by the World Bank, which subsequently supported some 75 major projects and studies on the proposed Meso-American Biological Corridor.

21. For more information about SICA, see http://www.sicanet.org.

22. Discussions are underway, for example, between Chile and CARICOM, Chile and CACM, between CACM and Mercosur, between the Andean Community and Mercosur, between Venezuela and Mercosur, between Mexico and Mercosur, between Mexico and the "northern triangle" countries in Central America, between Mexico and Nicaragua, between Mexico and CACM as a whole, between Mexico and Peru, and between Mexico and Ecuador.

23. The full text of the Canada-Chile Free Trade Agreement can be found at http://www.dfait-maeci-gc.ca/tna-nac/bilateral-e.asp. Information about the environmental side agreement can be found at http://can-chil.gc.ca/English/Resource/Agreements/AECCC/Default.cfm.

24. The Integrated Border Environmental Plan had six workgroups (Air, Water, Hazardous and Solid Wastes, Pollution Prevention, Contingency Planning and Emergency Response, and Enforcement) to deal with environmental issues.

25. For further information on this program, see http://www.epa.gov/usmexicoborder/et.htm.

26. For more information about the bilateral agreement and the BECC and NADBank, see http://www.becc.org and http://www.nadbank.org.

27. For the full texts of these agreements, see http://www.sice.oas.org/tradee.asp#cancr_CR.

28. The recent side agreement between Canada and Costa Rica also warrants exploration.

29. See Economic Commission for Europe Committee on Environmental Policy Fourth Ministerial Conference "Environment For Europe," Aarhus, Denmark, 1998 Convention On Access To Information, Public Participation in Decision-Making And Access to Justice in Environmental Matters.

30. For further information, see http://www.caricom.org/chartercivilsoc.html.

31. For more information, see http://www.caricom.org.

32. The following suggestions are based on the Trade Rules and Sustainability in the Americas project of the International Institute for Sustainable Development, the International Centre for Trade and Sustainable Development, and the United Nations Environment Program. It sought to apply the Winnipeg Principles for Trade and Sustainable Development to the process of Western hemispheric integration, and specifically to the FTAA process, by comparing examples from the existing subregional trade regimes.

33. See e.g. UNEP 2001; WWF and CIPMA 2001; OECD 2000; CEC 1996, 1999.

34. This initiative brings together ECOS in Uruguay, the Mexican Environmental Law Center, the International Institute for Sustainable Development in Canada, the Centro Internacional de Política Económica para el Desarrollo Sostenible and the Centro de Investigación y Planificación del Medio Ambiente, Fundación Futuro Latinoamericana in Ecuador, and the Global Environment and Trade Study and the National Wildlife Federation in the US.

35. One strong model for such an effort is the International Centre for Trade and Sustainable Development in Geneva for WTO issues. See http://www.ictsd.org.

36. The Grupo Zapallar is a Latin American network on trade and sustainability created in 2000 by the Fundación Futuro Latinoamericana. For further information, see http://www.ffla.org. NGOs and policy centers are also working to foster collaborative research on FTAA and the environment. A joint project of the Yale Center for Environmental Law and Policy, the Centro Internacional de Política Económica para el Desarrollo Sostenible, and the Centro de Investigación y Planificación del Medio Ambiente is one example. A 1999 symposium hosted by the Miami-based North-South Centre and other partners also developed a series of recommendations for environmentally sound trade expansion in the Americas (Rosenberg 2000).

37. For detailed information about the OAS strategy for soliciting input from civil society, see http://www.oas.org.

References

Andean Community. 2000. *Indicadores de la Communidad Andina, Mercosur y Chile*. http://www.communidadandina.org.

CACM. 2000. "Indicadores." Central American Common Market. Available at http://www.cacm.org.

Caldwell, J. 1999. NWF's Communication to the Committee of Government Representatives for the Participation of Civil Society's (CGR). Washington: National Wildlife Federation.

CARICOM. 2000a. Caribbean Trade and Investment Report 2000. CARICOM.

CARICOM. 2000b. Background Document prepared for the Second Special Consultation on the CARICOM Single Market and Economy, Barbados, 20–21 November. CARICOM.

CEC. 1996. *A Survey of Recent Attempts to Model the Environmental Effects of Trade: An Overview and Selected Sources*, Montreal: Commission for Environmental Cooperation.

CEC. 1999. *Assessing the Environmental Effects of the NAFTA: An Analytical Framework and Issue Studies*, Montreal: Commission for Environmental Cooperation.

Charnovitz, S. 1997. "Two Centuries of Participation: NGOs and International Conferences." *Michigan Journal of International Law* 18, no. 2: 183–286.

Esty, D. 1994. *Greening the GATT: Trade, Environment and the Future.* Washington: Institute for International Economics.

IADB. 1996. Annual Report on the Environment and Natural Resources. Inter-American Development Bank.

ICTSD. 1999. "Comercio y medio ambiente en los acuerdos regionales." *Puentes Entre el Comercio y el Desarollo Sostenible* 2, no. 1: 1–15.

Iglesias, E. 1997. "El nuevo rostro de la integracion regional en America Latina y el Caribe." In *Acuerdos de comercio e integracion en las Americas: Un compendio analitico.* Washington: Inter-American Development Bank and Organization of American States.

IISD. 1994. *Principles for Trade and Sustainable Development.* Winnipeg: International Institute for Sustainable Development.

Inman, S. 1965. *Inter-American Conferences 1826–1954.*

Inter-American Dialogue. 1997. *Integrating the Hemisphere: Perspectives from Latin America and the Caribbean,* ed. A. Jatar and S. Weintraub.

Lucas, N. 2000. "Some Issues for Consideration on Participation." Paper presented at Washington conference on "The FTAA and the Environment: What Can We Learn from the NAFTA Model?"

Mann, H., and K. von Moltke. 1999. *NAFTA's Chapter 11 and the Environment: Addressing the Impacts of the Investor-State Process on the Environment.* Winnipeg: International Institute for Sustainable Development.

Mercosur. 2001a. Acuerdo Marco sobre Medio Ambiente del Mercosur. Approved Text from XX Reunión del Consejo Mercado Común, Mercosur/CMC/DEC. no. 2/01.

Mercosur. 2001b. Resolución no. 38/95 of the Grupo Mercado Común and Recomendación no. 01/01 of Sub-Grupo no. 6 "Environment." Available at http://www.mercosur.org.

Morton, C. 1998. *Progress Toward Free Trade in the Western Hemisphere since 1994.* La Jolla: Institute of the Americas.

Murillo, C. 1998. *Trade and Environment: Opening the Dialogue.* San José: National Audubon Society and Intel.

OAS. 1999. Regular Meetings of the Inter-American Committee on Sustainable Development. Ser. W/XIII.2.2, CIDI /CIDS/ doc.9/99.

OECD. 2000. *Assessing the Environmental Effects of Trade Liberalisation Agreements.* Paris: OECD

Robins, N., and Roberts, S. 1997. *Unlocking Trade Opportunities: Case Studies of Export Success from Developing Countries.* London: International Institute for Environment and Development.

Rodriguez-Mendoza, M. 1997. "The Andean Group's Integration Strategy." In *Integrating the Hemisphere-Perspectives from Latin America and the Caribbean*. Bogota: Inter-American Dialogue.

Rosenberg, R., ed. 2000. *Environmentally Sound Trade Expansion in the Americas: A Hemispheric Dialogue*. North-South Center Press.

Scott, J. 1916. *Second Pan American Scientific Congress: The Final Act and Interpretive Commentary 7*.

Segger, M., M. Munos, P. Meirles, J. Taurel, and V. Paul. 2000. *Trade Rules and Sustainability in the Americas*. Winnipeg: International Institute for Sustainable Development.

Segger, M. et al.. 2001. *Ecological Rules and Sustainability in the Americas*. Winnipeg: International Institute for Sustainable Development.

UNCTAD. 1997. "Decision 285 of the Commission of the Cartagena Agreement 1991." In *World Investment Report 1997: Transnational Corporations, Market Structure and Competition Policy*. Geneva: UNCTAD.

UNEP. 2000. Notes from 08 October 2000, 5th Inter-sessional meeting of the Latin America and Caribbean Environment Ministers Forum. Doc. UNEP/LAC-IC.5/8, Anexo I, Parte 1: Proceso de preparacion regional para Rio +5, from the Comite Tecnico Interagencial del Foro de Ministros de Medio Ambiente de America Latina y el Caribe.

UNEP. 2001. *Reference Manual for the Integrated Assessment of Trade-related Policies*. Geneva: United Nations Environment Program, Division of Technology, Industry and Economics (Economics and Trade Unit).

WWF (World Wildlife Fund) and CIPMA (Centro de Investigación del Medio Ambiente). 2001. *Sustainability Assessments in the Americas*. Gland: WWF.

Zadek, S., and P. Tiffen. 1997. *Fair Trade: Paths to Sustainability*. London: IIED, New Economics Foundation, and Twin Trading.

18

Trade and Environment: Reflections on the NAFTA and Recommendations for the Americas

Carolyn L. Deere and Daniel C. Esty

Negotiations for a new trade agreement for the Americas offer governments a unique opportunity to adopt an approach to economic integration that advances development and environmental protection. A hemispheric trade agreement that expressly addresses the link between trade and the environment would be more economically sound and politically durable than one that does not. It would also be more consistent with the vision of hemispheric peace and prosperity articulated at the 1994 Summit of the Americas in Miami, where the 34 participating heads of state called for progress on three policy dimensions: social, economic, and environmental.[1]

At present, the FTAA negotiating structure does not provide any focus on environmental issues.[2] The United States, Canada, and (more recently) Chile stand alone in their desire to consider such issues in the negotiations.[3] But the past hostility to an environmental focus may be changing. In Mexico, long the country most opposed to any trade-environment linkage, the appointment of Victor Lichtinger—former Executive Director of the North American Commission for Environmental Cooperation (NACEC)— as Environmental Minister bodes well.[4] In any case, Mexico's opposition to incorporating environmental considerations into the architecture of trade agreements makes no sense. While there are real issues to address in ensuring that protectionists do not misuse environmental provisions, avoiding trade and environment abuse is easier where these issues are taken head on. Trade ministers in Latin America and the Caribbean would therefore do better to follow the leadership of Chile in adopting a forward-looking and constructive approach to the trade and environment agenda.

The argument that trade objectives can best be advanced without a link to issues of pollution control and natural-resource management has no analytic basis and does not withstand scrutiny.[5] To deny the existence of a trade and environment relationship is not just descriptively off base, it is normatively wrong. Uninternalized externalities, such as those that arise from trans-boundary pollution spillovers or overexploited shared natural resources, threaten market failures that diminish the efficiency of international economic exchanges, reduce the potential gains from trade, lower social welfare, and generate unnecessary environmental degradation (Esty 2001).

Denial of any trade-environment connection also threatens to impede progress toward a viable regional trade agreement. In the United States, forward momentum in the trade realm is difficult to sustain (Bergsten 1992; Aaronson 2001; Esty 1998a). Advocates of hemispheric economic integration need the support of the pro-trade and pro-environment "swing vote" senators and representatives (Audley 2001; Mayer 2002). In view of the importance of environmental considerations in consolidating US political support for trade agreements, recalcitrance on the environment front lengthens the odds of an FTAA's success (Destler and Balint 1999). The idea that one could win ratification of a regional trade agreement without an environmental element is misguided, never mind that such a narrowly focused approach to hemispheric integration would almost certainly accomplish less than a broader agreement that addresses the inescapable set of environmental issues that arise. Taking environmental issues seriously in the FTAA context is thus both a practical reality and a political necessity.

Understanding the opposition to efforts to connect trade policy and environmental policy is a fundamental objective of this book. We have focused on the NAFTA because that agreement is the key to understanding Mexico's hostility to any "trade and environment" agenda and because it provides a significant base of experience on which to draw in structuring a hemisphere-wide trade area.

A careful review of the NAFTA record reveals no evidence that the NAFTA-inspired environmental provisions that Mexico, Canada, and the United States put in place have had any negative effect. On the contrary, the NAFTA experience suggests that governments can address environmental concerns in a carefully tailored, thoughtful, and systematic manner both within and alongside a trade agreement. Despite Mexico's fears, the NAFTA's environmental provisions have not harmed trade or investment flows to Mexico. Rather, they have offered Mexico important economic and environmental benefits (Miller 2002; Torres 2002).

The NAFTA's environmental approach therefore offers a useful starting point for western hemisphere integration. Nevertheless, the NAFTA model is not perfect, nor are its institutions. Indeed, the recognized shortcomings—insufficient funding for environmental cooperation, provisions for trade sanctions, certain elements of the controversial chapter 11 investment regime, weak links between the NAFTA's trade bodies and its environmental organizations—must not be repeated. But the environmental provisions of the agreement and the North American Agreement on Environmental Cooperation (NAAEC, or the "environmental side agreement") offer an important foundation for efforts to advance a regional trade agreement for the Americas.

Lessons from the NAFTA

The preceding chapters have identified a great many lessons that can be drawn from the NAFTA. In the pages that follow, we highlight those that are most relevant to the prospects for successful trade policies and environmental protection in the FTAA context.

Lessons from the Negotiations of NAFTA's Environmental Provisions

The NAFTA experience makes clear that negotiating dynamics matter. While Mexican officials continue to complain about the effects of the environmental provisions of the NAFTA (Mancera 2002), several contributors to this book have shown that the Mexican resistance to linking trade with environmental issues has no substantive or empirical foundation (Araya 2002; Miller 2002; Torres 2002). Rather, Mexican opposition derives from a range of historic, procedural, and emotional factors. It reflects, in part, lingering unhappiness about the tuna-dolphin dispute and fear of protectionist environmental measures that could undermine Mexican exports. Unhappiness with the NAFTA negotiation process—especially a feeling of having been abused by the Clinton administration's last-minute additional demands, some of which (e.g., the provisions for trade sanctions for environmental shortcomings) were rather outrageous—also fuels the Mexican stance.[6] Mexico's environmental stance may stem as well from a desire to obtain negotiating leverage with the United States in the FTAA context.

The fundamental lesson here is that *negotiation* of the NAFTA frustrated and insulted Mexican officials—many of whom went on to hold influential positions in the Zedillo administration—and bolstered Mexico's resolve to deflect legal and policy measures to address trade-environment links (Araya 2002). Mexico's distress over the NAFTA process has a valid foundation. Bullying by the United States during the final stages of the NAFTA negotiations and the manner in which US domestic political considerations encroached on the international negotiating arena (especially the eleventh-hour efforts of the Clinton administration to appease US environmental groups) were objectionable. And Mickey Kantor's high-handed negotiating style is cited by many as a special source of antagonism. The depth of the resentment can be seen in the fact that the Mexican government's attitude persists despite its recognition of the demonstrable economic and environmental benefits Mexico has obtained from the NAFTA (Mancera 2002).

In sum, trust and good will are important ingredients in successful negotiations, especially with regard to contentious issues. The perceived fairness of negotiating processes can affect the long-term willingness of countries to cooperate on the environment and on other issues. In the coming years, the United States and Canada—the main proponents of trade-environment linkage in the FTAA—will have to find ways to raise issues and advance their own political agendas without provoking the antipathy that emerged in the NAFTA negotiations. Forcing countries to the environmental negotiating table hurts the prospects for long-term cooperation and undermines the independent validity of the environmental aspects of a trade agreement.

Participation and Transparency in Trade Policy Making

A second lesson from the NAFTA experience is that broad public participation in trade policy making generates a more robust decision-making process and ultimately produces a better agreement. Participation of non-governmental organizations and a full spectrum of civil-society groups broadens the debate and helps build political understanding and support for the agreements that emerge (Esty 1998b). In the United States, the relatively inclusive approach taken by the US Trade Representative during NAFTA negotiations helped to focus attention on a range of pollution-control and natural-resource-management issues, which broadened the political coalition backing the NAFTA and made congressional approval possible (Audley 1997; Esty 1994b; Mayer 2002).

Perhaps these lessons have been overshadowed by the World Trade Organization's bitter episode in Seattle in 1999. Indeed, the sight of some environmental advocates walking arm in arm with avowed protectionists could have been taken as confirmation that environmentalists are simply protectionists. But such a sweeping generalization would be unfounded. Moreover, the tactics of some environmental groups should not be used to discredit the underlying logic of many environmental groups and proposals. Nevertheless, many FTAA negotiators remain skeptical about the need for greater transparency or about the benefits of engaging NGOs in the trade policy process. Opening up the FTAA negotiations would, however, generate a range of benefits:

• richer and deeper public discourse, ensuring that critical issues are addressed more thoughtfully and systematically
• increased public understanding and appreciation of the potential benefits and costs of new trade agreements and of the tradeoffs that must be made to achieve these benefits
• vigorous intellectual "competition" in policy making, with the interplay among different visions strengthening the ultimate agreement
• more data-driven analysis, ultimately promoting more careful and effective policy proposals.

In sum, the NAFTA experience suggests that inclusiveness and transparency enhance the credibility and legitimacy of the negotiations and the trade agreement that ultimately emerges (Esty 2002). Perceived fairness depends on "due process" and on a sense that all voices were heard. An open process is especially important where some interests will lose, or perceive themselves to have lost, from the agreement. In the end, while public scrutiny may make government officials uncomfortable, openness and procedural inclusiveness are essential to good public decision making.

Environmental Impacts and Reviews of Trade Agreements

Consistent with economic theory, this volume demonstrates that trade liberalization can have positive and negative effects on the environment. Environmental impact assessments or reviews of trade agreements thus represent an important tool for identifying trade-environment links (Audley 2001).[7] Whether the issue at hand is pollution levels, risks to human health, or the sustainable management and use of natural resources, an environmental review can help a government to identify the potential issues and opportunities for environmental gains. Such reviews, if done well, throw a spotlight on the potential environmental implications of freer trade, identify and analyze options for mitigating these effects, and avert economic inefficiency and harms to trade flows by anticipating and preventing potential trade-environment conflicts (de Almeida 2001; Jha et al. 1999).

As this volume makes clear, some of the dirtiest industries are actually cleaner in Mexico than in the United States because the process of liberalizing trade and investment created incentives for the construction in Mexico of new plant and equipment, which is almost always less polluting (Gallagher 2002). Steel stands out as an example in this regard. In other Mexican industries, pollution levels may be higher than in the United States or in Canada, and environmental performance may worsen under trade-induced competitive pressures (Esty 1996). This risk looms largest in sectors that rely on end-of-the-pipe pollution-control approaches and which face cost pressures that can lead companies to try environmental corner cutting. The Mexican paper industry appears to be one industry in which the "race toward the bottom" dynamic can be found (Gallagher 2002).

Steady progress is being achieved in the methodologies for trade-related environmental reviews.[8] More sophisticated future environmental reviews may be able to highlight sectors and regions of special sensitivity that deserve extra policy attention in the context of trade liberalization. The rapid transformation of Mexico's agricultural sector, with growth in some segments (e.g., tomatoes) and rapid decline in others (e.g., corn), has created environmental and social stresses that have not been managed effectively (Nadal 2002). Similarly, in the FTAA context, proper forecasting might enable participating countries to better protect natural resources from unsustainable rates of exploitation (driven in part by export incentives) and to minimize and manage environmental challenges that arise within and along their borders.

Some Environmental Provisions Must Be Incorporated into the Trade Agreement

There are several reasons to fold environmental provisions into the text of a trade agreement. First, some links between trade and environmental policy are inescapable. The failure to deal with issues of this type explicitly just invites trouble down the road as disputes break out over how to balance competing trade and environmental policies. Second, placing language about sustainable development in the trade agreement itself helps to demonstrate that the underlying economic integration strategy is broadly conceived. Such provisions make it clear that drafters recognize that other values, such as protection of the environment, must be respected in the pursuit of trade goals. Finally, environmental provisions help to clarify expectations, increase predictability, and thus smooth the implementation of the agreement.

The NAFTA demonstrates that trade goals and environmental objectives can be pursued in tandem. Indeed, the environmental provisions incorporated directly into the NAFTA text have generated little controversy.[9] The emphasis on sustainable development in the NAFTA's preamble and a number of environmental elements throughout the agreement itself represent a real breakthrough.

The NAFTA goes a considerable distance to reassure environmental advocates and the public at large that trade goals will not mindlessly trump important environmental values or policies. It affirms the provisions of multilateral environmental agreements—such as the Montreal Protocol on Substances That Deplete the Ozone Layer, the Convention on International Trade in Endangered Species, and the Basel Convention on the Control of Transboundary Movements of Hazardous Wastes—will not be overridden by trade commitments. The commitment of NAFTA parties to strive to achieve full enforcement of their respective environmental laws also serves as a major step toward reducing trade-environment tension.

As this volume demonstrates, whether the NAFTA goes far enough in addressing systemic environmental concerns associated with trade remains a matter of some controversy and warrants further exploration. One thing is clear: The explicit environmental provisions in the NAFTA's text have not caused economic harm to Mexico. Indeed, environmental performance in Mexico, at least in some areas, seems to have improved as a result of NAFTA-related environmental initiatives (Hufbauer et al. 2000).

Ironically, other (non-environmental) provisions have generated serious controversy, including an environmental backlash. Notably, the unanticipated negative environmental consequences of the NAFTA's investment chapter have drawn significant attention (Mann and von Moltke 1999; CIEL et al. 1999).

Several other textual issues should be addressed in future trade agreements.

First, imprecise and vague terminology has produced unnecessary controversy. Words such as "legitimate" and "necessary" present a considerable hurdle to environmental policies and standards under traditional trade-law jurisprudence. What constitutes a legitimate trade measure? What constitutes a necessary regulation? Who should decide? Both of these expressions have peculiar "term of art" interpretations in the trade domain. In both instances, environmentalists argue that the NAFTA officials, trade negotiators, and dispute panels should be more deferential to other non-trade experts and policy makers in their interpretations.[10] In future agreements, governments can avoid generating anxiety about the environmental impacts of trade and investment rules by clarifying terminology. The emerging World Trade Organization jurisprudence provides useful guidance on some of these points (Wofford 2000).

Second, the NAFTA did not adequately address the important issue of standards relating to production and processing methods (PPMs). In an ecologically interdependent world, *how* things are produced is often more important than *what* is produced. In determining whether PPMs are legitimate, analysis should focus on the scope of the potential harm and whether the provisions in question are derived through a scientific process, are proportionate to the environmental harm in question, and are applied in a non-discriminatory manner to both domestic and imported products. Measures that address trans-boundary effects or which derive from international environmental agreements are most likely to meet these conditions. Consistent with this approach to PPMs, the NAFTA and future trade agreements should explicitly affirm the compatibility with trade rules of eco-labels (including eco-labels awarded based on PPM criteria) so long as these schemes apply equally to both domestic and imported products (Deere 1999). The underlying challenge here is to find approaches that accommodate developing countries' legitimate concerns about de facto discrimination against their products.

Third, governments should carefully consider the environmental opportunities and challenges presented by the incorporation of investment provisions into trade agreements. To avoid negative effects on environmental law and policy processes, the investment provisions in future trade agreements will have to be constructed carefully. Governments should include specific provisions to ensure environmental objectives are advanced, not undermined. Base-line environmental requirements (perhaps incorporating already established environmental standards, such as World Bank criteria for project loans) should be adopted for all projects supported by foreign investment. Negotiators should not overlook the risk of causing a "regulatory chill" in domestic regulatory processes because of poorly drafted and overly broad provisions protecting foreign investors.

Lessons from NAFTA's Environmental Side Agreement and Institutions
There is now widespread agreement that the NAFTA environmental side agreement provision permitting trade sanctions (the withdrawal of tariff benefits) to be imposed on any NAFTA party that derogates from its own environmental standards was a mistake. Canada was right to reject this provision. The United States was wrong to have forced Mexico to accept it. Although it is enforceable only after an arbitration panel has concluded— through an elaborate and convoluted process that has never been applied— that a party has persistently failed to enforce its environmental laws effectively, the possibility of "snap-back" tariffs makes no sense. This single provision has engendered high levels of ill will and suspicion, especially among Mexican trade officials. It has done nothing to promote environmental progress.

This sop to environmental advocates in the United States was advanced by the Clinton administration without sufficient thought or analysis.[11] Indeed, many of the one-time supporters of this provision have now come to view it as a hindrance to environmental cooperation.[12] In retrospect, where an issue of environmental derogation arises within the context of a trade agreement, a process of consultation and cooperative efforts to improve performance (as originally contemplated in the pre-Clinton version of the NAFTA) seems a better mechanism for ensuring environmental compliance.

The NAFTA also shows the importance of institutional reinforcement of environmental commitments. While flawed in serious ways, the NACEC, BECC, and NADBank have contributed—through financial and technical assistance—to improvements in Mexico's environmental performance.[13] With proper funding and better political support, these institutions would have been able to do even more to ensure that environmental progress and the construction of pollution-control infrastructure kept pace with the changes brought about by economic integration. Even without adequate backing, these bodies have had some important successes—for example, NACEC's watershed management in the San Pedro River area, which helped leverage outside funding and establish partnerships to address a pollution problem. More creative funding for environmental infrastructure should be considered in the FTAA context.

The presence of NACEC has also helped to invigorate collaborative environmental efforts among the three NAFTA countries. NACEC-led initiatives to increase information flows, spur the transfer of environmentally efficient technologies, expand the scope of environmental cooperation and financing, add to the stock of environmental data and analysis, stimulate improvements in the way that Mexican business approaches environmental practice and compliance, and implement mechanisms to promote civil society's participation in environmental decision making have begun to pay dividends.

The presence of a tri-national environmental institution has proved essential in another respect. Specifically, the NACEC provisions for citizen "submissions" (complaints about government malfeasance or nonfeasance) and independent reports (under articles 14 and 15 of the NAAEC) have been useful mechanisms for keeping the NAFTA governments on their environmental toes. Several of the complaints have brought serious problems to light. Some cases have also stimulated unprecedented cooperation by government, industry, experts, and NGOs to resolve disputes and address potential crises.

The NAFTA's environmental institutions are, however, severely constrained by inadequate funding, narrow mandates, and weak political support. Isolation from the NAFTA Free Trade Commission remains a major problem for the NACEC. The two bodies have not worked together on any matter of substance in the seven years since the NACEC's inception. Ironically, eleven NAFTA working parties and committees refer to environmental issues in their work plans. How can it be that the NACEC has never been invited to meet with any of them?

Finally, the NACEC suffers from a narrow mandate. It lacks authority to address many environmental issue that arises in the NAFTA context, such as the environmental impacts of subsidies. Moreover, it might have been given a formal role in advising the NAFTA's Free Trade Commission on the chapter 11 controversy. The NACEC might also have been used to provide both impartial advice and a forum for governments, industry, scientific experts, environmental experts, and communities to use to settle disputes.

The Border Environmental Cooperation Commission (BECC) and the North American Development Bank (NADBank) have provided important technical advice to communities and local authorities (Abel 2000; Rodríguez 2000), but their success has been constrained by limited funding and political disinterest. Needed institutional and procedural changes in how the NADBank and BECC conduct their business and how they approach the trans-national nature of border environmental infrastructure challenges before them have been slow in coming (Abel 2000). Without these changes, the ability of both agencies to provide the kind of support that border communities actually need to respond to environmental challenges is severely restricted (Carlsen and Salazar 2002; Hufbauer et al. 2000).[14]

The experience of the NADBank and BECC also demonstrate the value of—and real limits to—community-based efforts to address environmental challenges.[15] Communities need to be engaged in the process of setting priorities, but their capacity limitations must be acknowledged. Neither the NADBank, the BECC, nor other philanthropic and civil-society initiatives will ever have the capacity to address the full range of infrastructure needs on the border. National governments and local authorities on both sides need to undertake concrete action to finance and implement infrastructure projects for the benefit of the border communities.

The NAFTA's environmental institutions were not designed to rectify all the outstanding environmental issues in the NAFTA region. However, to achieve even their limited objectives and mandates they need more consistent political support, better integration with the structure of trade policy, better financing, and clearer authority to address environmental problems. The experience under the NAFTA environmental side agreement suggests that the creation of a "parallel" track to promote environmental cooperation can generate numerous benefits without economic harm. Environmental issues cannot, however, be isolated from the day-to-day business of trade agreements. Appropriate institutional reinforcement of environmental efforts facilitates economic integration.

Recommendations for Greening Trade in the Americas

Finding ways to manage trade-environment links is not a matter of choice but a requirement of economic and political reality. The question for trade policy makers is not whether to respond to environmental challenges but how to do so. In the FTAA context, the question is: Will governments address the trade-environment relationship in an open and systematic manner, or on an untransparent and ad hoc basis? What follows is a set of recommendations for a serious but carefully circumscribed strategy of folding environmental considerations into regional trade arrangements. The approach advanced here focuses on the FTAA but would also be applicable, with minor modifications, to other trade initiatives.

As a starting point, we identify the set of issues that governments must address within the trade agreement. Other issues are best taken up on a parallel track designed to promote environmental cooperation among the parties. NAFTA provides one source of lessons for the Americas, much can also be learned from other emerging subregional agreements (Segger and Borregaard 2002). Because the NAFTA involved three countries with shared borders, some of the specific provisions that were relevant and even necessary in North America will be less essential or desirable in the hemispheric context. But the essential features of the NAFTA's success—an express ex ante approach to environmental issues, a commitment to dealing with core trade-environment links in the trade treaty, a focus on institutionalizing environmental cooperation, creation of mechanisms to promote broad public dialogue, funding for environmental infrastructure investments—should be replicated.

The recommendations that follow fall into three broad categories:

• a *process* that promotes transparency, public participation, and consideration of environmental issues—and in doing so enhances the legitimacy of and builds support for the final trade agreement
• a tightly defined, *substantive* trade and environment agenda to be addressed within the regional trade agreement
• a "parallel track" of *environmental activities* designed to promote hemispheric cooperation on pollution control and management of shared natural resources but with clear points of intersection with the trade negotiations, the ultimate trade agreements, and their implementation of the economic-integration strategy.

Process-Oriented Commitments
The governments involved in the FTAA negotiations should make a number of procedural commitments to ensure that environmental issues are fully considered in the course of hemispheric economic integration. An emphasis on open dialogue and thorough debate should cover not only the FTAA but also the parallel negotiations for managing environmental cooperation

and the implementation of both the ultimate trade treaty and any environmental agreements. In particular, FTAA governments should

• commit to addressing trade and environmental issues expressly

• abolish the FTAA's existing but entirely ineffective Committee on Civil Society and, in its place, create a Regional Civil Society Forum engaging local, national, and regional NGOs, civil society groups, and representatives of business in an ongoing process of evaluating trade and environment issues and devising policy proposals and alternatives. The forum would host periodic consultations and dialogues with government officials on substantive topic areas and an annual Forum.

• create a transparent process to fold environmental sensitivity into the FTAA analysis and negotiations

• maintain regular contact with civil-society organizations and the business community at the national level. Trade ministries should also engage in consultations with non-trade ministries at the national level in the development of trade policy.

• conduct regular briefings on issues under negotiation to inform the public and to elicit analyses and options on how best to address these issues

• establish a high-level Trade and Environment Advisory Group that would meet regularly with representatives of the nine FTAA negotiating groups and a proposed Environmental Negotiating Group. The Advisory Group would consist of a rotating group of around 30 advisors selected from the countries of the hemisphere.

• conduct environmental reviews of the potential effects from a movement toward more open markets across the hemisphere

• establish an Environment Negotiating Group composed of qualified government officials that would ensure that pollution and resource management issues are systematically addressed in the FTAA negotiation process. Drawing from environmental reviews, the advice of a Trade and Environment Advisory Group, and input from civil-society organizations, and supported by the proposed Environment Negotiating Group, each of the FTAA's nine existing negotiating groups should take up environmental converns. For example: The Market Access Negotiating Group should negotiate toward zero tariffs on environmental goods and services to promote trade in pollution-control and clean-energy technologies and would pursue the elimination of trade-distorting and environmentally damaging subsidies in natural-resource sectors such as energy, water, agriculture, fisheries, and forest products.[16] The Agriculture Negotiating Group should advance measures to coordinate health, plant-health, and environmental standards and rules in order to facilitate agricultural trade flows and would develop transition strategies for the communities and the ecological resources likely to be affected by freer trade and mechanisms for conserving genetic diversity (e.g., native crop varieties). The Investment Negotiating Group should structure investment rules to stimulate efforts to control pollution and to manage natural resources sustainably and would stipulate a series of base-line environmental requirements to be met by foreign investors in all projects. The Dispute Settlement Negotiating Group should (1) establish a structure for settling environment-related disputes that would build on and improve the NAFTA model, (2) ensure access to scientific and technical expertise, and (3) commit to balancing trade goals and the environmental aims and principles enumerated in international environmental agreements as well as domestic laws.

Environmental Provisions within the FTAA

Certain elements of the trade-environment relationship are so tightly intertwined with the process of trade liberalization that they should be dealt with in the main text of any regional trade agreement. Several NAFTA-like provisions should be explored for the Americas, including provisions that do the following:

• Make clear that trade commitments and rules do not overrule existing international environmental agreements. Standards, including provisions for trade measures as enforcement mechanisms in international agreements such as the Montreal Protocol, the Convention on International Trade in Endangered Species, and the Basel Convention as well as regional and bilateral environmental agreements, should be expressly acknowledged.

• Provide for deference to national environmental standards, as long as they reflect legitimate environmental policy making and are applied non-discriminatorily to both domestic and foreign products.[17] Those who challenge environmental policies under trade agreements should be required to demonstrate that the rules or regulations to which they object are a masked barrier to trade, are not based on a scientific process, or are protectionist in intent.

• Discourage countries from lowering environmental standards or relaxing enforcement as a mechanism for achieving competitive advantage or attracting foreign investment.

• Place the burden of proof on the party that challenges another party's environmental or health measures.

• Establish a dispute-settlement process that reprovides ready access to environmental, scientific, and technical expertise.

In other critical regards, the NAFTA presents a starting point but should not be replicated in the hemispheric context. In particular, FTAA negotiators should refine NAFTA provisions and move beyond them to do the following:

• Eliminate the threat of environment-based trade sanctions. As was discussed extensively in earlier chapters, the NAFTA provision in this regard makes no sense. One alternative option to consider is the process established in the Canada-Chile Free Trade Agreement's parallel environmental agreement whereby accusations of environmental non-compliance are first addressed through dispute resolution and dialogue. Another model to consider would be the NAFTA's draft environmental provisions as of 1993 (before the Clinton administration's sanctions provisions were added) emphasizing dialogue and disclosure of weak performance.

• Eliminate investment provisions that permit expropriation claims by investors when environmental standards are realigned.[18] The general principle should be that governments are free to revise their regulatory standards without compensating those whose economic position is adversely affected. Claims for compensation should be permitted only when consistent with the stringent requirements of US "takings" jurisprudence.

A regional trade agreement for the Americas could also be strengthened by adding new environment provisions that were not in the NAFTA but which address sources of ongoing trade environment tension. Specifically, it would be useful to do these things:

• Refine the environmental "exceptions" clause that is found in the NAFTA (and in article XX of the General Agreement on Tariffs and Trade). The word "necessary" should not be interpreted in ways that present an insurmountable hurdle to legitimate environmental policies and standards.[19] In this regard, the WTO Appellate Body's recent Shrimp-Turtle decision provides a much more carefully balanced approach to weighing competing trade and environmental values than earlier trade cases.[20]

• Recognize that environmental standards related to production processes and methods (PPMs) can be legitimate. If they derive from international, regional, or bilateral environmental agreements, or if they address issues that have trans-boundary effects with measures that are proportionate to the environmental harm in question and are applied nondiscriminatorily, PPMs should be considered legitimate.

• Declare eco-labels to be consistent with the terms of the trade agreement subject to specific disciplines so long as such labels are not arbitrary (i.e., lacking a scientific basis or applied discriminatorily) or a disguised barrier to trade, they should be considered FTAA consistent even if they address production processes and methods.

Environmental Parallel Track

There is no substitute for strong national environmental programs as a way to reduce trade-environment tension. With this reality in mind, the FTAA negotiators should ensure that environmental officials are part of a collaborative effort to strengthen environmental performance within each country and across shared borders. Specific environmental finance and cooperation initiatives and institutions in the spirit of those that accompany NAFTA are necessary counterparts to the commitments being made for deeper economic integration.

In keeping with the 1994 Miami Declaration of the Summit of the Americas, we believe the FTAA governments should commit to a parallel track of environmental negotiations (led by environmental agencies with the participation of trade officials) to establish a Hemispheric Environmental Commission (HEC). The HEC would institutionalize attention to environmental issues arising in the context of regional economic integration and support environmental cooperation for improved national and regional environmental results. Its institutional form should build on but be more flexible than that of the NACEC. Its responsibilities should include the following:

• Environmental data gathering and analysis. Providing sound environmental data and analysis should be a central role of the HEC. Difficulties comparing existing data across countries combined with methodological challenges of linking environmental data with economic data to show cause-effect relationships limit governmental capacity to identify optimal strategies for addressing environmental risks. A commitment of modest resources across the hemisphere would permit the shifting of environmental decision making from its current less-than-rigorous approach to a much more analytically driven foundation.

• Environmental reviews of the trade agreement. The HEC should be asked to provide technical expertise and advice as well as financial assistance to governments as they undertake both prospective and retrospective environmental reviews of trade policy options.

• Capacity building, coordination, policy exchange, and sharing of "best practices." There are a great many ways for the countries of the hemisphere to improve cooperation among national and inter-governmental environmental institutions and initiatives. The progress of environmental cooperation in North America provides a good example of what can be done. Additionally, the US-Mexico model for border cooperation on environmental issues provides useful lessons (positive and negative) for collaboration by pairs or groups of countries that share borders. Integrated common border plans designed to jointly manage shared natural resources (especially fisheries and forests) and to prevent and address pollution spillovers should be undertaken wherever trans-boundary issues arise in the Americas.

• Environmental compliance and public participation. To support governments' efforts to monitor compliance with environmental legislation, the FTAA countries should also provide a NAFTA-like procedure for citizen submissions and independent investigations relating to compliance with the environmental provisions of the FTAA. In designing public participation strategies, governments can seek guidance from 1996 Inter-American Strategy for the Promotion of Public Participation in Decision-Making for Sustainable Development (developed by the OAS in response to a mandate from the 1996 Bolivia Summit of the Americas on Sustainable Development).

• Financing for the environment. Ensuring that adequate resources are available to build the necessary environmental infrastructure at the local, provincial/state, and national levels across the hemisphere will also require concerted efforts. Environmental bonds or other innovative financing would be an important addition to an overall infrastructure initiative. Governments should also work to leverage private-sector financing for the environment through strategic partnerships.

• Private-sector environmental cooperation and environmental technology transfer. Intergovernmental cooperation to promote the transfer of environmental technologies can be developed using a number of existing models. The work of the US Environmental Training Institute provides one example of how international environmental collaboration can be undertaken.[71] Efforts to strengthen national systems for environmental certification, accreditation, and implementation of environmental management systems (such as ISO 14000) would also be useful.

In its institutional form, we see a Hemispheric Environmental Commission as a lean, largely "virtual" mechanism for coordinating national and regional environmental efforts through a small secretariat, perhaps based at the Organization of American States. A modest commitment of resources and official time and energy could be structured to pay significant dividends, leveraging a decentralized public policy "network" structure. We envision this commission working closely with the secretariats of existing multilateral, regional, and bilateral environmental agreements and other environmental institutions and organizations. In particular, we see real opportunities for the new commission to work with the Economic

Commission for Latin America and the Caribbean, the NACEC, the OAS's
Trade Unit, UNEP's Regional Office for Latin America, and the Division
of Integration, Trade and Hemispheric Issues of the Inter-American
Development Bank.

The work of the HEC should be shaped and reviewed by an annual meeting of regional environmental ministers (building on existing annual
meetings of the region's environmental ministers hosted by UNEP—see
UNEP 2000). The HEC should also meet on a periodic basis on substantive
issues with whatever secretariat emerges for the FTAA. Both groups should
be charged with jointly gathering and responding to input, issue identification, and analysis from the aforementioned high-level Trade and Environment Experts Advisory Group and ongoing Western Hemisphere Civil
Society Forum. This forum would involve designated civil society representatives drawn from national consultation processes and would include
formal mechanisms for dialogue with relevant government officials (Lucas
2000).

An Action Plan

Trade policy can and must address environmental issues in concrete ways.
To this end, the debate on trade and the environment should be grounded
in empirical evidence and sound analysis, not in emotion and rhetoric. The
central lesson of the NAFTA is that trade-environment links can be constructive. Economically, countries have little to fear from NAFTA-like environmental provisions. The negotiation and implementation of NAFTA's
environmental provisions have fostered useful models, processes, precedents, and lessons for environmental cooperation among the regional trading partners.

The opportunities presented by expanded trade across the western hemisphere loom large for many countries in the Americas. Strengthening pollution-control and natural-resource-management programs represents a
parallel policy priority. There is no reason why economic and environmental performance cannot advance in tandem. Building on the foundation laid by the NAFTA, regional trade negotiators should commit to an
environmental action plan that might do the following:

• Launch environmental reviews of the FTAA, nationally and regionally,
aiming to identify critical issues that negotiators should address.

• Create an FTAA Environment Negotiating Group to coordinate the integration of environmental considerations into the FTAA negotiation process and initiate parallel track of negotiations for a regional environmental agreement.

• Advance a "win-win" trade and environment agenda focused on reducing trade-disrupting and environmentally damaging subsidies in the fisheries, timber, water, energy, and agricultural sectors. A commitment to zero tariffs on environmental goods and services to maximize transfers of pollution-control technology should also be adopted.

• Commit to addressing all directly trade-related environment issues in a series of NAFTA-style environmental provisions within the FTAA. Investment provisions deserve special attention, with the shortcomings from NAFTA's chapter 11 addressed.

• Open parallel negotiations on a Regional Environmental Agreement establishing a Hemispheric Environmental Commission to foster cooperation and to manage issues related to the implementation and the effects of the trade agreement.

• Replace the FTAA's Committee on Civil Society with a standing Trade and Environment Experts Advisory Group and an ongoing Civil Society Forum to facilitate public participation in the negotiations and implementation of the regional trade area.

FTAA negotiators have a clear opportunity to reverse the course of a hemispheric integration journey that openly ignores the environment. The trade community of the Americas could build on—and improve—the foundation laid in the NAFTA, in Mercosur, and in other regional agreements, thus creating a trade agreement for the western hemisphere that promotes economic prosperity, environmental vitality, and a better quality of life for all its citizens.

Notes

1. The Miami Declaration specifically states: "Social progress and economic prosperity can be sustained only if our people live in a healthy environment and our ecosystems and natural resources are managed carefully and responsibly. . . . We will advance our social well-being and economic prosperity in ways that are fully cognizant of our impact on the environment." For the complete text of this declaration, see http://www.ftaa-alca.org.

2. The FTAA's Trade Negotiations Committee (TNC) has divided negotiations among nine FTAA negotiating groups that have specific mandates from ministers and the TNC to negotiate text in their subject areas. They were established for market access; investment; services; government procurement; dispute settlement;

agriculture; intellectual property rights; subsidies, antidumping and counter-vailing duties; and competition policy. The negotiating groups meet regularly throughout the year. For further information on the FTAA negotiating process, see http://www.ftaa-alca.org.

3. Among Latin America countries, Chile has distinguished itself by taking posi-tive leadership on the trade-environment linkage (Matus and Rossi 2002). It remains unclear how strongly the Bush administration will take up and advocate environ-mental considerations in the trade context. But the history of NAFTA, WTO, and fast track negotiations in the US strongly signals that no US administration will succeed in expanding external trade relations without significant concessions to domestic environmental constituencies.

4. On Mexico's traditional stance on trade and environment issues, see de la Calle 1999. With the arrival of Vicente Fox as the president of Mexico, the dynamics of the trade and environment debate may change. Fox's appointment of Victor Lichtinger (with his experience on trade and environment affairs at the NACEC) as Minister of the Environment may bode well for a more sympathetic Mexican dis-position toward environmental issues. Already, Fox has stressed support for greater public participation, transparency, and accountability in NAFTA affairs. Fox has suggested that the NACEC does play a valuable role in the NAFTA region, but that it has not always been used to its full potential. He has advocated strengthening cooperation to address this shortcoming.

5. Some environmental organizations, business groups and policy makers in the US and Canada joined the Mexican government's opposition to the incorporation of environmental provisions in the NAFTA. Many business groups and policy makers consider environmental considerations extraneous to their primary goal of expand-ing trade and argue that such issues should be dealt with external to trade arrange-ments. Some environmental groups share opposition to a trade-environment linkage, but their rationales differ. They fear that environmental considerations could simply serve as a "greenwash" of trade arrangements that they consider inher-ently unsustainable.

6. While the tuna-dolphin case played out in the multilateral trading regime (GATT and WTO) rather than the NAFTA regime, and Mexico ultimately won the case, Mexico has good reason to guard against similar trade and environment cases in the regional context. For more information on the tuna-dolphin case, see Kingsbury 1994.

7. The EU, Canada, and the US have each developed methodologies for, and imple-mented, environmental reviews of several trade policy initiatives. For more infor-mation on the European efforts, see the European Commission's official page on trade issues: http://europa.eu.int/comm/environment/eia/sea-studies-and-reports/sea_approach.pdf. The Canadian Ministry of Foreign Affairs and International Trade has released several documents regarding its methodologies and conclusions (2001). A useful starting point for information on the US process is the Clinton administration's Executive Order 13141 on environmental reviews of trade agree-ments. The NACEC is also conducting considerable analytical work on method-ologies and has commissioned a number of reports analyzing the environmental

impacts of NAFTA (NACEC 1999). Finally, the World Wildlife Fund has played an important role in stimulating discussion of methodologies for assessment. The WWF recently released its report of an International Experts Meeting on Sustainability Assessments of Trade Liberalisation (2000). In 1999, the NACEC released three case studies of environmental impacts of the NAFTA. A further 14 case studies were commissioned and presented at a NACEC Trade and Environment Symposium held in October 2000. The NACEC's work takes both a sectoral approach (e.g., agriculture, electronics, energy, tourism) as well as an environmental media approach (e.g., water, forests).

8. The NACEC, a direct institutional product of the NAFTA, has generated significant research on methodologies for environmental reviews. In 1999, NACEC officials commissioned a series of trade and environment case studies to evaluate the analytical framework it developed (NACEC 2000). A landmark North American symposium in 2000 on the NACEC's methodology underscored the need to make progress in the fine-tuning of the analytical framework, but also highlighted the potential of environmental reviews as a tool for integrating environmental sensitivity into the trade policy making process.

9. For an overview of the provisions, see appendix A.

10. Environmental measures that restrict trade have to be justified in terms of whether they are "necessary" and "legitimate." The term "tantamount to expropriation" arises in the context of NAFTA's chapter 11 investment provisions. For more information about chapter 11, see CIEL et al. 1999, Mann and von Moltke 1999, and Mann and Araya 2002.

11. This conclusion does not imply, however, that trade measures for environmental purposes are necessarily bad. The utility of trade measures to promote compliance with the provisions of several multilateral environmental agreements has been repeatedly demonstrated (Brack 2001).

12. John Audley's (2001) presentation at the April 2000 conference on which this volume builds stands as a leading example of the change of heart that many environmentalists have undergone since the adoption of the NAFTA.

13. For an overview of the extensive environmental challenges that remain in Mexico, see Barkin 1999 and Mumme 1999. The US-Mexico border in particular still faces immense environmental problems. Border communities continue to lack basic environmental and sanitary infrastructure. As the negative impacts of air and water pollution on local communities and their health continues, increased trade and investment across the border region is intensifying environmental pressures. Many of the border communities that are most impacted lack the mechanisms and means to draw state or national attention to their plight. See Weintraub 2000, Public Citizen and RMALC 1996; Carlsen and Salazar, 2001. Two excellent sources of information about the environmental problems along the border and the struggles for environmental justice are the web site of the Environmental Health Coalition (http://www.environmentalhealth.org) and that of *Borderlines*, a monthly publication of the Interhemispheric Resource Centre (http://www.us-mex.org/borderlines/).

14. To address border environmental needs, the NADBank needs to be a development institution as much as a bank. In response, a "giant window" has now been

opened. Because NADBank is considered a foreign lender, Mexican communities have to work with Mexican federal agencies as intermediaries to access NADBank monies. And the NADBank has experienced considerable difficulties transferring funds from the NADBank to local municipalities (Faulkner 2000).

15. The BECC experience highlights the important of technical assistance to help communities design, build and sustain environmental projects. BECC officials have learnt about the particular barriers that small, lower-income communities face in accessing project assistance and funding (Faulkner 1999; Kelly et al. 2001). These communities tend to lack the institutional capacity to design, implement and maintain projects that the meet the BECC and NADBank technical and financial criteria. Again, partnerships have proved very important. BECC officials have worked with the US Environmental Protection Agency to deliver grants through a Project Development Assistance Program. However, BECC certification of a project still does not guarantee NADBank or funds from third-party lenders.

16. Interest in the environmental and trade impacts of "perverse" subsidies is high. For a review of the scale and scope of perverse subsidies in agriculture, fossil fuels/nuclear energy, road transportation, water and fisheries, see Myers 1998. In working to reduce harmful subsidies, governments must bear in mind that the removal of subsidies may be a necessary but not sufficient condition for improved environmental management, that some subsidies may help promote transitions to more environmentally sound modes of production, and that some subsidies may be important in developing countries for food security and livelihoods, particularly on a short-term basis.

17. The emerging GATT article jurisprudence (Wofford 2000) and the existing work on "legitimacy" (Esty 1994a) provide a foundation.

18. To the extent that environmental standards target a single company or a small class of companies, consideration should be given to whether this limited set of actors is bearing an unfair or undue burden from a revised regulatory structure. For specific recommendations for the FTAA investment regime, see Mann and Araya 2001.

19. For a discussion of the term "necessary" in the NAFTA context, see Mann 2000. For a discussion of the issue in the GATT realm, see Esty 1994a.

20. For information about the WTO shrimp-turtle decision, see Wofford 2000, Stilwell and Arden-Clarke 1998, and WTO 1998.

21. The US Environmental Training Institute is a private, non-profit organization, established by the US Environmental Protection Agency in 1992, that promotes sustainable development by addressing the specific environmental training needs of industrially developing countries worldwide, see http://www.ctc.com/about/rpv/pubs/display/international.pdf.

References

Aaronson, S. 2001. *Redefining the Terms of Trade Policymaking*. Washington: National Policy Association.

Abel, A. 2000. "NAFTA's Institutional Framework: Lessons Learned and Relevance to the FTAA: Some Issues to Consider Regarding the BECC and NADBank Models." Paper presented at Washington conference on "The FTAA and the Environment: What Can We Learn from the NAFTA Model?"

Araya, M. 2002. "Mexico's NAFTA Trauma: Myth and Reality." In this volume.

Audley, J. 1997. *Green Politics and Global Trade: NAFTA and the Future of Environmental Politics.* Georgetown University Press.

Audley, J. 2001. "Back to the Bad Old Days of NAFTA." Paper presented at NWF/Yale Center for Environmental Law and Policy conference on "Trade and Environment in the Americas: Lessons Learned from the NAFTA," April 2000; revised April 2001. Available at http://www.ycelp.org.

Brakin, D. 1999. "Free Trade and Environmental Policymaking in Mexico." *Borderlines* 7, no. 9: 14–15.

Bergsten, F. 1992. The Primacy of Economics. *Foreign Policy* 87, summer: 3–24.

Brack, D. 2001. "Environmental Treaties and Trade: Multilateral Environmental Agreements and the Multilateral Trading System." In *Trade, Environment and the Millennium,* ed. G. Sampson and W. Chamber. Tokyo: UN University Press.

Canadian Ministry of Foreign Affairs and International Trade. 2001. Framework for Conducting Environmental Assessments of Trade Negotiations. Ottawa: DFAIT.

Carlsen, L., and H. Salazar, eds. 2001. Respuestas Sociales ante la integración: Impacts socioambientales del TLCAN. Mexico City: RMALC.

Carlsen, L., and H. Salazar. 2002. "Limits to Cooperation: A Mexican Perspective on the NAFTA's Environmental Side Agreement and Institutions." In this volume.

CIEL et al. 1999. *Investment Agreement of the Americas: Environmental, Economic and Social Perspectives.* Center for International Environmental Law, Instituto del Tercer Mundo, and Preamble Center.

de Almeida, L., ed. 2001. Trade and Environment: A Positive Agenda for Sustainable Development. Preliminary Document for the XIII Meeting with the Latin American and Caribbean Environment Ministers, Brazilian Ministry of Environment, Secretariat of Policies for Sustainable Development.

Deere, C. 1999. *Ecolabelling and Sustainable Fisheries.* Rome and Washington: FAO and IUCN.

de la Calle, L. 1999. Presentation at WTO High Level Symposium on Trade and Environment. Available at http://www.wto.org.

Destler, I., and P. Balint. 1999. *The New Politics of American Trade: Trade, Labor, and the Environment.* Washington: Institute for International Economics.

Esty, D. 1993. "Integrating Trade and Environment Policy Making: First Steps in the North American Free Trade Agreement." In *Trade and Environment,* ed. D. Zaelke et al. Island.

Esty, D. 1994a. *Greening the GATT: Trade, Environment and the Future.* Washington: Institute for International Economics.

Esty, D. 1994b. "Making Trade and Environmental Policies Work Together: Lessons from NAFTA." *Aussenwirtschaft* 49: 59–79.

Esty, D. 1996. "Revitalizing Environmental Federalism." *Michigan Law Review* 95: 570–653.

Esty, D. 1998a. "Environmentalists and Trade Policymaking." In *Constituent Interests and US Trade Policies*, ed. A. Deardorff and R. Stern. University of Michigan Press.

Esty, D. 1998b. "Non-Governmental Organizations at the World Trade Organization: Cooperation, Competition or Exclusion." *Journal of International Economic Law* 1, no. 1: 123–148.

Esty, D. 2001. "Bridging the Trade-Environment Divide." *Journal of Economic Perspectives* 15, no. 3: 113–130.

Esty, D. 2002. "The World Trade Organization's Legitimacy Crisis." World Trade Review 1, no. 1.

Faulkner, T. 1999. "BECC Technical Assistance to Small Communities." *Borderlines* 7, no. 2: 11.

Gallagher, K. 2002. "Industrial Pollution in Mexico: Did the NAFTA Matter?" In this volume.

Hufbauer, G., D. Esty, D. Orejas, L. Rubio, and J. Schott. 2000. *NAFTA and the Environment: Seven Years Later*. Washington: Institute for International Economics.

Jha, V., A. Markandya, and R. Vossenaar. 1999. *Reconciling Trade and the Environment*. Elgar.

Kelly, M., C. Reed, and L. Taylor. 2001. The Border Environmental Cooperation Commission (BECC) and the North American Development Bank (NADBank): Achieving Their Environmental Mandate. Austin: Texas Center for Policy Studies.

Kingsbury, B. 1994. "The Tuna-Dolphin Controversy, the WTO and the Liberal Project to Reconceptualize International Law." *Yearbook of International Environmental Law* 5: 1–40.

Lucas, N. 2000. "Some Issues for Consideration on Participation." Paper presented at Washington conference on "The FTAA and the Environment: What Can We Learn from the NAFTA Model?"

Magraw, D. 1995. *NAFTA and the Environment: Substance and Procedure.* American Bar Association.

Mancera, J. 2002. "A Mexican View on Trade and Environment." In this volume.

Mann, H. 2000. "Assessing the Impact of NAFTA on Environmental Law and Management Processes." Paper presented at First North American Symposium on Understanding the Linkages between Trade and Environment. Available at http://www.cec.org/symposium.

Mann, H., and M. Araya. 2002. "An Investment Regime for the Americas: Challenges and Opportunities for Environmental Sustainability." In this volume.

Mann, H., and K. von Moltke. 1999. NAFTA's Chapter 11 and the Environment: Addressing the Impacts of the Investor-State Process on the Environment. Winnipeg: International Institute for Sustainable Development.

Matus, M., and E. Rossi. 2002. "Trade and the Environment in the FTAA: A Chilean Perspective." In this volume.

Mayer, F. 2002. "Negotiating the NAFTA: Political Lessons for the FTAA." In this volume.

Mayer, F. 1998. *Interpreting NAFTA: The Science and Art of Political Analysis.* Columbia University Press.

Miller, E. 2002. "Did Mexico Suffer Economically from the NAFTA's Environmental Provisions?" In this volume.

Mumme, S. 1999. "NAFTA and the Environment." *Foreign Policy in Focus* 4, no. 26: 1–4.

Myers, N. 1998. Perverse Subsidies: Tax $s Undercutting Our Economies and Environments Alike. Winnipeg: International Institute for Sustainable Development.

NACEC. 1999. Assessing Environmental Effects of the North American Free Trade Agreement (NAFTA). An Analytic Framework (Phase II) and Issue Studies. Environment and Trade Series no. 6, North American Commission for Environmental Cooperation.

NACEC. 2000. *First North American Symposium on Understanding the Linkages between Trade and Environment.* Papers available at http://www.cec.org/symposium.

Nadal, A. 2002. "Zea Mays: Effects of Trade Liberalization of Mexico's Corn Sector." In this volume.

OECD. 2000. *Assessing the Environmental Effects of Trade Liberalisation Agreements.* Paris: OECD.

Public Citizen and RMALC. 1996. *NAFTA's Broken Promises: The Border Betrayed, US-Mexico Border Environment and Health Decline in NAFTA's First Two Years.* Washington: Public Citizen Publications.

Public Citizen and RMALC. 1999. *Foreign Direct Investment and the Environment.* Paris: Organization for Economic Cooperation and Development.

Rodríguez, R. 2000. "The North American Development Bank: Lessons for the FTAA." Presented at Washington conference on "The FTAA and the Environment: What Can We Learn from the NAFTA Model?"

Segger, M., and N. Borregaard. 2002. "Sustainability and Hemispheric Integration: A Review of Existing Approaches." In this volume.

Segger, M., M. Munos, P. Meirles, J. Taurel, and V. Paul. 2000. *Trade Rules and Sustainability in the Americas.* Winnipeg: International Institute for Sustainable Development.

Stilwell, M., and C. Arden-Clarke. 1998. *Dispute Settlement in the WTO: A Crisis for Sustainable Development.* Center for International Environmental Law, WWF International, Oxfam-GB, and Community Nutrition Institute.

Torres, B. 2002. "The North American Agreement on Environmental Cooperation: Rowing Upstream." In this volume.

UNEP. 2000. Final Report: XII Meeting of the Forum of Ministers of the Environment of Latin America and the Caribbean, Bridgetown, Barbados. UNEP/LAC-IG.XII/4. Nairobi: United Nations Environment Program.

Weintraub, S. 2000. "Economic Impacts of NAFTA's Environmental Provisions." Presented at Washington conference on "The FTAA and the Environment: What Can We Learn from the NAFTA Model?"

Wofford, C. 2000. "A Greener Future at the WTO: The Refinement of WTO Jurisprudence on Environmental Exceptions to GATT." *Harvard Environmental Law Review* 24, no. 2: 563ff.

WTO. 1998. United States Import Prohibition of Shrimp and Shrimp Products: Report of the Panel. Geneva: World Trade Organization.

WWF. 2000. Report of International Experts Meeting on Sustainability Assessments of Trade Liberalisation. Gland: World Wildlife Fund.

Appendix A
Environmental Provisions of the North American Free Trade Agreement

Environmental Provisions in the Text

In scope and in nature, the environmental provisions included in the NAFTA's main text represent an unprecedented insertion of environmental issues in a trade agreement. Environmental considerations are incorporated in the following ways:

Preamble

The preamble contains hortatory language that plays an important role in setting the tone and the framework for the agreement. Four of the paragraphs discuss environmental matters, stating that the NAFTA should promote sustainable development, preserve the flexibility of governments to safeguard public welfare, strengthen the development and enforcement of environmental laws and regulations, and pursue trade goals in a manner "consistent with environmental protection and conservation."

General Exceptions

For environmental measures in the area of Trade in Goods and Technical Barriers to Trade,[1] the NAFTA incorporates the General Agreement on Tariffs and Trade's article XX provisions on measures necessary to protect human, animal, or plant life or health, or relating to the conservation of living and non-living exhaustible natural resources. Article 2101 of the NAFTA states that, provided that measures are not applied in a manner that would constitute a means of arbitrary or unjustifiable discrimination between countries where the same conditions prevail or a disguised restriction on trade between the parties, nothing in the NAFTA's chapters on

Goods, Technical Barriers to Trade,[2] Cross Border Trade in Services, or Telecommunications shall be construed to prevent adoption or enforcement by any party of measures necessary to secure compliance with laws or regulations that are not inconsistent with the provisions of the agreement, including those relating to health and safety and to consumer protection.

International Environmental and Conservation Agreements

In article 104, the NAFTA provides that, in the event of conflict between the NAFTA and the Convention on International Trade in Endangered Species, the Montreal Protocol on Substances That Deplete the Ozone Layer, or the Basel Convention on the Movement of Hazardous Wastes, the provisions in these agreements shall prevail to the extent of the inconsistency (provided that where a party has a choice among equally effective and reasonably available means of complying with such obligations, the party chooses the alternative that is least inconsistent with other provisions of the NAFTA). In the annex to article 104.1, the same principle is applied to several bilateral agreements between the parties and countries are given the option of adding agreements to the annex.

Sanitary and Phytosanitary Measures (Chapter 7)

The NAFTA's articles most relevant to SPS measures and the environment are article 712, stating the right to adopt SPS measure more stringent than international standards; article 713, establishing international standards as the basis for the SPS regime with no "downward harmonization"; article 714.1, committing the parties to seek equivalence of SPS measures among countries without reducing level of environmental protection; and articles 718–719, requiring transparency in any efforts to modify domestic SPS rules. Article 712 also requires states to ensure that SPS measures are not used to discriminate arbitrarily or unjustifiably between one country's goods and like goods of the other countries. In addition, governments must ensure that any sanitary or phytosanitary measure adopted, maintained or applied is (i) based on scientific principles, taking into account relevant factors including, where appropriate, different geographic conditions; (ii) not maintained where there is no longer a scientific basis for the measure; and (iii) based on a risk assessment, as appropriate to the circumstances.

Standards-Related Measures (Chapter 9)

The NAFTA affirms the existing rights and obligations of countries with respect to the GATT Agreement on Technical Barriers to Trade and other agreements, including environmental and conservation agreements (article 903). Article 904 affirms the right of each country to adopt, maintain, and apply standards related to safety, the protection of human, animal, and plant life and health, the environment, and consumers (as well as measures to ensure their enforcement or implementation) at a level that it considers most appropriate. The same article also permits countries to prohibit importation of goods that do not comply with such standards. Article 905 affirms the right of countries to adopt standards-related measures that are more stringent than the international standard. The NAFTA also encourages countries to seek equivalence among their respective standards-related measures to the greatest extent possible, without reducing the level of protection of the environment (article 906). Article 909.1 calls for transparency in the modification of national rules related to standards-related measures.

Dispute Settlement

A country that objects to an action taken by another NAFTA member to adopt, maintain, or protect the environment (taken under article 104 (Environmental Agreements), subchapter 7[b] (Sanitary and Phytosanitary measures), or chapter 9 (Standards-Related measures) may request recourse to the NAFTA's dispute-settlement procedures. The specific environment-related articles of chapter 20 on dispute settlement are article 2005, stating the right to request that any dispute related to SPS measures, Standards-Related measures or agreements set forth in article 104 be addressed exclusively under the NAFTA (as opposed to less-specific GATT); article 2015, enabling the dispute-settlement panel to request a written report from an independent Scientific Review Board on any issue of fact concerning the environment and obligating the panel to take it into account when rendering its decision; article 2017, establishing that the aforementioned written report and any comments from parties on it shall be appended to final report of the panel when it is made public. The NAFTA also contains articles that place the burden of proof the government challenging the legitimacy of SPS or standards-related measures (articles 723.6 and 914.4, respectively).

Public Consultation
Article 2002 states that the NAFTA Free Trade Commission may seek the advice of non-governmental persons or groups and that it may establish, and delegate responsibilities to, ad hoc or standing committees, working groups, or expert groups.

Investment
Article 1114 (1) permits governments to adopt, maintain, or enforce any measure otherwise consistent with the chapter that it considers appropriate to ensure that investment activity in its territory is undertaken in a manner sensitive to environmental concerns. Article 1114 (2) discourages parties from working to establish, acquire, retain, or expand investment by relaxing, waiving, or derogating from domestic health, safety, or environmental measures. If one NAFTA party considers that another party has offered such an encouragement, it may request a consultation with the other country with a view to avoiding such behavior. In terms of Performance Requirements, the NAFTA secured the right of countries to require that an investment use a technology to meet generally applicable measures related to health, safety, or environmental standards (article 1106).

The North American Agreement for Environmental Cooperation

The North American Agreement for Environmental Cooperation (NAAEC) is commonly known as "the environmental side agreement." The objectives of the environmental side agreement are as follows:
• foster the protection and improvement of the environment in the territories of the parties for the well-being of present and future generations
• promote sustainable development based on cooperation and mutually supportive environmental and economic policies
• increase cooperation among the parties to better conserve, protect, and enhance the environment, including wild flora and fauna
• support the environmental goals and objectives of the NAFTA
• avoid creating trade distortions or new trade barriers
• strengthen cooperation on the development and improvement of environmental laws, regulations, procedures, policies, and practices
• enhance compliance with and enforcement of environmental laws and regulations

• promote transparency and public participation in the development of environmental laws, regulations, and policies
• promote economically efficient and effective environmental measures
• promote pollution-prevention policies and practices.

The North American Commission for Environmental Cooperation

The North American Commission for Environmental Cooperation (NACEC) is an international organization created by Canada, Mexico and the United States under the North American Agreement on Environmental Cooperation. The NACEC was established to address regional environmental concerns, to help prevent conflicts between trade goals and environmental objectives, and to promote effective enforcement of environmental law.

The NACEC is organized as follows:

• The Council, the governing body of the NACEC, is composed of the environmental ministers (or the equivalent) of the three countries. It meets at least once a year to discuss NACEC programs and activities.

• The Joint Public Advisory Committee (JPAC) is composed of fifteen ministers, five from each of the three countries, who are appointed by their respective governments. Its members act independently. Their responsibility is to advise the Council on all matters within the scope of the North American Agreement on Environmental Cooperation.

• The secretariat is composed of professional staff that implement initiatives and conduct research in core program areas on topics pertaining to the North American environment, environmental law and standards, and other environment and trade issues. The secretariat also processes citizens' submissions on enforcement matters. Located in Montreal, with a liaison office in Mexico City, the secretariat provides technical and operational support to the Council, as well as to committees and groups established by the Council.

The NACEC's work program has four central components, with the following associated projects:

Environment, Economy and Trade
　　Emerging Environmental Trends in North America
　　Assessing Environmental and Trade Relationships
　　Supporting Biodiversity Conservation through Green Goods and Services (e.g., sustainable tourism, shade coffee)
　　Market and Financial Mechanisms in Support of the Environment

Conservation of Biodiversity
    Strategic and Cooperative Action for the Conservation of Biodiversity in North America
    North American Bird Conservation Initiative
    Species of Common Conservation Concern
    Mapping of Marine and Estuarine Ecosystems of North America
    North American Marine Protected Areas Network
    Global Program of Action for the Protection of the Marine Environment from Land-based Activities in North America
    Closing the Pathways of Aquatic Invasive Species across North America
    North American Biodiversity Information Network

Pollutants and Health
    Facilitating Trinational Coordination in Air Quality Management
    Developing Technical and Strategic Tools for Improved Air Quality in North America
    Trinational Air Quality Improvement Initiative: North American Trade and Transportation Corridors
    Sound Management of Chemicals
    North American Pollutant Release and Transfer Register
    Capacity Building for Pollution Prevention
    Children's Health and the Environment in North America

Law and Policy
    Comparative Report on Environmental Standards
    North American Regional Enforcement Forum
    Enforcement and Compliance Capacity Building
    Enforcement/Compliance Reporting
    Environmental Management Systems to Promote Compliance and Environmental Performance

North American Fund for Environmental Cooperation

The NACEC created the North American Fund for Environmental Cooperation (NAFEC) in 1995 as a means to fund community-based projects in Canada, Mexico, and the United States that promote the NACEC's goals and objectives.

    NAFEC supports projects that

• are community-based (involve a clearly defined community of stakeholders who actively participate in the design and implementation of the project)
• respond to a specific issue or problem and lead to concrete results
• reflect cooperative and equitable partnerships between or among organizations from different sectors and/or countries within North America

- meet the objectives of the NACEC (by complementing the current NACEC program)
- strengthen and build the capacities of local people, organizations, and institutions
- emphasize sustainability and link environmental, social, and economic issues
- leverage additional support, but are unlikely to obtain full funding from other sources.

Article 13 Initiatives
NAAEC's article 13 allows the secretariat to prepare a report to the Council on any matter within the scope of its annual work program. To date, the secretariat has produced reports on electricity and the environment, on preserving trans-boundary migratory bird habitat on the upper San Pedro River, on continental pollution pathways, and on the death of migratory birds at the Silva Reservoir.

Border Environmental Cooperation Commission and North American Development Bank

The Governments of Mexico and the United States agreed in November 1993 on the creation of two institutions—the Border Environment Cooperation Commission (BECC) and the North American Development Bank (NADBank)—to assist communities on both sides of the US-Mexico border in coordinating and carrying out environmental infrastructure projects. The agreement was designed to further the goals of the North American Free Trade Agreement and the North American Agreement on Environmental Cooperation.

Border Environment Cooperation Commission
Headquartered in Ciudad Juarez, Chihuahua, the BECC was created to provide technical assistance to border communities and to certify environmental infrastructure projects in the border region for financing consideration by the NADBank and other sources. Certification is based on a set of environmental, health, technical, financial, community participation, and sustainable development criteria through a process that ensures extensive public and local input. The goal of the BECC's technical assistance is to

ensure technically sound and feasible projects, master plans, project design, environmental assessment, and local institutional capacity building.

North American Development Bank

Headquartered in San Antonio, Texas, and capitalized in equal shares by the United States and Mexico, the NADBank was created to facilitate financing for the development, execution, and operation of environmental infrastructure projects along the US-Mexico border. All environmental projects financed by this bank must be certified by the Border Environment Cooperation Commission; related to potable water supply, wastewater treatment, or municipal solid waste management; and located in the border region. In early 2001, the NADBank made preliminary moves to broaden the scope of its activities along the border.

The NADBank provides three primary services to border communities:

• Advisor and financial strategist. The NADBank can provide basic guidance to smaller communities that may require assistance with comprehensive long-term infrastructure planning and development.

• Investment banker. The NADBank works to structure the most affordable financial package possible for its clients and seeks ways to reduce project costs by securing funding from both private and public sources.

• Lender. The NADBank provides financing for project costs not covered by other funding sources and guarantees to encourage financing from other lenders. The NADB also administers US Environmental Protection Agency grant resources.

Notes

1. Except to the extent that a provision in Part Two (Trade in Goods) applies to services or investment or Part Three (Technical Barriers to Trade) applies to services.

2. To the extent that a provision of Part Two (Trade in Goods) or Part Three (Technical Barriers to Trade) applies to services.

Appendix B
Overview of the FTAA Process

First Summit of the Americas (Miami, December 9–11, 1994)

Heads of state and government of 34 countries in the region agreed to "promote prosperity through economic integration and free trade." The participants resolved to begin immediately to construct the Free Trade Area of the Americas (FTAA), in which barriers to trade and investment would be progressively eliminated, and to complete negotiations for the agreement by 2005.

First Western Hemisphere Trade Ministers Meeting (Denver, June 30, 1995)

Trade ministers of the 34 signatories to the Miami Declaration met to initiate negotiations for the FTAA.

Second Western Hemisphere Trade Ministers Meeting (Cartagena, March 21, 1996)

Trade ministers met to review the efforts of the vice-ministers at preliminary discussions and to ask the vice-ministers to recommend timing for a lunch of the FTAA negotiations.

Summit of the Americas on Sustainable Development (Santa Cruz de la Sierra, Bolivia, December 1996)

Elected heads of state and government affirmed their "determination to move forward toward sustainable development and to implement the decisions and commitments set forth in the Rio Declaration and Agenda 21." They also approved a Plan of Action for Sustainable Development of the Americas.

Third Western Hemisphere Trade Ministers Meeting (Belo Horizonte, Brazil, May 16, 1997)

Trade ministers reviewed the progress that had been achieved in trade liberalization in the hemisphere since the Miami Summit of the Americas. The ministers agreed that the FTAA negotiations should be initiated at the Second Summit of the Americas in 1998.

Fourth Western Hemisphere Trade Ministers Meeting (San José, Costa Rica, March 19, 1998)

Trade ministers reviewed the results of the preparatory work for the negotiations of the FTAA. They recommended to their heads of state and government the initiation of negotiations and set out the structure and general principles and objectives to guide the negotiations.

Second Summit of the Americas (Santiago, Chile, April 19, 1998)

On the basis of the San José Declaration, the FTAA negotiations were launched formally at this summit. Heads of state and governments direct their ministers responsible for trade to begin negotiations for the FTAA, with the goal of concluding the agreement no later than 2005. The leaders agreed that the FTAA negotiating process should be transparent and should take into account the differences in the levels of development and size of the economies in the Americas, in order to facilitate full participation by all countries.

Fifth Western Hemisphere Trade Ministers Meeting (Toronto, November 4–5, 1999)

Ministers instructed the negotiating groups to prepare draft texts of their respective chapters and to discuss the modalities and procedures for negotiations in their respective areas. Ministers also approved several business-facilitation measures designed to facilitate commercial exchange in the hemisphere, particularly in the area of customs procedures.

Sixth Western Hemisphere Trade Ministers Meeting (Buenos Aires, April 7, 2000)

FTAA ministers resolved that the agreement should be finalized no later than January 2005 and fully operational no later than December 2005. The ministerial declaration that emerged from the meeting also states that draft texts of the trade agreement will be made available after each subsequent Summit of the Americas.

Third Summit of the Americas (Quebec City, April 2000)

All 34 heads of state signed a declaration that included a pledge to conclude FTAA negotiations on the FTAA no later than January 2005 for implementation by December 2005.

Seventh Western Hemisphere Trade Ministers Meeting (Quito, October 2002)

Deadline for second draft of the text of an agreement

January 2005

Deadline for completion of FTAA negotiations

December 2005

Agreement scheduled to enter into force.

Appendix C
Official Web Resources

Government of Canada

Department of Foreign Affairs and International Trade (DFAIT): www.dfait-maeci.gc.ca

National Advisory Committee on the NAAEC: www.naaec.gc.ca/english/nac

Council of Ministers of the Environment: www.mbnet.mb.ca/ccme

Environment Canada: www.ec.gc.ca

Government of Mexico

Secretaría de Economía: www.economia.gob.mex

Secretaría de Medio Ambiente y Recursos Naturales: www.semarnat.gob.mx

Comisión Nacional Para el Conocimiento y uso de la Biodiversidad: www.conabio.gob.mx

Instituto Nacional de Ecología: www.ine.gob.mx

NAFTA Office: www.naftaworks.org

Procuraduría Federal de Protección al Ambiente: www.profepa.gob.mx

Government of the United States

Trade Representative: www.ustr.gov

Governmental Advisory Committee to the US Representative to the CEC / National Advisory Committee to the US Representative to the CEC: www.epa.gov/ocempage/reports

Environmental Protection Agency: www.epa.gov

Regional Agencies

Organization of American States: www.oas.org
Economic Commission for Latin America and the Caribbean: www.eclac.org
Inter-American Development Bank: www.iadb.org

NAFTA and FTAA Sites

NAFTA Secretariat: www.nafta-sec-alena.org
North American Commission for Environmental Cooperation: www.cec.org
Border Environment Cooperation Commission: www.cocef.org
North American Development Bank: www.nadbank.org
Commission for Labor Cooperation: www.naalc.org
Free Trade Area of the Americas:www.ftaa-alca.org

Contributors

Gustavo Alanis-Ortega is President of the Mexican Environmental Law Center, a fellow of the Leadership for Environment and Development Program, and a columnist for the newspaper *Reforma* in Mexico City. He teaches environmental law at the Universidad Iberoamericana in Mexico City and is co-director of a diploma course on environmental law and policy.

Mónica Araya is Director of the Sustainable Americas Project at the Yale Center for Environmental Law and Policy. She has worked as a trade and environment advisor in the Department of International Trade Negotiations at Costa Rica's Ministry of Foreign Trade. She is a doctoral candidate in environmental policy at the Yale School of Forestry and Environmental Studies.

Nicola Borregaard, Executive Director of Chile's Recursos e Investigación para El Desarrollo Sustentable (RIDES), is a former director of the economic unit of CONAMA (Chile's environmental agency). She has written or edited several national and international publications, including *Mercosur and the Environment* (with Hernán Blanco).

Douglas Jake Caldwell is Program Coordinator for Trade and Environment at the National Wildlife Federation in Washington. He recently served as a member of the North American Commission for Environmental Cooperation's Advisory Committee to the North American Symposium on Trade and Environment.

Laura Carlsen is a researcher, a writer, and a member of the Mexican Action Network on Free Trade (RMALC). She has published extensively on Mexican economy, society, and environment in US and Mexican publications.

Marie Claire Cordonnier Segger is Project Director of the Americas Portfolio at the International Institute for Sustainable Development. She chairs the Hemispheric Working Group on Trade and the Environment and has been the lead author of several IISD books on trade and sustainability in the Americas, including *Trade Rules and Sustainability in the Americas*. She is also Director of the Centre for International Sustainable Development Law Commission.

Carolyn Deere is Assistant Director in the Global Inclusion theme of the Rockefeller Foundation. She has worked at the World Conservation Union, the International Center for Trade and Sustainable Development, and the Overseas Development Council. Her publications include *Globalization and Grantmakers, Eco-labelling* *and Sustainable Fisheries*, and *Net Gains: Linking Fisheries Management, International Trade and Sustainable Development*.

Daniel Esty is Director of the Yale Center for Environmental Law and Policy and Associate Dean of the Yale School of Forestry and Environmental Studies. From 1989 to 1993 he served as Special Assistant to EPA Administrator William Reilly, then as the EPA's Deputy Chief of Staff and its Deputy Assistant Administrator for Policy. He is the author, editor, or co-editor of a number of books, including *Greening the GATT*, *Thinking Ecologically*, and *Sustaining the Asia Pacific Miracle*.

Richard Fisher is Managing Partner of Fisher Family Fund L.P. He served in the Clinton administration (with the rank of ambassador) as Deputy US Trade Representative, with primary responsibility for Latin America, Canada, and Asia. As the chief operating officer for NAFTA, he had oversight responsibilities for bilateral trade issues in Canada, Mexico, and the countries of the Western Hemisphere, and for the development of the FTAA.

Kevin Gallagher is Research Associate at the Global Development and Environment Institute at Tufts University, where he works on GDAE's Sustainable Hemispheric Integration Project. His publications include *International Environmental Negotiations* (edited with Lawrence Susskind and Willliam Moomaw) and *International Trade and Sustainable Development* (with Jacob Werksman).

Eduardo Gitli was a professor at the International Center on Economic Policy for Sustainable Development (CINPE). He consulted for UNCTAD, ECLAC, FAO, and various national governments on trade-related issues and was the author of three books and numerous papers. He died as this volume was being finalized.

Ana Karina González-Lützenkirchen is Coordinator of the Program in Trade and Environment at Centro Mexicano de Derecho Ambiental (CEMDA). She has collaborated on several publications on trade and the environment in Mexico and the Western Hemisphere.

Javier Mancera is Director of the Mexican Ministry of Economy's NAFTA Office at the Embassy of Mexico in Washington. Previously he served as Economic Counselor at the Mexican embassy in Washington and was the liaison with the private sector and labor unions at Ministry of Economy in Mexico City.

Howard Mann is an Ottawa-based lawyer and policy consultant specializing in international environmental and trade law. Before opening his own practice, he worked as a lawyer for the Canadian government, often representing Canada in international environmental negotiations such as the 1992 Climate Change Convention and the environmental side agreement to the NAFTA.

Mario Matus is Director for Multilateral Economic Affairs in Chile's Ministry of Foreign Affairs. He has served as negotiator and legal counsel for negotiating teams on many Chilean trade agreements, including the Chile-Mexico Free Trade Agreement, the Chile-Canada Free Trade Agreement, Chile's accession to the NAFTA, and the GATT's Uruguay Round negotiations in Geneva. He also headed the Trade Negotiation Division's Services and Investment Department and served as chairman of the FTAA's Negotiating Group on Dispute Settlement.

Frederick W. Mayer is an associate professor of public policy studies and political science and the Director of the Center for North American Studies at Duke University. His book *Interpreting NAFTA* explores the nature of the political processes that created the agreement. He served as a senior international trade and foreign policy advisor to former US Senator Bill Bradley and as a policy analyst at the US Environmental Protection Agency.

Carlos Murillo is a professor at the International Center on Economic Policy for Sustainable Development (CINPE). He has headed the School of Economics at the Universidad Nacional in Costa Rica and CINPE. From 1997–98, he was Costa Rica's Vice-Minister of Foreign Trade. He is currently coordinating Costa Rica's Integration, Trade and Environment (INCA) project.

Alejandro Nadal is a professor of economics at the Center for Economic Studies and Coordinator of the Science and Technology Program at the Colegio de Mexico. On behalf of eight Mexican civic organizations and networks, he coordinated a team of economists that developed an alternative national economic plan for Mexico. He is the author of numerous articles exploring the origins of the NAFTA and its social and environmental impacts.

Edda Rossi heads the Trade and Sustainable Development Department of the Trade Negotiation Division of Chile's Ministry of Foreign Affairs. She formerly worked in the Trade Negotiation Division's FTAA/North America Department, where she was responsible for issues related to the process of the creation of the Free Trade Area of the Americas. She also served as a member of Chile's negotiation team.

Hilda Salazar is Director of Mujer y Medio Ambiente in Mexico City, a non-governmental organization that is a member of the Mexican Action Network on Free Trade (RMALC). She was among the founding members of RMALC, and she is an active participant in its activities. She has written widely on environment and environmental politics in Mexico.

Marianne Schaper is Economic Affairs Officer in the Environment and Development Division of the United Nations Economic Commission for Latin America and the Caribbean (ECLAC). She has also worked in ECLAC's Division of Industrial Development and its International Trade Division, and in UNCTAD's Commodities Division.

Index